SIMPLE-G

Iman Haqiqi • Thomas W. Hertel

Editors

SIMPLE-G

A Gridded Economic Approach
to Sustainability Analysis of the Earth's
Land and Water Resources

Springer

Editors
Iman Haqiqi
Center for Global Trade Analysis,
Department of Agricultural Economics
Purdue University
West Lafayette, IN, USA

Thomas W. Hertel
Center for Global Trade Analysis,
Department of Agricultural Economics
Purdue University
West Lafayette, IN, USA

ISBN 978-3-031-68053-3   ISBN 978-3-031-68054-0   (eBook)
https://doi.org/10.1007/978-3-031-68054-0

This Springer imprint is published by the registered company Springer Nature Switzerland AG
The registered company address is: Gewerbestrasse 11, 6330 Cham, Switzerland

If disposing of this product, please recycle the paper.

The authors dedicate this book to their families, in the hope for a sustainable future: Monireh Aqababaei, Sana and Yassna Haqiqi, Adriela Fernandez, Alexander and Sarah Hertel Fernandez.

Foreword

Imagine this setting: A workshop at the Massachusetts Institute of Technology in September 2002. A man standing at the blackboard, surrounded by a group of economists, holding forth on some new idea. Ideas are flying as much as the chalk dust. It was difficult for me not to get drawn in. This was how I first met Tom Hertel. He was the man at the chalkboard. I am a global environmental scientist and was possibly the only noneconomist in the group; whatever Tom was saying went over my head, but it was clearly exciting to the group. Tom noticed me at some point and said something along the lines of "Navin, this is where your data would come in and be most useful." This has been the story of our relationship since: Tom sees connections that few others—including myself—identify.

Since that first meeting more than two decades ago, Tom Hertel and I have collaborated on multiple papers and projects. I am trained as an earth and environmental scientist. I used to work on the global climate problem, but these days I focus on the sustainability of land use and food systems. Tom Hertel is an economist whose work has focused on the intersections of trade, environment, and development; he is particularly prominent for his work in global economic modeling. Our long-term collaboration has built on our mutual interest in global problems, our commitment to strongly empirical work, our deep respect for each other's intellect and integrity, and our desire to develop new insights from leveraging the power of economic thinking to addressing global sustainability challenges. I learned nearly everything I know of economic theory from working with Tom.

This book is another example of Tom's ability to see the big picture and seize opportunities for advancing the field. The book is fundamentally about the challenge of scale: How one can represent and analyze processes that interact between scales? For example, global models are great at capturing interactions between different countries and regions of the world but are very poor at representing local-scale processes. Local models are excellent at representing local processes in great detail but not how they influence, or are influenced by, other locations nor how they may influence the globe. This challenge of representing cross-scale interactions plagues nearly all academic disciplines. This book presents a modeling framework for representing and simulating such cross-scale interactions in the realm of global sustainability challenges.

But how did Tom get here? His pioneering work has been at the global scale, especially with the GTAP (Global Trade Analysis Project) model. GTAP is a model of the world's economy that represents global interactions among countries, sectors, and factors of production and has been used to analyze the economic and welfare implications of different policy interventions. But over time, Tom was drawn to address our pressing global environmental and sustainability challenges. His 2010 presidential address to the Agricultural and Applied Economics Association discussed the multiple pressures on the global agricultural resource base, including feeding the world's growing population, environmental sustainability, and capacity to withstand shocks from extreme weather events or governmental interventions.

Over the last decade, Tom's focus has shifted to addressing the role of global agricultural land use in achieving the United Nations Sustainable Development Goals. In 2013, Tom developed a much simpler version of GTAP, a model appropriately called SIMPLE, that allowed one to make much faster calculations than GTAP and that was also more transparent and accessible to noneconomists. SIMPLE allowed one to include richer representations of the environmental system and to run many more simulations than ever before. While SIMPLE was simple, it was still coarse, considering seven world regions and their interactions. While it was possible to increase the number of regions, doing so still didn't capture granularity at the local scale. A conversation between Tom and I at a meeting of the American Geophysical Union led to the "A-ha!" moment for Tom—SIMPLE is in fact simple enough that we can do the economics on the grid, at the local scale! Voilà—SIMPLE-G!

SIMPLE-G is a gridded, meaning geographically disaggregated, version of SIMPLE that can consider local biophysical characteristics and institutions. With SIMPLE-G, one can simulate how global policies will influence local land-use decision-making and responses and how these local responses will aggregate up and feed back to the global scale, what Tom calls the global-to-local-to-global framework. This book presents SIMPLE-G, including the economic theory (Part II), the nuts and bolts (Part III), and several illustrative applications (Part IV). The book illustrates how SIMPLE-G could be used to investigate policy interventions to address different UN Sustainable Development Goals and evaluate potential spillovers and trade-offs.

As one example of the power of SIMPLE-G, take the case of nitrate pollution in the US Mississippi Basin as a result of excessive fertilizer application in this agricultural region. Liu and coauthors (Chap. 14) consider four interventions to address this problem—a national tax on nitrogen losses, improving nitrogen use efficiency, drainage water management, and wetland restoration. They find that spatially targeted policies (drainage management, wetland restoration) are the most effective but result in spatial spillovers; that is, the increased crop prices and lower fertilizer prices resulting from the policies cause farmers in neighboring (nontarget) regions to increase crop production. The novel insight is that policies that may seem most effective at first blush may not be once cross-scale interactions are considered; ultimately, a mix of synergistic targeted and across-the-board policies are most effective. Part IV of this book is full of such interesting and insightful examples: addressing other environmental issues such as climate change and groundwater depletion, the role of different policy interventions such as investment in agricultural R&D or limiting environmentally harmful agricultural land use practices, multiple stressors such as climate change and the pandemic, and investigating the influence of critical meso-level mediators such as labor mobility.

So why should you read this book? First, if you are an environmental scientist interested in bringing economic insights into your work, this is a great place to start. SIMPLE-G might just be the tool you've been looking for. You can also have the confidence that this tool has been developed by one of the world's leading agricultural economists and his team. This book will also be useful to those economists who are interested in tailoring their work to address the world's environmental sustainability challenges. While the economic theory may be old hat for you, I am sure you will find the applications in Part IV enlightening.

The holy grail for global change research evolves over time. Decades ago, after the first models of the global atmospheric circulation were developed, the challenge was to couple different parts of the biophysical Earth system, including the oceans, the cryosphere, and terrestrial vegetation. Enormous progress has been made on those fronts in recent decades. The holy grail now is to represent human behavior in these biophysical models. Tom Hertel's efforts with GTAP, SIMPLE, and now SIMPLE-G allow us to better represent human responses, even if economic theory is only one framework for representing human behavior. Another holy grail, as I mentioned earlier, is scale, to represent the suite of processes between the local and the global and their interactions. This book presents a global-to-local-to-global modeling framework that addresses this scale challenge as well. The last chapter of the book (Part V) presents a comprehensive and thoughtful discussion of future directions for this type of work. As an earth and environmental scientist, I am excited to see how far we get in the coming decades to fully representing humans in Earth system models. We do live in interesting times.

Canada Research Chair in Data Science
for Sustainable Global Food Systems,
Professor and Director, Institute for
Resources, Environment and
Sustainability, Professor, School of
Public Policy and Global Affairs,
University of British Columbia,
Vancouver, Canada

Navin Ramankutty

Preface

The genesis of this book goes back to a conversation between one of the co-authors (Hertel) and Navin Ramankutty, his long-time friend and collaborator, while attending the meetings of the American Geophysical Union (AGU) in San Francisco. Hertel and Ramankutty had collaborated on a global land use database for use in conjunction with the Global Trade Analysis Project (GTAP) model, and the ensuing Agro-Ecological-Zones framework (GTAP-AEZ) was being widely used to analyze land-based climate mitigation policies as well as the indirect land use impacts of biofuels. However, it was frustrating to aggregate the rich, gridded data developed by Ramankutty and colleagues to match the aggregate regions used in the economic models rather than using it directly at the grid cell level. This aggregation greatly diluted the connection to the rich biophysical heterogeneity required for credible analysis of issues ranging from biodiversity to terrestrial carbon storage, water scarcity, and pollution. As a result, we found that physical scientists tended to lose interest when analyses were conducted at the regional or country level using models like the GTAP framework.

About this time, Hertel, Ramankutty, and Purdue University colleague Uris Baldos were collaborating on a paper revisiting the question of how advances in agricultural technology affect land use. Instead of using the economywide, GTAP framework for the analysis, the authors opted to use SIMPLE (a Simplified International Model of Prices Land use and the Environment), which Hertel and Uris Baldos had developed for use in the classroom. Unlike GTAP, which seeks to cover the entire economy, SIMPLE is a partial equilibrium model that focuses on a single sector or commodity, which facilitates its use in interdisciplinary graduate courses; the simplified structure greatly facilitates analysis and interpretation of results. In their paper, Hertel, Ramankutty, and Baldos were able to explain their novel findings about the impact of a prospective African Green Revolution on land use and terrestrial carbon emissions using simple graphical analysis in conjunction with analytical expressions derived from a condensed version of SIMPLE. This allowed them to clearly reconcile what had previously appeared to be irreconcilable positions of two groups of researchers: those arguing, along with Borlaug, that improved technology would be land-sparing, and those appealing to Jevons' Paradox, which suggests that improved technology would simply result in more land being devoted to agriculture. As demonstrated in their 2014 paper, both groups had ground to stand on: The correct answer depends on the underlying data and parameters.

Following this success, Hertel and Ramankutty looked for the next potential collaboration. They also concluded it was time to stop aggregating the biophysical data. At this point, Hertel realized that the equations in SIMPLE were simple enough, and the data and parameter requirements sufficiently modest, to permit implementation at the level of individual grid cells. Rather than determining commodity supply via national-level equations, these could be determined at the local level, with national-level outcomes determined by summation across grid cells within a particular market. This concept was subsequently implemented by Uris Baldos, and the first published application was led by another Purdue collaborator, Jing Liu, working with a group of global hydrologists. The success of SIMPLE-G can be attributed in large part to Iman Haqiqi's contributions, bringing new economic perspectives, enabling large model computations, improving parameter estimation, and implementing the first comprehensive model validation. Iman Haqiqi further developed the SIMPLE-G framework, and the SIMPLE-G model was officially christened with a short course in 2019 and a publication in 2020.

SIMPLE-G has formed a core element of several large interdisciplinary grants from the US Department of Agriculture, the National Science Foundation, and the Department of Energy. It is also playing an integral role in GLASSNET (glassnet. net), an NSF-funded network of networks aimed at promoting global-local-global analysis of sustainability. The theme of GLASSNET is that, while sustainability analyses focusing on land and water resources are inherently local, it is also important to connect these local analyses with the global changes that drive many of these stresses. Furthermore, responses to local stresses can spill over to other locations and have global impacts. This global-local-global theme is explored in depth in a special issue of the interdisciplinary journal, *Environmental Research Letters*. Several contributions to this special issue used the SIMPLE-G framework, and we have included them in the applications section of this book (Part IV).

By keeping the economics simple, we have found far greater scope for transdisciplinary collaboration. Throughout our work with SIMPLE and SIMPLE-G, we have found that the simplicity of these models allows for far richer collaborations, as noneconomist collaborators can readily grasp the key economic concepts. This permits them to engage with the research in a meaningful way and identify new avenues for enriching the analysis and parameterization of SIMPLE-G. The structure of this book is informed by the example of the so-called *GTAP Book*, published in 1997 and edited by Tom Hertel, one of the co-authors of this book, and cited over 5,000 times, at the time of this publication, according to Google Scholar. The design of that book was similar, with theory, data, and parameters preceding a set of applications of the model authored by a variety of contributors. These applications formed the basis for the first several annual offerings of the GTAP short course. We are following a similar path with SIMPLE-G. Much of the material in this book was developed in the context of the first two SIMPLE-G courses, during which we refined our approach to teaching SIMPLE-G to economists as well as noneconomists seeking to address the world's land and water sustainability challenges (https://mygeohub.org/groups/glassnet/learning-hub/courses_page/simpleg).

As with the GTAP example, wherein the book and database gave rise to a remarkably large network of researchers and decision-makers (at publication, more than 28,000 members, spread across more than 175 countries: www.gtap.org), we hope that this new book will facilitate the development of a community producing many new versions of SIMPLE-G. These may include new focus regions (e.g., SIMPLE-G-EU) or a different sectoral focus (e.g., SIMPLE-G-Livestock), or it may even involve the integration of SIMPLE-G and GTAP. Eventually, we hope to provide a common global repository for model files, data, parameters, and model specifications to which the community can contribute and from which collaborators can draw to address new sustainability policies.

West Lafayette, IN, USA   Thomas W. Hertel
                                                                                        Iman Haqiqi

Acknowledgments

This book is the product of nearly a decade of collaborative work based in the Center for Global Trade Analysis at Purdue University and led by the GLASS (Global to Local Analysis of Systems Sustainability) lab group, which was formed to share the knowledge and techniques necessary to implement this novel form of economic modeling. Key collaborators in this lab group—including Uris Baldos, Alfredo Cisneros-Pineda, Elizabeth Fraysse, Iman Haqiqi, Thomas Hertel, Jing Liu, Srabashi Ray, and Zhan Wang—have each led or contributed to different sections of this book. This book would not have been possible without their sustained contributions to the development and application of SIMPLE-G.

We also acknowledge the contributions of our many collaborators who introduced us to the nuances of the sustainability challenges faced by our planet and how they can be better understood and addressed within the SIMPLE-G framework. These include Laura Bowling, Sylvie Brouder, Maksym Chepeliev, Kelsi Ferin, Karen Fisher-Vanden, Steve Frolking, Danielle Grogan, Alexandra Hill, Matthew Huber, David Johnson, Justin Johnson, Christopher Kucharik, Richard Lammers, Emiliano Lopez-Barrera, Dominique van der Mensbrugghe, Stephen Polasky, Carol Song, Shanxia Sun, Ed Taylor, Brayam Valqui Ordonez, Nelson Villoria, Jeffrey Volenec, Mort Webster, Wil Wollheim, Jungha Woo, Lan Zhao, and Shan Zuidema.

This research would not have been possible without the efforts of Mark Horridge and Ken Pearson, developers of the GEMPACK software suite, which is central to the implementation of SIMPLE-G.

Completion of this book also owes a great debt to Amy Bekkerman of Precision Edits, who tirelessly edited our work and made valuable suggestions about how to better organize and communicate this vast amount of material.

Funding for this research has come from a great variety of sources. Purdue University's Big Idea Award through Discovery Park helped to launch GLASS at Purdue.

Support from the US Department of Energy's Office of Science (DOE Cooperative Agreement #DE-SC0016162) allowed for the development and application of SIMPLE-G to hydrological and energy challenges and the development of coupled models (DOE Cooperative Agreement #DE-SC0022141).

Competitive grants from USDA/NIFA allowed us to engage with a variety of domain experts in studying a range of sustainability challenges facing the food system (grants include #2016-67007-24957 "Assessing the Long Run Sustainability of US Agriculture in an Integrated Global Economy," #2019-67023-29679 "Economic Foundations of Long Run Agricultural Sustainability," and #2022-67023-36403 "Labor Markets and the Impacts of Environmental Stresses and Conservation Policies on US Agriculture").

Competitive grants from the US National Science Foundation facilitated novel applications of SIMPLE-G in the context of the food–energy–water nexus (NSF-CBET grant #1855937), telecoupling analysis (NSF-DEB-1924111), and networking international researchers from diverse disciplines (AccelNet NSF-OISE-2020635).

Support from US National Science Foundation, Institute for Geospatial Understanding through an Integrative Discovery Environment (I-GUIDE, NSF award #2118329) allowed for the development of cyberinfrastructure for sustainability analysis as well as the development and application of validation methods for SIMPLE-G-US.

Funding from the Alexander von Humboldt Foundation and from the USDA Hatch Project (#1003642) helped support PI Hertel's work on this book.

Finally, a grant from Purdue's Office of the Vice President for Research and the College of Agriculture supported the editing and open access provision of this book.

Contents

Part I   Introduction

1   **Introduction**   3
    Thomas W. Hertel

Part II   Theory

2   **SIMPLE Economic Theory**   11
    Thomas W. Hertel

3   **Grid-Level Analysis Using SIMPLE-G**   23
    Srabashi Ray and Thomas W. Hertel

Part III   Model

4   **Equilibrium Conditions and General Assumptions
    for a Quantitative Geospatial Economic Model**   35
    Iman Haqiqi

5   **SIMPLE-G Model Specification: Mathematical Equations
    in a Multiscale Market Equilibrium Model**   51
    Iman Haqiqi

6   **Benchmark Data: Integrating Biophysical and Economic
    Information in a Consistent Geospatial Dataset**   81
    Iman Haqiqi and Uris Lantz C. Baldos

7   **Behavioral Parameters: Capturing Geospatial Heterogeneity
    in Economic Decisions and Responses**   93
    Iman Haqiqi

8   **Computation and Baseline: Efficient Methods for Solving a Large
    System of Equations for Projection and Scenario Analysis**   103
    Iman Haqiqi and Uris Lantz C. Baldos

9   Model Validation: Comparing Gridded and Regional
     Simulations to Observations   113
     Iman Haqiqi, Zhan Wang, and Uris Lantz C. Baldos

Part IV   Applications

10   The R&D Cost of Climate Mitigation in Agriculture   135
     Keith Fuglie, Srabashi Ray, Uris Lantz C. Baldos,
     and Thomas W. Hertel

11   Gridded Implications of Total Factor Productivity Growth   159
     Elizabeth A. Fraysse, Thomas W. Hertel, and Srabashi Ray

12   Local Groundwater Sustainability Policies and Global Spillovers . .   173
     Iman Haqiqi, Laura Bowling, Sadia Jame, Uris Lantz C. Baldos,
     Jing Liu, and Thomas W. Hertel

13   The Role of Labor Markets in Determining the Efficacy
     and Distributional Impact of Sustainability Policies   199
     Srabashi Ray, Iman Haqiqi, Alexandra E. Hill, J. Edward Taylor,
     and Thomas W. Hertel

14   Tackling Policy Leakage and Targeting Hot Spots Could Be Key
     to Addressing the "Wicked" Challenge of Nutrient Pollution
     from Corn Production in the United States   217
     Jing Liu, Laura Bowling, Christopher Kucharik, Sadia Jame,
     Uris Lantz C. Baldos, Larissa Jarvis, Navin Ramankutty,
     and Thomas W. Hertel

15   The Role of Transportation Infrastructure Expansion
     in the Transmission of Global Crop Price Shocks
     to the Brazilian Agriculture   235
     Zhan Wang

16   Global Groundwater Sustainability and Virtual Water Trade   253
     Iman Haqiqi, Chris J. Perry, and Thomas W. Hertel

17   Interplay Between the Pandemic and Environmental Stressors   283
     Iman Haqiqi, Danielle S. Grogan, Marziyeh Bahalou, Jing Liu,
     Uris Lantz C. Baldos, Richard Lammers, and Thomas W. Hertel

Part V   Future Directions

18   Future Directions: Policy Implications, Model Extensions,
     and Institutional Innovation   307
     Iman Haqiqi, Thomas W. Hertel, Zhan Wang, Uris Lantz C. Baldos,
     Alfredo Cisneros-Pineda, and Jing Liu

Index   325

List of Figures

Fig. 1.1   United Nations Sustainable Development Goals, representing a blueprint for a better future 4

Fig. 2.1   The unified global economy 12
Fig. 2.2   Impact of a technological improvement in Region A on the world market and on the rest of the world 16

Fig. 3.1   Cropland area response under varying land supply conditions .. 28
Fig. 3.2   Input substitution determines the magnitude of the intensive margin of supply 29

Fig. 4.1   Overview of the SIMPLE-G model 37
Fig. 4.2   Schematic representation of local-to-global supply linkages in SIMPLE-G 39
Fig. 4.3   Spillover effects (**a**) to similar crop-producing locations, (**b**) across subregions, and (**c**) across countriesLeakage and spillover effects depend on the degree of mobility of inputs and the degree of product differentiation 41

Fig. 5.1   Basic gridded production structure of a simple two-input model and associated markets 54
Fig. 5.2   Basic production structure of a SIMPLE-G model with three-input nested CES 58
Fig. 5.3   Basic production structure with four inputs 62
Fig. 5.4   Basic production structure with land allocation 65
Fig. 5.5   Structure of crop production at each grid cell 71

Fig. 6.1   Overview of main data and parameter processing method for SIMPLE-G-US 86

Fig. 7.1   (**a**) Economic supply for water with maximum availability determined by asymptote C. (**b**) Estimated groundwater elasticities in the continental United States 98

Fig. 8.1   Linear relationship between solution time for SIMPLE-G-US
            and number of grid cells 106
Fig. 8.2   Sample window from the SIMPLE-G web application
            at https://mygeohub.org/tools/simpleus 107

Fig. 9.1   SIMPLE model global validation 115
Fig. 9.2   SIMPLE model regional validation 116
Fig. 9.3   Validation of crop output and cropland for Brazil 117
Fig. 9.4   Crop supply by market in Brazil, calculated using market share
            from GTAP database 118
Fig. 9.5   Grain price index and its linear regression model 119
Fig. 9.6   Steps in the validation of the SIMPLE-G-US-Allcrops model,
            2001–2002 to 2016–2017 121
Fig. 9.7   Share of cropland in total grid cell area calculated from the
            National Land Cover Database, 2016 121
Fig. 9.8   Observed and simulated cropland area in the United States in
            2016–2017 based external drivers, 2001–2002 to 2016–2017 .. 125
Fig. 9.9   Observed and simulated cropland area by major farming states,
            2016–2017 126
Fig. 9.10   Observed and simulated cropland area by USDA Farm Resource
            Regions in 2016–2017 126
Fig. 9.11   Supply and demand drivers of change in cropland area by USDA
            Farm Resource Regions, 2016–2017 127
Fig. 9.12   Cumulative distribution of change in cropland (ha) by grid cell,
            2001–2002 to 2016–2017 128

Fig. 10.1   Agricultural greenhouse gas emissions and output in world
            regions, 1990 and 2019 138
Fig. 10.2   Areas targeted for cropland restrictions in the policy simulations
            (5% of global cropland area) 147
Fig. 10.3   Effect of research and development and land policy scenarios on
            farmgate greenhouse gas emissions intensity (kg CO2e per
            constant 2017 USD) 150
Fig. 10.4   Change in (a) crop prices, (b) undernutrition, and (c) cropland
            use as a result of global policy scenarios: differences between
            2017 and 2050, by region 151

Fig. 11.1   Map of selected grid cells used in this chapter 162
Fig. 11.2   Decomposition of changes in US crop price by ten sources of
            change, 2017–2050 168
Fig. 11.3   Decomposing the global and local drivers of crop output in grid
            WA 170

Fig. 12.1   Structure of crop production at each grid cell 177
Fig. 12.2   Change in (a) crop production by region and (b) US
            groundwater withdrawals, 2010–2050 183

Fig. 12.3   Change in (**a**) US irrigated land and (**b**) US groundwater
               withdrawals in response to a population increase of one million,
               by region   185
Fig. 12.4   Percentage change in (**a**) groundwater withdrawals and (**b**) value
               (shadow price) of groundwater to producers by 5 arcmin grid
               cells in response to a policy restricting groundwater withdrawals
               to the level of average annual recharge   187
Fig. 12.5   Percentage change in (**a**) irrigation surface water withdrawals
               and (**b**) total cropland area by 5 arcmin grid cells in response
               to groundwater restrictionsAreas in *white* are not cultivated   188
Fig. 12.6   Percentage change in (**a**) global production and (**b**) land
               use in response to a US groundwater sustainability constraint
               in 2050   189
Fig. 13.1   Structure of crop production in SIMPLE-G-CZ   205
Fig. 13.2   Grid-cell level response to a permanent 10% price hike as a
               percentage change in (**a**) production, (**b**) employment, and (**c**)
               land use under a perfectly elastic labor market contrasted with
               responses in (**d**) production, (**e**) employment, and (**f**) land use
               under restricted labor mobility   209
Fig. 13.3   Changes in groundwater use due to a global price shock in (**a**)
               perfectly elastic and (**b**) "sticky" labor markets   209
Fig. 13.4   Impact of crop price shock on wages at the commuting zone
               level (restricted labor mobility model)   210
Fig. 13.5   Impact of US sustainable groundwater policy shock as a
               percentage change in (**a**) crop production, (**b**) employment, and
               (**c**) wages in the western United States under perfect mobility
               versus impact on (**d**) crop production, (**e**) employment, and (**f**)
               wages under restricted labor mobility   211

Fig. 14.1   Schematic of the SIMPLE-G-US-CS
               modelSIMPLE-G-US-CS is a modification of the standard
               SIMPLE-G model (Fig. 4.1)   220
Fig. 14.2   Connections between Agro-IBIS and SIMPLE-G-US-CS and
               nitrogen (N) loss mitigation policies   221
Fig. 14.3   Changes in nitrogen (N) loss under (**a**) an N loss tax, (**b**) split N
               application, (**c**) controlled drainage, (**d**) wetland restoration, and
               (**e**) combined strategies of tax, split N, and wetland
               restorationUnits are tons of N loss per 5 arcmin grid cells   224
Fig. 14.4   N loss reduction by (**a**) state and mitigation strategy and (**b**)
               accumulated percentage   226

Fig. 15.1   Role of logistics costs in Brazilian crop exports   236
Fig. 15.2   Economic framework of national and subnational responses to a
               world price shock (*dotted line*) and transportation infrastructure
               expansion (*dashed line*)   237

Fig. 15.3   Overview of the SIMPLE-G-Brazil model for the
             transportation infrastructure application. (Adapted
             from Wang et al. (2024)) 239
Fig. 15.4   Railway extension plan from PNL2035 and its contribution to
             reducing transportation costs 245
Fig. 15.5   (a) Percentage change in national crop price levels under the
             business-as-usual (BAU), low, and high transportation
             expansion scenarios. Gridded farm-gate crop price received
             under (b) the BAU scenario, (c) the low scenario, and (d)
             the high scenario 246
Fig. 15.6   Simulated changes in crop production and the decomposition
             by extensive margins, intensive margins, and their interactions at
             the state and national level under (a) the BAU scenario, (b) the
             low scenario, and (c) the high scenario 248
Fig. 15.7   Simulated changes in greenhouse gas emissions and the
             decomposition by drivers 249

Fig. 16.1   Overall gridded production structure
             of the SIMPLE-G model 259
Fig. 16.2   Employing hydroclimatic information to inform the economic
             model about irrigation water availability and nonrenewable
             groundwater irrigation 261
Fig. 16.3   Spatial distribution of irrigation showing the share of blue water
             in total gridded crop water consumption around the year 2017 . 263
Fig. 16.4   Spatial distribution of unsustainable groundwater illustrating the
             share of unsustainable groundwater in total gridded irrigation
             water consumption around the year 2017 264
Fig. 16.5   Decomposition of global crop response to global groundwater
             sustainability restrictionsThe leftmost (red) bar shows the
             immediate/direct impact prior to adaptation 266
Fig. 16.6   Percentage change in (a) irrigated and (b) rainfed cropped areas
             in response to global groundwater sustainability restrictions
             produced by the water balance model 268
Fig. 16.7   Global changes in surface water irrigation in response to global
             groundwater sustainability restrictions 269
Fig. 16.8   Percentage change in (a) irrigated and (b) rainfed crop
             production in response to global groundwater sustainability
             restrictions 270
Fig. 16.9   Global changes in agricultural employment in (a) irrigated and
             (b) rainfed crop production (%) in response to global
             groundwater sustainability restrictions 275

Fig. 17.1   Method overview 285
Fig. 17.2   Grid cell shock to (a) rainfed crop yields and (b) surface water
             supply due to weather impacts 289

Fig. 17.3   Percentage change in (**a**) greenhouse gas emissions from the
              agricultural sector and (**b**) undernourished populations in
              developing nations due to the compound shocks, with and
              without adaptations 291
Fig. 17.4   Decomposition of the drivers of change in undernourished
              populations under the (**a**) no-adaptation scenario and (**b**)
              adaptation scenario 293
Fig. 17.5   Decomposition of adaptation contributions to changing
              undernourished populations compared with the no-adaptation
              scenario in (**a**) percentage change and (**b**) millions
              of people 295
Fig. 17.6   The interaction effect of weather and pandemic on (**a**) crop
              prices and (**b**) undernourishment outcomes in Sub-Saharan
              Africa 298

Fig. 18.1   Nesting SIMPLE-G within the GTAP-AEZ framework.
              Livestock and forestry demand for land are derived from a
              national production function, while crop production is modeled
              at the grid-cell level 316

List of Tables

Table 4.1	Market equilibrium conditions in the SIMPLE-G model	47
Table 4.2	Major assumptions in the SIMPLE-G model	47
Table 5.1	SIMPLE-G models in Part III	52
Table 5.2	Summary of gridded equations for SIMPLE-G-1.2.1	53
Table 5.3	Gridded equations for SIMPLE-G-1.3.1 with three inputs	60
Table 5.4	Gridded equations for SIMPLE-G-1.4.1 multiscale markets	63
Table 5.5	Gridded equations for SIMPLE-G-2.4.1	68
Table 5.6	Gridded equations for SIMPLE-G-2.6.1	72
Table 5.7	Major equations for the demand side of SIMPLE-G	73
Table 6.1	Gridded data for SIMPLE-G-US	85
Table 7.1	Parameters in the gridded model	95
Table 9.1	Global drivers of land-use change, 2002–2017	123
Table 10.1	Population and income growth assumptions (2017–2050)	140
Table 10.2	Elasticities of agricultural research and development (R&D) capital stock, by region and technology provider (Fuglie 2018)	144
Table 10.3	Research and development (R&D) and environmental policy scenarios for the simulations	146
Table 10.4	Effects of research and development (R&D) and environmental policy scenarios on global agricultural greenhouse gas (GHG) emissions, prevalence of malnutrition, food prices, cropland, and agricultural production	149
Table 10.5	Marginal costs of agricultural greenhouse gas (GHG) emissions abatement costs from the global policy scenarios	153
Table 10.6	Effects of regional European Union (EU) policy scenarios on agricultural productivity, cropland, and greenhouse gas (GHG) emissions	154

Table 11.1   SIMPLE parameters for 12 selected grid cells for in-depth
              analysis of model results 162
Table 11.2   Simulated and calculated impacts of a 1% TFP shock to 12
              selected grids in SIMPLE-G1 164
Table 11.3   Decomposition of total change in crop production as a
              percentage change 165
Table 11.4   Minimodel analysis of grid-cell impacts of the baseline
              scenario, incorporating population, income, biofuels, and
              technology growth worldwide, 2017–2050 166

Table 12.1   Projected percentage changes in population, income, and
              productivity, 2010–2050 180
Table 12.2   Projected changes in cropland area, water withdrawal, fertilizer
              applications, and crop productions in response to US
              groundwater restriction 190

Table 13.1   National change (%) in crop production, employment,
              wages, land use, groundwater use, and crop price
              in the United States 208

Table 14.1   Nitrogen (N) loss reduction outcomes, impacts on crop output
              and price, and mitigation efficiency across management
              strategies 222

Table 15.1   Data sources for SIMPLE-G-Brazil transportation
              application 241

Table 16.1   Total crop water demand and contribution of nonrenewable
              resources for major groundwater users circa 2017 263
Table 16.2   Long-term regional impacts of restricting groundwater
              irrigation to renewable resources 272
Table 16.3   Trade impacts of restricting groundwater irrigation
              (percentage change in supply to the domestic market, exports,
              and imports) 273
Table 16.4   Impact of groundwater sustainability policy on virtual blue
              water exports by source 273
Table 16.5   Impact of groundwater sustainability policy on employment in
              crop production by practice and region 276

Table 17.1   Pandemic and weather shock scenario design 288
Table 17.2   Undernourished population by region in the initial condition
              (no disaster) and change in undernourished population
              due to the compound disaster with and without adaptation 292

Table 18.1   Gridded (SIMPLE-G) and subregional (SIMPLE-S) versions
              in the case of the US-focused model from Chap. 12 320

Contributors

Marziyeh Bahalou Department of Industrial Engineering, Purdue University, West Lafayette, IN, USA

Uris Lantz C. Baldos Center for Global Trade Analysis, Department of Agricultural Economics, Purdue University, West Lafayette, IN, USA

Laura Bowling Department of Agronomy, Purdue University, West Lafayette, IN, USA

Alfredo Cisneros-Pineda Center for Global Trade Analysis, Department of Agricultural Economics, Purdue University, West Lafayette, IN, USA

Elizabeth A. Fraysse Center for Global Trade Analysis, Department of Agricultural Economics, Purdue University, West Lafayette, IN, USA

Keith Fuglie Economic Research Service, US Department of Agriculture, Washington, DC, USA

Danielle S. Grogan Earth Systems Research Center, University of New Hampshire, Durham, NH, USA

Iman Haqiqi Center for Global Trade Analysis, Department of Agricultural Economics, Purdue University, West Lafayette, IN, USA

Thomas W. Hertel Center for Global Trade Analysis, Department of Agricultural Economics, Purdue University, West Lafayette, IN, USA

Alexandra E. Hill Department of Agricultural and Resource Economics, University of California, Berkeley, CA, USA

Sadia Jame Department of Agricultural and Biological Engineering, Purdue University, West Lafayette, IN, USA

Larissa Jarvis McGill Sustainability Systems Initiative, McGill University, Montreal, QC, Canada

Christopher Kucharik Department of Agronomy, University of Wisconsin, Madison, WI, USA

Richard Lammers Earth Systems Research Center, University of New Hampshire, Durham, NH, USA

Jing Liu Center for Global Trade Analysis, Department of Agricultural Economics, Purdue University, West Lafayette, IN, USA

Chris J. Perry Independent Researcher, London, UK

Navin Ramankutty School of Public Policy and Global Affairs, University of British Columbia, Vancouver, BC, Canada

Srabashi Ray Center for Global Trade Analysis, Department of Agricultural Economics, Purdue University, West Lafayette, IN, USA

J. Edward Taylor Department of Agricultural and Resource Economics, University of California, Davis, CA, USA

Zhan Wang Center for Global Trade Analysis, Department of Agricultural Economics, Purdue University, West Lafayette, IN, USA

Part I
Introduction

Chapter 1
Introduction

Thomas W. Hertel

The world faces significant sustainability challenges in the decades ahead (United Nations 2019). Of the 17 UN Sustainable Development Goals, nine bear directly on the world's land and water resources (Fig. 1.1). These goals cannot be attained without prudent management of these natural resources, yet growing populations and rising incomes place unprecedented stress on them. Of the four planetary boundaries identified as being at risk, three—genetic diversity, land system change, and nitrogen and phosphorous flows—are directly linked to land use—and the fourth—climate change—is also substantially driven by developments in global land use (Steffen et al. 2015). The challenge posed in assessing potential sustainability solutions is that these stresses do not respect disciplinary boundaries. And while stresses are often highly localized, the drivers of these stresses are often global (e.g., Haqiqi et al. 2023), and local responses can influence national and international outcomes. For this reason, underlying risks, as well as solutions, are typically assessed using a suite of models with increasingly complex approaches. Unfortunately, this complexity often renders replication and use by researchers outside the core group impossible (Obersteiner et al. 2016; Springmann et al. 2018).

To date, only a few open-source, bottom-up, economic-environmental modeling frameworks have been capable of analyzing global sustainability at the resolution of individual grid cells (Lotze-Campen et al. 2008; Valin et al. 2013). There is clearly a trade-off between complexity and accessibility. Models used in teaching and

This chapter draws heavily on a paper originally published as Baldos, Uris Lantz C., Iman Haqiqi, Thomas W. Hertel, Mark Horridge, and J. Liu. 2020. SIMPLE-G: A multiscale framework for integration of economic and biophysical determinants of sustainability. *Environmental Modelling & Software* 133: 104805. https://doi.org/10.1016/j.envsoft.2020.104805.

T. W. Hertel (✉)
Center for Global Trade Analysis, Department of Agricultural Economics, Purdue University, West Lafayette, IN, USA
e-mail: hertel@purdue.edu

I. Haqiqi, T. W. Hertel (eds.), *SIMPLE-G*,
https://doi.org/10.1007/978-3-031-68054-0_1

Fig. 1.1 United Nations Sustainable Development Goals, representing a blueprint for a better future. Of the 17 goals, nine are related to food, water, and land – the focus of this book. Image source: United Nations Sustainable Development Goals (https://www.un.org/sustainabledevelopment/). The content of this publication has not been approved by the United Nations and does not reflect the views of the United Nations or its officials or Member States.

academic research are generally simpler than those developed by national and international laboratories and research institutions. A relatively simple, global, grid-resolving sustainability framework that could also be run "in-cloud" would allow wider participation in sustainability discussions and facilitate more crowdsourcing of new ideas, data, and parameters to enrich the representation of local stresses, policies, and adaptations. This book documents one such modeling framework: SIMPLE-G, a Simplified International Model of agricultural Prices, Land use, and the Environment, Gridded version.

The SIMPLE-G framework allows researchers to analyze the interplay between economic and environmental systems. In doing so, it accounts for the actions of local agricultural producers with respect to land and water use within the context of regional and global commodity markets. This model integrates economic theories with environmental sciences to analyze the biophysical and economic impacts at different geospatial scales. The model's economic supply of land and water accounts for information about local institutions, biophysical characteristics, sustainability criteria, and physical availability. Consequently, heterogeneous local constraints lead to different rates of change in land and water use. On the demand side, growth in income and population lead to changes in food consumption baskets and changes in agricultural trade patterns.

Integrating human and earth system analysis within a single global economic framework is a challenging task and often focuses on one-way linkages, such as those used in downscaling regional results to a grid cell level (Reilly et al. 2012). It has also been common practice to extrapolate from sophisticated grid-level analyses to the national scale, assuming that the share of grids in national aggregate of production or land use are unchanging (Schlenker and Roberts 2009). Bridging local, national, and global scales within a single, integrated framework is essential if we wish to consider the influence of local decision makers' behavior on global sustainability outcomes. In the SIMPLE-G framework, these local decisions are endogenous to the model and account for local biophysical characteristics and institutions as well as the simultaneous determination of local input prices and globally determined commodity prices.

This book features eight applications of the SIMPLE-G framework to illustrate its versatility in sustainability analyses. Chapter 10 introduces the reader to the SIMPLE model (nongridded) through an exploration of the role of agricultural productivity growth in ensuring both environmental sustainability and food security. Chapter 11 builds on Chap. 10 by introducing the simplest possible representation of economic and biophysical processes, dubbed SIMPLE-G1-US: two inputs (land and nonland), with just one region (i.e., the United States) represented at a fine scale. Subsequent applications gradually introduce greater complexity, including groundwater and surface water withdrawals (Chap. 12), labor markets (Chap. 13), and nitrogen fertilizer applications (Chap. 14), represented at a fine scale. Some examples of sustainability challenges that have been assessed using these versions of SIMPLE-G range from the implications of public research and development for agricultural productivity and greenhouse gas emissions (Chap. 10) to the consequences of groundwater conservation policies for employment, production, and trade (Chap. 16). Another application (Chap. 14) focuses on mitigating agricultural nonpoint source water pollution and one other (Chap. 15) focuses on the spatial environmental impacts of infrastructure development (i.e., road and rail construction) in Brazil. A final application (Chap. 17) aims to understand the global interplay between future pandemics and adverse climate shocks. In short, this book offers a rich array of SIMPLE-G applications for the reader to replicate and build upon in their own future research. All of the files required for replication are available at https://gtap.agecon.purdue.edu/simple-g/.

To take full advantage of the SIMPLE-G framework and resources, readers need deeper insights into the theory, data, parameters, and analysis tools underpinning these models. In Part II, we review the economic theory underlying the SIMPLE-G model. Using this theory, we can obtain important insights about the likely consequences of sustainability policies (e.g., changes in agricultural productivity or the withdrawal of natural resources for conservation purposes). We encourage model users to go beyond simply reporting results and instead provide in-depth analysis of the channels through which these external shocks to food and environmental systems affect key social and environmental outcomes. The analytical results developed in Part II also highlight the critical parameters that drive these findings, which are typically obtained from statistical studies or from transfer functions developed through simulations of biophysical systems models, as described in Part III. These parameters are inherently uncertain; understanding how parameter uncertainty translates into uncertainty in key metrics is an important part of any SIMPLE-G analysis. Part IV illustrates these principles in a series of applications.

An important and novel innovation documented in this book is the use of what we refer to as minimodels, which we have developed to address the challenge of understanding the critical drivers of local responses in a clear and transparent manner. This concept represents our effort to take the reader beyond the maps that are typically used to report results from global gridded models. Of course, the problem with in-depth analysis of any individual grid cell/activity lies in the fact that it is connected, in some way, to all the other grid cells in this type of bottom-up, multiscale analysis. The trick is to find a way to isolate the grid cell from its neighbors to permit a more in-depth analysis of local changes.

The benefit of the minimodel concept comes into play once the user has identified certain grid cells that respond in an extreme or perhaps even counterintuitive manner. These grid cells can be extracted for deeper analysis: This extracted minimodel consists of all the data, parameters, and equations describing behavior in a given grid cell. The minimodel also includes the solution values for all communicating variables for the focus grid cell, solved using the full model. The *communicating variables* are those variables that are determined in concert with the other grid cells and other world regions. In the simplest application (SIMPLE-G1-US), this is just the crop commodity price. In that model—illustrated in Chap. 11— the crop price is a sufficient statistic summarizing all of the developments in the national and global markets. In the minimodel, this price is treated as an exogenous variable and is shocked by the solution value of said variable, emerging from the full model simulation. When this communicating variable is shocked, along with any other grid-specific exogenous variables being perturbed in the full model simulation (e.g., total factor productivity in the case of Chap. 11), the user should obtain identical solutions from both the minimodel and the full SIMPLE-G model simulations. In this manner, a pared-down minimodel opens a way into in-depth analysis of behavioral responses in a particular grid cell.

More complex applications of the model contain more communicating variables across grid cells. The other applications in Part IV include wage changes determined within a commuting zone, national fertilizer prices, and natural resource shadow prices, among others. Therefore, the structure of the minimodel will vary across the applications. Once the particular minimodel is obtained and the equivalence of the two solutions is verified, the user can employ the powerful AnalyseGE software, embedded in the GEMPACK software suite, to analyze grid cell behavior, focusing on one equation at a time. This allows the user to isolate the source of unexpected or counterintuitive results as well as the key parameters driving these outcomes. Upon identifying the source of the surprising result, the researcher can scrutinize the reason for this outcome—possibly reevaluating the underlying data or parameters—or simply register this as a new finding.

To enhance the reader's experience and provide further motivation and insights into the SIMPLE-G framework, all of the chapters in this book are accompanied by voice-over PowerPoint presentations available here: https://gtap.agecon.purdue.edu/simple-g/. Readers may wish to view these short lectures before reading the associated chapter. These presentations also facilitate the use of this book as a text for courses utilizing SIMPLE-G.

We believe that this book will facilitate widespread use of the SIMPLE-G framework. By discussing the issues of database construction and model parameterization as well as model implementation and validation (Part III) in detail, along with providing a range of diverse applications for readers to replicate, we anticipate the construction of many new versions of SIMPLE-G. As this work develops, we plan to assemble a common global repository for data, parameters, and model specifications. This repository will allow researchers to draw on previous work to develop their own variants of SIMPLE-G or to use existing versions in new ways to assess the sustainability dimension of policies bearing on the world's land and water resources. Further discussion of these future directions for SIMPLE-G may be found in Chap. 18.

1 Summary

The SIMPLE-G framework represents a significant step forward in the computational modeling of human and environmental systems. This model incorporates market-mediated effects and feedback from human systems to natural systems and informs spatially heterogeneous economic decisions. It provides a framework for measuring spillover effects transmitted from one location to another location through markets and accounts for telecoupled distant regions, enabling analysis of interregional dependencies at a global scale. The model is particularly useful for evaluating the potential impacts of changes in government policies, technological improvements, and climate change on food production, consumption, and prices. With its ability to address the spatial distribution of activities and explicitly model the spatial heterogeneity associated with biophysical and socioeconomic systems, SIMPLE-G offers a valuable tool for designing effective conservation strategies and sustainability policies.

References

Baldos, Uris Lantz C., Iman Haqiqi, Thomas W. Hertel, Mark Horridge, and Jing Liu. 2020. SIMPLE-G: A multiscale framework for integration of economic and biophysical determinants of sustainability. *Environmental Modelling & Software* 133: 104805. https://doi.org/10.1016/j.envsoft.2020.104805.

Haqiqi, Iman, Laura Bowling, Sadia Jame, Uris Baldos, Jing Liu, and Thomas Hertel. 2023. Global drivers of local water stresses and global responses to local water policies in the United States. *Environmental Research Letters* 18: 065007. https://doi.org/10.1088/1748-9326/acd269.

Lotze-Campen, Hermann, Christoph Müller, Alberte Bondeau, Stefanie Rost, Alexander Popp, and Wolfgang Lucht. 2008. Global food demand, productivity growth, and the scarcity of land and water resources: A spatially explicit mathematical programming approach. *Agricultural Economics* 39: 325–338. https://doi.org/10.1111/j.1574-0862.2008.00336.x.

Obersteiner, Michael, Brian Walsh, Stefan Frank, Petr Havlík, Matthew Cantele, Junguo Liu, Amanda Palazzo, et al. 2016. Assessing the land resource–food price nexus of the Sustainable Development Goals. *Science Advances* 2: e1501499. https://doi.org/10.1126/sciadv.1501499.

Reilly, John, Jerry Melillo, Yongxia Cai, David Kicklighter, Angelo Gurgel, Sergey Paltsev, Timothy Cronin, Andrei Sokolov, and Adam Schlosser. 2012. Using land to mitigate climate change: Hitting the target, recognizing the trade-offs. *Environmental Science & Technology* 46: 5672–5679. https://doi.org/10.1021/es2034729.

Schlenker, Wolfram, and Michael J. Roberts. 2009. Nonlinear temperature effects indicate severe damages to U.S. crop yields under climate change. *Proceedings of the National Academy of Sciences* 106: 15594–15598. https://doi.org/10.1073/pnas.0906865106.

Springmann, Marco, Michael Clark, Daniel Mason-D'Croz, Keith Wiebe, Benjamin Leon Bodirsky, Luis Lassaletta, Wim De Vries, et al. 2018. Options for keeping the food system within environmental limits. *Nature* 562: 519–525. https://doi.org/10.1038/s41586-018-0594-0.

Steffen, Will, Katherine Richardson, Johan Rockström, Sarah E. Cornell, Ingo Fetzer, Elena M. Bennett, Reinette Biggs, et al. 2015. Planetary boundaries: Guiding human development on a changing planet. *Science* 347: 1259855. https://doi.org/10.1126/science.1259855.

United Nations. 2019. *The future is now—Science for achieving sustainable development*, Global Sustainable Development Report. New York: United Nations, Department of Economic and Social Affairs.

Valin, Hugo, Petr Havlík, Niklas Forsell, Stefan Frank, Aline Mosnier, Daan Peters, Carlo Hamelinck, Matthias Spöttle, and Maarten van den Berg. 2013. *Description of the GLOBIOM (IIASA) model and comparison with the MIRAGE-BioF (IFPRI) model*. Laxenburg: International Institute for Applied Systems Analysis (IIASA), Ecofys and E4tech. https://previous.iiasa.ac.at/web/home/research/modelsData/GLOBIOM/Copy_of_Describing_GLOBIOM_and_comparison_with_MIRAGE-BioF_O.pdf.

Part II
Theory

Chapter 2
SIMPLE Economic Theory

Thomas W. Hertel

The SIMPLE modeling framework is firmly grounded in economic theory. As such, it is useful to review the foundational concepts before moving into the specification of the model itself, which is the subject of Part III of this book. This chapter introduces the theory of global land use change in the SIMPLE (global, non-gridded) framework, while Chap. 3 introduces the theory behind the gridded modeling. This economic theory part of the book is written for an interdisciplinary audience with an eye to making it broadly accessible. However, readers eager to get into the applications of SIMPLE-G are free to jump ahead, returning to this section as needed to understand the underlying theory, which will be referenced in many of the applications. Reading this chapter and Chap. 3 will enhance the readers' understanding of the economic mechanisms at work in SIMPLE-G.

1 One-Region Model: The Global Food System in a Nutshell

In the spirit of starting with the simplest possible economic model of global land use, we begin with a unified global agricultural economy (Fig. 2.1). We focus on cropland expansion in response to growth in demand for agricultural commodities (due to, e.g., growth in population, income, and biofuels). In this context, it seems intuitive that, if demand were to increase by 10% and farming techniques remained

Sections 1, 2, and 3 draw heavily on Hertel, Thomas W. 2018. Economic perspectives on land use change and leakage. *Environmental Research Letters* 13: 075012. https://doi.org/10.1088/1748-9326/aad2a4.

T. W. Hertel (✉)
Center for Global Trade Analysis, Department of Agricultural Economics, Purdue University, West Lafayette, IN, USA
e-mail: hertel@purdue.edu

© The Author(s) 2025
I. Haqiqi, T. W. Hertel (eds.), *SIMPLE-G*,
https://doi.org/10.1007/978-3-031-68054-0_2

11

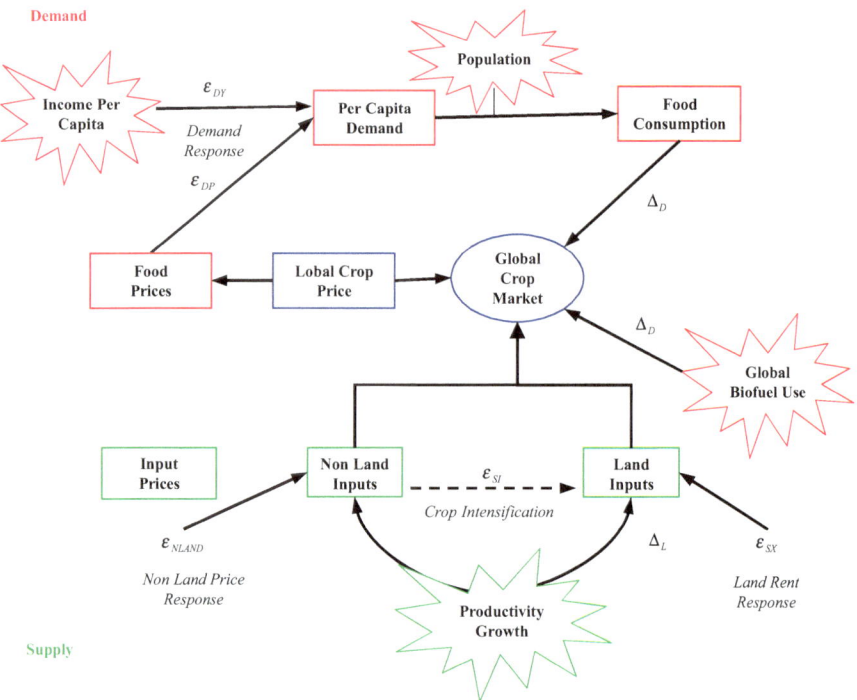

Fig. 2.1 The unified global economy. *Red outlines* denote demand factors while *green outlines* denote the supply side of the economy. Epsilon denotes elasticities while delta denotes shocks, or perturbations to the system

unaltered, the world would need 10% more land—of equivalent average quality—to meet the additional demand. However, historical data show that global production increased nearly fivefold over the five decades from 1961 to 2011, while the extent of global cropland increased only modestly (Fuglie et al. 2019). A large part of this difference can be explained by agricultural technologies, which experienced tremendous advancements over the same period. Among other things, many parts of the world have experienced the Green Revolution: Crop yields increased considerably through the introduction of new seed varieties and more intensive use of fertilizers and irrigation (Evenson and Gollin 2003). The intensity with which existing lands were cropped also increased significantly through double and even triple cropping (Bruinsma 2009). The modest net increase in cropland extent over this period is therefore an indication that, in the global footrace between growing demand and rising productivity, demand appears to have dominated. Although advances in technology have nearly kept pace with growing demand from an ever-richer global population.

The preceding narrative has abstracted from any economic factors. For example, we have not considered the response of consumers and producers to changes in the scarcity of cropland or food. As will be shown, these economic responses can be important in shaping the long-run evolution of land use. To see how these economic forces interact with biophysical factors (e.g., cropland extent, population growth, crop yield), consider Fig. 2.1, which introduces economic considerations (parameters shown in gray-shaded boxes and denoted with the epsilon symbol) into this discussion. On the demand side, it has been well documented that people tend to initially consume more food as their incomes rise from low levels. The parameter $\varepsilon_{DY}(y)$ captures this effect and is known to economists as the income elasticity of demand for food. In addition, as incomes rise, consumers also tend to shift their consumption patterns toward more livestock-based products (Pingali 2007), with important consequences for land use. Livestock—particularly cattle—are relatively poor converters of land into food (Eshel et al. 2014). In addition to responding to changing incomes, consumer demand is also price responsive. The absolute value of price responsiveness—also called the price elasticity of food demand, $\varepsilon_{DP}(y)$—tends to diminish as consumers become wealthier (Muhammad et al. 2011). This is reflected in the fact that the price elasticity is itself a function of per capita income. High-income consumers can largely ignore rising prices of a commodity (e.g., rice), which accounts for just a small share of their total budget, whereas low-income consumers must adjust by either consuming less food or switching to food sources that are cheaper per calorie provided.

Economic factors also play a role in the supply side of the global food economy, as illustrated in the lower half of Fig. 2.1. As land becomes increasingly scarce in the face of growing demand, land prices (and farmland rental prices) rise. During boom periods, these land rental responses can be very large (Henderson et al. 2011), and they send signals to farmers to conserve land by intensifying production. This response is captured by ε_{SI}, referred to as the price elasticity of supply at the intensive margin (i.e., the price responsiveness of crop yields). In the long run, research and development and the adoption of agricultural technologies have also been shown to respond to relative prices (Ruttan 1977), with the effect of further increasing the value of ε_{SI}.

The potential for intensification of crop production depends on a variety of agronomic and economic factors and on the current level of intensification. It is expected that a favorable development in prices (e.g., higher output prices or lower input prices) could induce a significant supply response at this intensive margin in parts of the world with very low levels of commercial inputs. This has been found to be the case in Malawi, where significant fertilizer subsidies were introduced to increase maize production (Ricker-Gilbert et al. 2011). However, the magnitude of this response has been vigorously debated and clearly depends on a variety of other local factors, including the extent of soil degradation (Messina et al. 2017). This raises broader questions about the extent to which biophysical and socioeconomic constraints might alter producers' ability to respond to higher prices at the intensive margin.

Location is clearly a critical factor in environmental policy analysis, but many global economic models of land use change do not differentiate yield response by location. The greater the yield response to scarcity, the less the need to convert natural lands to cropland in the face of shocks (e.g., the US biofuels boom). Golub and Hertel (2012) show that when yield responses in non-US regions are allowed to depart from US-based estimates in the widely used GTAP-BIO model, dramatically different estimates of global land use and terrestrial carbon emissions as a result of the US biofuels mandates are obtained. They conclude that more geographically specific estimates of the endogenous intensification parameter, ε_{SI}, are required to accurately estimate cropland expansion. Havlik et al. (2013) have sought to introduce regional heterogeneity into their global model (GLOBIOM) by explicitly modeling the biophysical and economic systems at the grid cell level and then aggregating these systems at the national and global levels. Their linear programming approach is an important starting point; however, there is little doubt that much more research is required to accurately estimate location-specific intensive margins of supply response across the globe.

The final piece of this economic puzzle is the responsiveness of cropland supply to increased returns in farming. This parameter, ε_{SX}, called the price elasticity of supply at the extensive margin, is shown at the bottom of Fig. 2.1. In this simplified, long-run, partial equilibrium framework, we assume that nonland inputs are in "perfectly elastic" supply. This means that the nonland input prices are unaffected by developments in the farm sector and are instead determined by developments in the rest of the economy. We will relax this assumption in Chap. 3, showing that the responsiveness of the price of other inputs, such as labor, to developments in the farm sector can be very important for sustainability and equity outcomes.

The economic elements introduced above can be combined to allow for an analytical, partial equilibrium solution to this unified agricultural model. The resulting analytical solution (Eq. 2.1) expresses the long-run change in cropland, q_L^*, as being dependent on the exogenous shocks to demand, Δ_D; exogenous trends in yields due to improvements in agricultural technology, Δ_L, and the three economic margins (elasticities) of response to price:

$$q_L^* = (\Delta_D - \Delta_L)/(1 + \varepsilon_{SI}/\varepsilon_{SX} + \varepsilon_{DP}/\varepsilon_{SX}) \qquad (2.1)$$

Equation (2.1) reveals several important points about long-run global cropland change. First, since all price elasticities are defined as positive, their presence in the denominator of Eq. (2.1) serves as a shock absorber to global land use changes in response to changes in net demand for cropland ($\Delta_D - \Delta_L$). If the world finds itself in a period in which demand for cropland is outstripping yield-improving innovations, then prices will rise, households will curb consumption or shift to less land-intensive diets (e.g., less beef), and farmers will intensify production. All these responses will lead to a moderation in the amount of additional land actually brought under cultivation.

A second point to note from Eq. (2.1) is that what matters for land conversion is not the absolute size of the price elasticities of demand and supply but rather the relative size of the demand and intensive (yield response) supply elasticities compared to the extensive (area response) supply response. There are many examples of biophysical models of global cropland change that do not incorporate the demand

and intensive supply margins into their long-run projections. Historically, the GCAM (Wise and Calvin 2011), IMAGE (PBL Netherlands Environmental Assessment Agency 2015), and PIK models (Lotze-Campen et al. 2008) all fell into this category, although more recent versions have sought to remedy this limitation in various ways. In Eq. (2.1), ignoring these demand and supply intensive margins is equivalent to assuming that $\varepsilon_{SI} = \varepsilon_{DP} = 0$. In this case, when net demand rises by 10%, (productivity-adjusted) global cropland must also rise by 10%.

Baldos and Hertel (2013) explore the implications of ignoring these economic margins of response in the context of the 1961–2006 period for the global food economy (1961 is when the FAO data series begins, and 2006 is the year before a major global food price crisis). They first introduce the SIMPLE model (a Simplified International Model of Prices, Land use, and the Environment) and validate it against global data over this period. SIMPLE is a numerical implementation of the framework shown in Fig. 2.1 (see Part III for more details). Supply and demand in SIMPLE are developed at the level of geographic regions, and growth in population, income, and productivity are specified exogenously. The authors find that, when they run the model over this historical period with the demand and intensive supply margins eliminated (i.e., $\varepsilon_{SI} = \varepsilon_{DP} = 0$), the model overpredicts historical land use change nearly threefold. This illustrates the point, already evident from Eq. (2.1), that purely biophysical models will overstate cropland changes in response to exogenous shocks. This follows directly from the missing adjustments in consumer demand and producer yields in response to higher prices. It also helps to explain why some of the most prominent Integrated Assessment Models predict considerable expansion in cropland over the twenty-first century (Schmitz et al. 2014) despite slowing population growth (UN) and robust growth in agricultural productivity (Fuglie 2010). In short, land use change scientists ignore economic responses to scarcity at their peril when undertaking long-run projections of land use change.

One of the important conservation policies that we will consider in this book involves the withdrawal of land and/or water from agriculture, in favor of environmental uses. In our stylized framework, this introduces a backward shift in cropland supply, represented by Δ_S. This results in a modified version of Eq. (2.1):

$$q_L^* = \{(\Delta_D + \Delta_S - \Delta_L)/(1 + \varepsilon_{SI}/\varepsilon_{SX} + \varepsilon_{DP}/\varepsilon_{SX})\} - \Delta_S \qquad (2.2)$$

In the absence of any demand or intensive margin response (i.e., $\varepsilon_{SI} = \varepsilon_{DP} = 0$), this new term cancels out and does not influence the total land use outcome. In other words, if land is withdrawn in one location, it must expand somewhere else in order to satisfy a fixed level of global food demand. However, if the economic margins of demand and supply are price responsive, then this conservation policy will have a more nuanced impact, making land scarcer and raising prices. For example, if all three price elasticities were equal in absolute magnitude, then only one-third of the land set-aside would be replaced by expansion elsewhere. In this context, the conservation policy is more effective, as it now has a chance to operate on the intensive margin of supply as well as on the demand margin. In the—admittedly naïve—case of equal elasticities, two-thirds of the land conservation target would be effectively achieved—in equal parts—by reductions in demand and expansion in yields. Clearly, the success of environmental policies aiming to remove sensitive natural resources from agricultural production depends on economic responses to the conservation measures.

2   Two-Region Model: International Trade Creates Opportunities for Land Use Spillovers to Other Regions

Thus far in our model of global land use change, we have abstracted from international trade by assuming a unified global economy in which any shock (e.g., an improvement in technology or a conservation policy) applies worldwide. However, most of the empirical literature on land use spillovers and leakage focuses on cases in which producers in one region are "treated" with technological change or conservation set-asides and the remainder of the global economy is "untreated" (Angelsen and Kaimowitz 2001; Gasparri et al. 2016; le Polain de Waroux et al. 2019). Here, economic theory can again provide useful insights by shedding light on the conditions under which improvements in agricultural technology will lead to land expansion in the treated region.

In this analysis, we start with the simplest case—again, admittedly naïve—in which world markets for agricultural products are fully integrated. This specification—which might be termed trade economist nirvana—reflects a global economy in which the farm commodity can be purchased for the same price regardless of country or region. Therefore, we are abstracting from all trade frictions, such as transport costs, tariffs, and other government policies aimed at insulating domestic producers and/or consumers from developments in world markets. The three-panel diagram in Fig. 2.2 depicts this situation. The left-hand panel portrays an increase in agricultural commodity supply, for any given price, in innovating Region A, via an outward shift in the supply curve. The supply curve in the

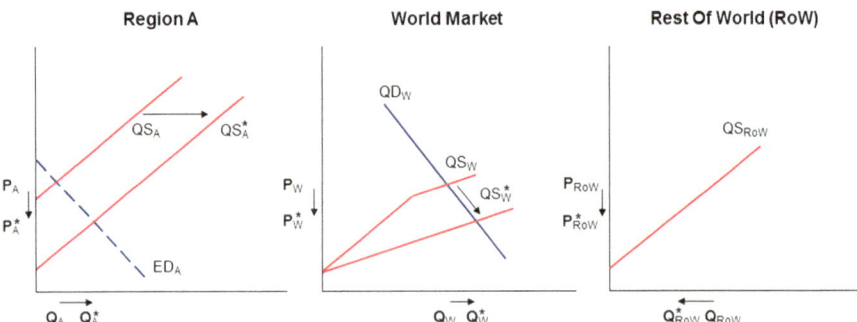

Fig. 2.2 Impact of a technological improvement in Region A on the world market and on the rest of the world

non-innovating rest of the world (RoW) remains unchanged, as shown in the righthand panel. The middle panel portrays global market equilibrium under integrated markets, where Region A's supply is added to that from the RoW to obtain a global supply curve. The intersection of global supply and demand determines world price, which naturally falls in response to the increased product supply coming from Region A.

The productivity gain shifts the supply curve in Region A—and hence the world—to the right, thereby depressing the world price and, as a consequence, production in other regions.

In this framework, the sufficient statistic for determining the direction of land use change in a region in response to an improvement in agricultural technology is the absolute value of the price elasticity of excess demand facing producers in the technology-advancing region (Hertel et al. 2014). This elasticity is denoted ε_{DP}^T, where superscript T denotes its association with the technically innovating region. It is governed by the slope of the demand curve facing producers in the innovating region (dashed line, labeled EDA in Fig. 2.2). The expression for ε_{DP}^T is given by Eq. (2.3), which shows that the excess demand elasticity depends not only on global demand conditions (ε_{DP}^W) but also on the responsiveness of RoW producers $\left(\varepsilon_{SI}^{RoW} + \varepsilon_{SX}^{RoW}\right)$ to price changes emanating from the innovating region.

$$\varepsilon_{DP}^T = \left[\varepsilon_{DP}^W + (1-\alpha)\left(\varepsilon_{SI}^{RoW} + \varepsilon_{SX}^{RoW}\right)\right]/\alpha \qquad (2.3)$$

When ε_{DP}^T is large, producers in Region A can expand production without significantly affecting the world price for the crop in question. When ε_{DP}^T is small, the expansion drives prices down and limits the incentive for further expansion.

The share of global supply provided by the innovating region, denoted α, also plays a critical role. The smaller this share is, the larger the excess demand elasticity will be. As shown in Hertel et al. (2014), the critical value for determining the direction of land use change in the innovating region is $\varepsilon_{DP}^T = 1$. When $\varepsilon_{DP}^T > 1$, improvements in productivity will lead to cropland expansion in the innovating region. When $\varepsilon_{DP}^T < 1$, the price-depressing effect of output expansion will curb production (i.e., the efficiency gains of the new technology will outweigh the effect of increased output) and land use in the innovating region will contract.

This analytical framework is extremely useful for sorting out the literature that has emerged from the Borlaug–Jevons debate on the land use impacts of agricultural technology change (Hertel et al. 2014). First, it is clear from Eq. (2.3) that, contrary to assertions in much of the early literature on this topic (Angelsen and Kaimowitz 2001), the question cannot be answered simply by observing global demand conditions for the innovated product; the supply response in the rest of the world is also critically important. Second, given the important role of α in Eq. (2.3), if an innovating region has only a small share of the global market, it is more likely that $\varepsilon_{DP}^T > 1$ and land expansion will occur in the innovating region. This highlights the importance of being a small producer in the world market. Under these conditions, land use expansion in the innovating region becomes more likely. Villoria (2019) provides an empirical analysis, of the impact of national productivity improvements on domestic farmland expansion. He finds that technological improvements in most countries in the world lead to national land use expansion in the innovating region. However, in a few large and/or relatively closed economies (e.g., China and India), he finds that domestic productivity improvements have led to land conservation in that country.

To this point, we have not yet addressed the fundamental question of global land use, which was at the heart of Borlaug's (2007) assertion that the Green Revolution had spared land worldwide. Determining the conditions under which improved technology in one region will spare land globally is more complex than determining land use change in the innovating region alone. Economic theory dictates that with a lower world price, output and land use will fall in the non-innovating regions in the wake of the new technology, assuming that this is the only change occurring in the global economy. However, the decline in land use overseas could be offset by a rise in land use in the innovating region. Ultimately, Borlaug's hypothesis requires testing in an empirical context. However, if we assume that the supply elasticities are equal in both regions, we can derive Eq. (2.4) to highlight the critical role of relative yields in this debate. This equation states that Borlaug's hypothesis will be overturned when the price elasticity of world demand for the crop in question exceeds a weighted combination of relative yields in the innovating region versus globally (Y^A/Y^W) and the globally uniform total supply response to price (ε_S^W) under the condition of equal supply elasticities:

$$\varepsilon_{DP}^W > \left(Y^A/Y^W\right)\left(\varepsilon_S^W + 1\right) - \varepsilon_S^W \Rightarrow \text{rejectBorlaug} \qquad (2.4)$$

It is easy to see from Eq. (2.4) that the terms involving the supply elasticity would cancel if yields were the same between the treated and untreated regions. In this case, the critical value required for Borlaug's assertion to hold is that the global price elasticity of demand for the crop in question be less than 1. Since this is true for most staple foods in most regions (Muhammad et al. 2011), given the Green Revolution's emphasis on staple grains, Borlaug would seem to be strongly supported by the condition in Eq. (2.4).

However, what if yields in the two regions were unequal? In particular, what if yields in the innovating region are far lower than the world average (i.e., $Y^A/Y^W \ll 1$)? This opens the possibility of Jevons' paradox applying at global scale. Hertel et al. (2014) explore this possibility in greater detail using the SIMPLE model developed in Chap. 4 of this book and relax the restrictive assumptions about equal supply responses in the two regions. They find that, while the historical Green Revolution did indeed spare land globally, a prospective African Green Revolution might not have the same benefit due to very low relative crop yields in in sub-Saharan Africa. This is particularly likely if global markets were fully integrated, such that each commodity had a uniform price worldwide.

3   Segmented Markets: Product Differentiation Blunts the Cross-Border Transmission of Price Signals

The (increasingly challenged) efforts of the World Trade Organization (WTO) notwithstanding, the world is a long way from the stylized model of perfectly integrated global markets postulated in the foregoing analysis. This is particularly true for agricultural products for which government interventions remain pervasive. Indeed, agriculture is one of the primary reasons why the Doha Round of WTO trade negotiations was inconclusive (WTO 2018). In some regions—most notably parts of sub-Saharan Africa (SSA)—there is the added challenge of physical access to markets. Producers and consumers are isolated, not only from international markets but also from their own national markets (Porteous 2015). We adopt the term *market segmentation* to refer to a situation in which markets are not fully connected and international prices are imperfectly transmitted into national and local markets, due either to government policies or to poor infrastructure and weak logistics. This has the effect of reducing the excess demand elasticity in Eq. (2.3), thereby increasing the likelihood that an improvement in locally employed agricultural technology will reduce the extent of local cropland.

Regardless of the extent of market integration, we already know that if the technological innovation in the innovating region is the only perturbation to the system, then cropland in the rest of the world must fall (or at least not rise, relative to baseline) due to the expected decline in world prices. Therefore, a simultaneous decline in land use in the innovating region is a sufficient condition to satisfy the Borlaug hypothesis. By reducing the excess demand elasticity facing producers in the innovating region, market segmentation reinforces the Borlaug intuition. Hertel et al. (2014) confirm this point empirically by examining the impact of a prospective African Green Revolution in the presence of historically segmented agricultural markets. They conclude that, under this historical trade regime, such a future Green Revolution would indeed be land sparing, in contrast to the ambiguous outcome when markets are perfectly integrated. In summary, the extent of market integration is a key determinant of the sustainability of agricultural innovations as well as many other interventions in the global food system.

4   Summary

The one-region model of the global food system provides a simplified framework to understand the economic relationship between demand and supply factors and their impacts on land use. While the model abstracts from the complexities of the real world, it highlights the importance of technological change in increasing agricultural productivity and the role of economic responses to scarcity in shaping land use patterns. The model also suggests that intensification of crop production can be a viable strategy to meet growing demand, but its potential is constrained by a range of agronomic, environmental, and economic considerations. Understanding these factors is crucial for designing effective policies that promote sustainable land use and food security in the face of global challenges such as population growth, climate change, and resource scarcity. The two-region conceptual framework highlights the interplay between market developments in one part of the world and the consequences for outcomes elsewhere, as well as global environmental impacts. This framework highlights the role of market integration in determining the sustainability of agricultural innovations in the global food system.

References

Angelsen, A., and D. Kaimowitz, eds. 2001. *Agricultural technologies and tropical deforestation*. Wallingford: CAB International. https://doi.org/10.1079/9780851994512.0000.

Baldos, Uris Lantz C., and Thomas W. Hertel. 2013. Looking back to move forward on model validation: Insights from a global model of agricultural land use. *Environmental Research Letters* 8: 034024. https://doi.org/10.1088/1748-9326/8/3/034024.

Borlaug, Norman. 2007. Feeding a hungry world. *Science* 318: 359–359. https://doi.org/10.1126/science.1151062.

Bruinsma, Jelle. 2009. The Resource Outlook to 2050: By how much do land, water and crop yields need to increase by 2050? In *FAO Expert Meeting on How to Feed the World in 2050*. Rome: Food and Agriculture Organization of the United Nations Economic and Social Development Department.

Dietrich, Jan Philipp, Christoph Schmitz, Hermann Lotze-Campen, Alexander Popp, and Christoph Müller. 2014. Forecasting technological change in agriculture—An endogenous implementation in a global land use model. *Technological Forecasting and Social Change* 81: 236–249. https://doi.org/10.1016/j.techfore.2013.02.003.

Eshel, Gidon, Alon Shepon, Tamar Makov, and Ron Milo. 2014. Land, irrigation water, greenhouse gas, and reactive nitrogen burdens of meat, eggs, and dairy production in the United States. *Proceedings of the National Academy of Sciences* 111: 11996–12001. https://doi.org/10.1073/pnas.1402183111.

Evenson, R.E., and D. Gollin. 2003. Assessing the impact of the green revolution, 1960 to 2000. *Science* 300: 758–762. https://doi.org/10.1126/science.1078710.

Fuglie, Keith. 2010. Total factor productivity in the global agricultural economy: Evidence from FAO data. In *The shifting patterns of agricultural production and productivity worldwide*, 63–95. Ames: The Midwest Agribusiness Trade Research and Information Center, Iowa State University.

Fuglie, Keith, Madhur Gautam, Aparajita Goyal, and William F. Maloney. 2019. *Harvesting prosperity: Technology and productivity growth in agriculture*. Washington, DC: World Bank. https://doi.org/10.1596/978-1-4648-1393-1.

Gasparri, Nestor Ignacio, Tobias Kuemmerle, Patrick Meyfroidt, Yann le Polain, and de Waroux, and Holger Kreft. 2016. The emerging soybean production frontier in southern Africa: Conservation challenges and the role of south-south telecouplings. *Conservation Letters* 9: 21–31. https://doi.org/10.1111/conl.12173.

Golub, Alla A., and Thomas W. Hertel. 2012. Modeling land-use change impacts of biofuels in the GTAP-BIO framework. *Climate Change Economics* 3: 1250015. https://doi.org/10.1142/S2010007812500157.

Havlík, Petr, Hugo Valin, Aline Mosnier, Michael Obersteiner, Justin S. Baker, Mario Herrero, Mariana C. Rufino, and Erwin Schmid. 2013. Crop productivity and the global livestock sector:

Implications for land use change and greenhouse gas emissions. *American Journal of Agricultural Economics* 95: 442–448. https://doi.org/10.1093/ajae/aas085.

Henderson, Jason, Brent A. Gloy, and Michael D. Boehlje. 2011. Agriculture's boom-bust cycles: Is this time different? *Economic Review* 96: 81–103.

Hertel, Thomas W. 2018. Economic perspectives on land use change and leakage. *Environmental Research Letters* 13: 075012. https://doi.org/10.1088/1748-9326/aad2a4.

Hertel, Thomas W., Navin Ramankutty, Lantz C. Uris, and Baldos. 2014. Global market integration increases likelihood that a future African Green Revolution could increase crop land use and CO2 emissions. *Proceedings of the National Academy of Sciences* 111: 13799–13804. https://doi.org/10.1073/pnas.1403543111.

le Polain de Waroux, Yann, Rachael D. Garrett, Jordan Graesser, Christoph Nolte, Christopher White, and Eric F. Lambi. 2019. The restructuring of South American soy and beef production and trade under changing environmental regulations. *World Development* 121: 188–202. https://doi.org/10.1016/j.worlddev.2017.05.034.

Lotze-Campen, Hermann, Christoph Müller, Alberte Bondeau, Stefanie Rost, Alexander Popp, and Wolfgang Lucht. 2008. Global food demand, productivity growth, and the scarcity of land and water resources: A spatially explicit mathematical programming approach. *Agricultural Economics* 39: 325–338. https://doi.org/10.1111/j.1574-0862.2008.00336.x.

Messina, Joseph P., Brad G. Peter, and Sieglinde S. Snapp. 2017. Re-evaluating the Malawian Farm Input Subsidy Programme. *Nature Plants* 3: 17013. https://doi.org/10.1038/nplants.2017.13.

Muhammad, Andrew, James L. Seale Jr., Birgit Meade, and Anita Regmi. 2011. *International evidence on food consumption patterns: An update using 2005 international comparison program data*, Technical Bulletin 1929. Washington, DC: US Department of Agriculture, Economic Research Service.

PBL Netherlands Environmental Assessment Agency. 2015. Welcome to IMAGE 3.0 Documentation. https://web.archive.org/web/20151108083904/http://themasites.pbl.nl/models/image/index.php/Welcome_to_IMAGE_3.0_Documentation.

Pingali, Prabhu. 2007. Westernization of Asian diets and the transformation of food systems: Implications for research and policy. *Food Policy* 32: 281–298. https://doi.org/10.1016/j.foodpol.2006.08.001.

Porteous, Obie C. 2015. High trade costs and their consequences: An estimated model of African agricultural storage and trade. In *Joint Annual Meeting*. San Francisco: Agricultural and Applied Economics Association & Western Agricultural Economics Association. https://doi.org/10.22004/ag.econ.205776.

Ricker-Gilbert, Jacob, Thomas S. Jayne, and Ephraim Chirwa. 2011. Subsidies and crowding out: A double-hurdle model of fertilizer demand in Malawi. *American Journal of Agricultural Economics* 93: 26–42. https://doi.org/10.1093/ajae/aaq122.

Ruttan, Vernon W. 1977. Induced innovation and agricultural development. *Food Policy* 2: 196–216. https://doi.org/10.1016/0306-9192(77)90080-X.

Schmitz, Christoph, Hans Van Meijl, Page Kyle, Gerald C. Nelson, Shinichiro Fujimori, Angelo Gurgel, Petr Havlik, et al. 2014. Land-use change trajectories up to 2050: Insights from a global agro-economic model comparison. *Agricultural Economics* 45: 69–84. https://doi.org/10.1111/agec.12090.

Villoria, Nelson. 2019. Consequences of agricultural total factor productivity growth for the sustainability of global farming: Accounting for direct and indirect land use effects. *Environmental Research Letters* 14: 125002. https://doi.org/10.1088/1748-9326/ab4f57.

Wise, M., and K. Calvin. 2011. *GCAM 3.0 agriculture and land use: Technical description of modeling approach*, PNNL-20971. Richland: Pacific Northwest National Laboratory.

World Trade Organization. 2018. The Doha Round. https://www.wto.org/english/tratop_e/dda_e/dda_e.htm. Accessed 1 Feb 2024.

Chapter 3
Grid-Level Analysis Using SIMPLE-G

Srabashi Ray and Thomas W. Hertel

While it is instructive to engage in aggregate market analysis of the sort discussed above, in the end, most sustainability issues are local in nature and depend on local biophysical conditions (e.g., climate, soils, hydrology, and land cover) and socio-economic circumstances (e.g., local governance, household characteristics, and choice of technology). To incorporate these sources of heterogeneity into a global sustainability analysis, a finer spatial resolution is necessary. Typically, physical scientists characterize fine scales using grid cells, which are spatial units of equal length and width, with the area varying according to the latitude of the grid cell (larger at the equator and smaller moving toward the poles). The resolution of grid cells in SIMPLE-G and the construction of the underlying data are the subject of Part III of this book. At this point, it is sufficient to note that the grid cells in SIMPLE-G are small enough to allow the incorporation of available heterogeneity in data and parameters and small enough to not influence commodity prices, so they can be treated exogenously. In the SIMPLE-G-US model, for example, approximately 75,000 grid cells are involved in US crop production.

Taking global and national drivers of market outcomes as exogenously given allows us to focus on relationships within the grid cell. This approach greatly simplifies our gridded analysis and allows us to reach some important conclusions about the likely impact of a global change in commodity prices on local land and

This chapter draws on two previously published papers: Ray, Srabashi, Iman Haqiqi, Alexandra E. Hill, J. Edward Taylor, and Thomas W. Hertel. 2023. Labor markets: A critical link between global-local shocks and their impact on agriculture. *Environmental Research Letters* 18: 035007. https://doi.org/10.1088/1748-9326/acb1c9; Ray, Srabashi, and Thomas W. Hertel. 2023. Labor market rigidities mediate the effectiveness and distributional impacts of conservation policies. Working Paper 4427548. SSRN. https://doi.org/10.2139/ssrn.4427548

S. Ray (✉) · T. W. Hertel
Center for Global Trade Analysis, Department of Agricultural Economics, Purdue University, West Lafayette, IN, USA
e-mail: ray152@purdue.edu; hertel@purdue.edu

water use as well as labor demand. Of course, since every grid cell will likely respond to any global changes and may compete for common inputs (e.g., labor and fertilizer), we cannot understand the full impact of these market developments without considering all grid cells simultaneously. Since the combined responses of all the grid cells influence global supply and hence world prices, this model must ultimately be solved as a multiscale system. This will be discussed at length in Part III.

The specific design of the crop production functions in SIMPLE-G is also discussed at length in Part III. In general, design will vary by application. *SIMPLE-G is a framework*, not a single model. In this section, we illustrate the framework with a two-input specification. In the two-input case, we continue to group the natural resource-related inputs (e.g., land) into one aggregate, which will be termed "resources" and indexed with R. Resources include all inputs required for agricultural production that are sourced from nature, including cropland, groundwater, and surface water. In practice, we also include irrigation equipment in the resources category since it refers to investments needed to augment natural inputs and make them suitable for crop production. In the two-input case, the second aggregate of production factors encompasses all remaining inputs, including human labor and manufactured inputs (e.g., seeds, chemicals and fertilizers, and capital), and is indexed with H. As we will see, the economic distinction between these two input categories rests upon the underlying factor supply elasticities (i.e., how responsive these input supplies are to price changes). Given the geographic immobility of resources, we generally expect the factor supply elasticity for resource inputs to be lower than that for human and manufactured inputs, which we expect to be mobile across geographic locations.

Box 3.1 outlines the basic two-input, gridded model (Ray and Hertel 2023). The model uses a two-input CES production function, which allows for factor-neutral technical changes.[1] Since we are interested in studying changes in input use, input prices, and crop production as a result of an exogenous shock, Equations. 3.1–3.9 in Box 3.1 are presented in terms of percentage change in the corresponding variable. Equations 3.1–3.3 form the crux of the analytical framework for the two-input model. Equation 3.1, i.e., the conditional demand for either input, is derived from the usual producers' cost minimization problem (Gohin and Hertel 2003). Equation 3.2 is the constant price elasticity input supply equation for either inputs. It also includes a policy lever. For example, in case of resources, a positive value for the policy lever implies a backward shift in resource supply that can be implemented to model conservation policies that restrict the availability of resources. Equation 3.3 represents the zero-profit equation that enforces the free entry and exit assumption in

---

[1] The factor-neutral technical change parameter is useful in studying the impacts of improvements in total factor productivity on input use, input price, and total production. This is studied in depth in Chaps. 10 and 11.

the agricultural sector that ensures a competitive market structure, where any changes in crop price are attributed to both inputs in inverse proportion of their cost share.

Box 3.1: A Partial Equilibrium Two-Input Theoretical Model for Agricultural Production: Assessing the Impact of Conservation Policies

Agricultural production (Q) is a constant elasticity of substitution (CES) function of two inputs, natural resources (Q_R) and manufactured or human inputs (Q_H), augmented by any changes in total factor productivity (a), in a relevant spatial unit.

Percentage change in conditional demand of input j:

$$q_j^D = (q - a) - \sigma\{p_j - (p + a)\} \text{ for } j = R \text{ or } H \tag{3.1}$$

Percentage change in supply of input j with supply shifters:

$$q_j^S = v_j p_j - \phi_j \tag{3.2}$$

Zero-profit condition in terms of percentage change:

$$p + a = \sum_{j=R,H} \theta_j p_j, \text{ where } \sum_{j=R,H} \theta_j = 1 \tag{3.3}$$

Percentage change in optimized agricultural supply:

$$q^S \quad a - (p \mid a)c \quad \Gamma_R \phi_R, \tag{3.4}$$

where $\quad \varepsilon = \left[\dfrac{1}{\frac{\theta_R}{v_R + \sigma} + \frac{\theta_H}{v_H + \sigma}} - \sigma\right], \qquad \Gamma_R = \dfrac{\theta_R(v_H + \sigma)}{\text{denom}} \quad$ and

$\text{denom} = \sum_{j,k} \theta_j(v_k + \sigma), j, k = R, H.$

Using Eqs. 3.1–3.4, the equilibrium changes in input use and output for a given $p = p^*$ follow:

Percentage change in input H use (Q_H):

$$q_H = v_R \frac{(p^* + a)}{\theta_R} \Gamma_H + \sigma \frac{(p^* + a)}{\theta_R} \Gamma_H - \phi_R \Gamma_H. \tag{3.5}$$

Percentage change in input R use (Q_R):

$$q_R = (p^* + a) \frac{v_R}{\theta_R} \Gamma_R - \phi_R \Gamma_\phi. \tag{3.6}$$

(continued)

Box 3.1 (continued)

Percentage change in price of input H (P_H):

$$p_H = \frac{p^* + a}{\theta_H}(1 - \Gamma_R) - \frac{\theta_R \phi_R}{\text{denom}}. \tag{3.7}$$

Percentage change in price of input R (P_R):

$$p_R = \frac{p^* + a}{\theta_R}\Gamma_R + \frac{\theta_H \phi_R}{\text{denom}}. \tag{3.8}$$

Percentage change in production (Q):

$$q = a + (p^* + a)\frac{v_R}{\theta_R}\Gamma_R + (p^* + a)\frac{\sigma\theta_H}{\theta_R}\Gamma_\sigma - \phi_R\Gamma_R, \tag{3.9}$$

where $\Gamma_H = \frac{\theta_R v_H}{\text{denom}}$, $\Gamma_\phi = \frac{\theta_R(v_H + \sigma) + \theta_H \sigma}{\text{denom}}$, $\Gamma_\sigma = \frac{\theta_R(v_H - v_R)}{\text{denom}}$ for $v_R < v_H$, $0 < \Gamma_\sigma < \Gamma_H < \Gamma_R < 1$.

Notation: All price and quantity variables are denoted with lowercase representing percentage changes in the underlying indices.
$\sigma \geq 0$: Elasticity of substitution between two inputs.
$v_j \geq 0$, $\theta_j \geq 0$: Elasticity of supply to agriculture and cost share of input j.
$\phi_H = 0$, $\phi_R > 0$: Backward supply shifter for input R due to, for example, a conservation policy
Source: Ray and Hertel (2023).

Setting input demand equal to supply to clear the input markets (i.e., Eq. 3.1 = Eq. 3.2) and using Eq. 3.3, we arrive at Eq. 3.4, which is the change in optimized output supply (Ray and Hertel 2023). Changes in output supply could be due to changes in crop price, depending on the output supply elasticity (ε) and any policy shock that restricts the availability of resources. The supply elasticity is itself a function of the input supply response, cost shares, and input substitution parameters and therefore is endogenous to the model. The degree of responsiveness of changes in production to changes in output price depends on the fundamental parameters of the input markets and production function. Finally, Eqs. 3.5–3.8 show the changes in use and price of both inputs, and Eq. 3.9 gives the change in crop production.

Despite all the simplifying assumptions (i.e., a single composite crop, exogenous crop price, only two inputs), we see that the impact of a change in crop price (p) or a conservation policy, modeled as a backward shift in the resource supply curve (ϕ_R), can nonetheless be rather complex. Fortunately, we can manipulate and interpret these expressions to generate valuable insights using more complex versions of SIMPLE-G, developed in Part III and applied in Part IV of this book. In the remainder of this section, we discuss the impacts of an exogenous crop price change and a conservation policy shock on resource and augmented human input use, input prices, and crop production within this relatively simple production structure.

1 The Case of Perfectly Elastic Human and Manufactured Input Supplies

To begin with, we make the simplifying assumption that labor and manufactured inputs are available with perfectly elastic supply to producers in the grid cell.[2] The actual supply elasticity of manufactured inputs is likely to be high in the medium run since inputs such as seeds, fertilizer, and chemicals can be transported relatively easily over long distances. While the assumption of perfectly elastic supply is clearly an oversimplification in the case of agricultural labor, this assumption is commonly used to simplify analyses. To highlight the importance of explicit modelling of agricultural labor markets, we begin with this assumption and then contrast our findings with the case of inelastically supplied labor. The perfectly elastic supply of human inputs implies that additional workers, if required, can be hired at the current wage rate. It also implies that workers can move long distances in search of work at the going wage rate.

Taking the output price and wages as exogenous, we can solve for the grid cell output response under the perfect mobility scenario, obtaining Eq. 3.9 (see Box 3.1). Note that the scaling terms, Γ_R and Γ_σ, are all equal to 1 in the special case of perfectly elastic human and manufactured input supply. The first term on the righthand side of Eq. 3.9 reflects the impact of an increase in Total Factor Productivity (TFP) in the grid cell. Holding all else constant, a 10% increase in TFP generates a 10% increase in output. The second term on the righthand side of Eq. 3.9 refers to the extensive margin of supply response in the face of an output price change. A rise in output price encourages producers to bring more resources into crop production. The responsiveness of resource use to an output price change depends on the percentage change in land rents, $\left[\frac{p^* + a}{\theta_R}\right]$, as well as the resource supply elasticity, v_R. The fact that the crop price is "magnified" as it is translated into resource rents, $\left[\frac{1}{\theta_R} > 1\right]$, stems from the fact that wages do not change. This means that the increase in revenue is entirely capitalized into resource values. We expect both the cost share of resources and the resource supply elasticity to vary greatly by location, determined by both biophysical and socioeconomic factors. For example, in farmland areas adjacent to suburban and urban centers, we expect the cost share of resources to be high and the resource supply elasticity to be low, driven by the high

---

[2]Box 3.1 shows the analytical results when both inputs are inelastically supplied. The reader can derive the results when human inputs are perfectly elastically supplied by considering Eqs. 3.5–3.9 under the assumption that $v_H \to \infty$. When $v_H \to \infty$, each of the gamma terms tends to 1.

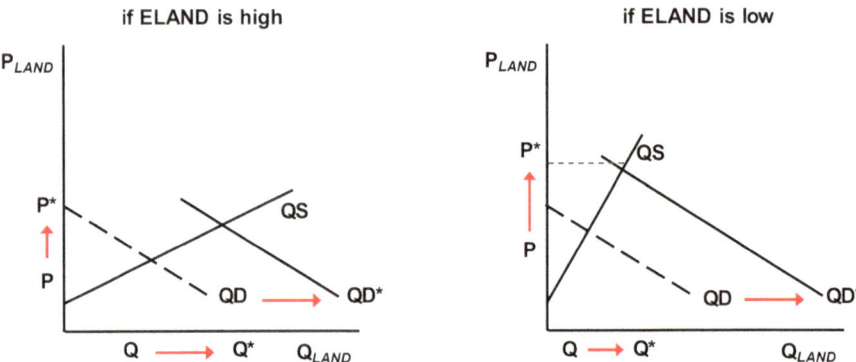

Fig. 3.1 Cropland area response under varying land supply conditions

rental value of land and relatively lesser availability of previously uncultivated land that can be brought under production in such areas. When combined, these factors result in a small extensive margin of supply in areas adjacent to urban centers. The opposite will be true in more remote locations where competition for land is more limited.

Figure 3.1 illustrates the differential responses of these two land supply situations in the face of increased crop demand. In the case of a remote location with a high land supply elasticity, an outward shift in demand translates mostly into cropland expansion, with only a minor impact on land returns. However, in regions where the land supply response is highly constrained, an outward shift in demand is primarily reflected in higher land prices.

The third term in Eq. 3.9 refers to the intensive margin of supply response to the output price change, $(p^* + a)\frac{\sigma\theta_H}{\theta_R}$. Once again, the magnification effect, $\left[\frac{1}{\theta_R} > 1\right]$, plays a role, as it translates the change in commodity prices into land prices. When human inputs are dominant, such that θ_H is large and θ_R is small, then this intensification effect will be more pronounced.

The elasticity of substitution between natural and human inputs, σ, also plays a key role in the intensive margin of supply in this model. A larger substitution elasticity indicates more potential for increasing crop yields in the face of heightened resource scarcity. Figure 3.2 illustrates the role of input substitutability in this framework in the context of a decline in human input prices facing producers in a given grid cell. When it is possible to vary the input mix (panel A), farmers' response to this price decline is to increase human inputs per hectare of land. The extent to which such substitution occurs—and hence the yield response to the price shock—depends on the curvature of the isoquant in panel A. The closer the isoquant is to a straight line, the higher the elasticity of substitution between resource and human inputs and the greater the impact on crop yields. Panel B illustrates the case where it is not possible to substitute human inputs for resource inputs; thus, yields do not change and resource and human inputs continue to be used in the same proportion.

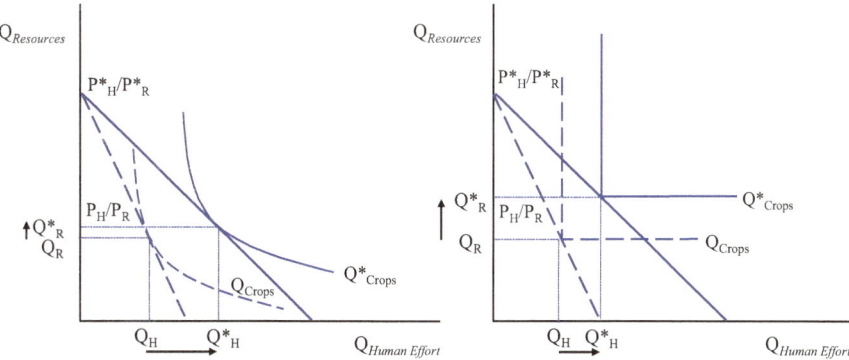

Fig. 3.2 Input substitution determines the magnitude of the intensive margin of supply

Of course, input usage still rises, as farming has become more profitable in the case of the human input price decline.

The final term in Eq. 3.9 relates to the effect of a conservation policy in which land is set aside, $\phi_R > 0$, to make room for nature and the delivery of other ecosystem services. In the absence of any impact on price, resource rents cannot change; with wages exogenously given (i.e., $\Gamma_R \rightarrow 1$), this translates into an equiproportionate decline in output. In reality, prices and resource rents will be affected, which will have consequences for potential "policy slippage," rendering the conservation policy less effective than initially expected for due to market-mediated responses by the affected producers.

The direction of change in human (Eq. 3.5) and resource (Eq. 3.6) input use depends on the total (extensive + intensive) supply response to the output price change (first term) net of the conservation policy shock (second term). If the conservation policy shock is dominant, $\left[(p^* + a) \frac{v_R + \sigma}{\theta_R} < \phi_R \right]$, then the total supply response only partially offsets job losses due to the conservation policy shock leading to a fall in employment and resource use. In contrast, if the total supply response is dominant, $\left[(p^* + a) \frac{v_R + \sigma}{\theta_R} > \phi_R \right]$, then there is an increase in employment and resource use in the grid (Ray and Hertel 2023). While wages remain constant by assumption in this first case, resource rental values increase unambiguously since scarcity in resources drives up their price.

2   The Case of Inelastically Supplied Human and Manufactured Inputs

When human inputs are also in less than perfectly elastic supply, the impacts of a commodity price increase and/or a conservation policy can be quite different from those of the preceding case. Here there is a possibility that workers might suffer a decline in wages, for example, in the context of a conservation policy. It also means that the magnification effect that boosted land rents so strongly in the context of a commodity price hike will no longer be so pronounced. In general, all the grid cell responses—extensive and intensive responses to output prices and the consequences of a conservation policy—are now more muted. This may be seen from Box 3.1 by the fact that the multiplicative scaling factors in Eqs. 3.1–3.5 (i.e., Γ_R, Γ_H, Γ_σ, and Γ_ϕ) lie between 0 and 1 as long as the resource input has the smaller supply response—an assertion that draws intuition from the fact that land is geographically immobile.

Under relatively inelastic labor market conditions, the factor supply elasticity determines the degree to which the production response to the conservation policy is dampened. This is because in a relatively tight labor market, it is more expensive to hire workers to manage the resources under production. Therefore, there is a smaller increase in resource use and employment when labor is inelastically supplied.

Further examination of the scaling terms reveals the role that labor supply elasticity plays in governing local responses to a commodity price boom (or bust). The difference in factor supply elasticities is clearly critical: If they are equal in value, then $\Gamma_\sigma = 0$, and there will be no incentive to intensify production in response to the commodity price hike. The rationale is that if both labor and the natural resource input are equally responsive to their respective factor prices, then these prices will respond symmetrically to the commodity price shock, increasing at the same rate. Since there is no incentive to substitute labor for resources, input use will also rise in lockstep with change in production.

Finally, if we consider the impact of a conservation policy on grid cell production, employment, and land use, we see that (holding output prices fixed) a policy reaches its maximum impact when labor is perfectly mobile since the conservation policy shock is premultiplied by the scaling factors. This is because the scaling factors reach their maximum value of 1 when labor is perfectly mobile. Therefore, as labor becomes less mobile, we expect that the impacts of a conservation policy, implemented as a shift in land supply, will become less effective.

3 Summary

The gridded theory provides a framework for understanding the complex relationships between market outcomes, land and water use, and labor demand at a local level. Given national drivers, the theory allows us to focus on relationships within individual grid cells. The two-input, gridded model uses a two-input production function, allowing for factor-neutral technical change. The economic distinction between the two input categories (resources and human/manufactured inputs) rests upon their availability to the agricultural sector, as captured by the underlying economic supply elasticities. These theoretical results highlight the fact that the impact of a change in crop price or a conservation policy can be complex, and a firm grasp of the underlying parameters is required to assess these impacts. This is as far as economic theory can take us. To say more about the potential impacts of a conservation policy, we need a quantitative model that can be parameterized and simulated in the face of a specific policy. Part III outlines the SIMPLE-G modeling framework as it is implemented numerically on a computer in order to assess the impacts of specific policies in specific contexts.

References

Gohin, Alex, and Thomas W. Hertel. 2003. *A note on the CES functional form and its use in the GTAP model.* West Lafayette, IN: Research Memorandum 2. Department of Agricultural Economics, Purdue University, Global Trade Analysis Project (GTAP). https://doi.org/10.21642/GTAP.RM02.

Ray, Srabashi, Iman Haqiqi, Alexandra E. Hill, J. Edward Taylor, and Thomas W. Hertel. 2023. Labor markets: A critical link between global-local shocks and their impact on agriculture. *Environmental Research Letters* 18: 035007. https://doi.org/10.1088/1748-9326/acb1c9.

Ray, Srabashi, and Thomas W. Hertel. 2023. *Labor market rigidities mediate the effectiveness and distributional impacts of conservation policies.* Working Paper 4427548. SSRN. https://doi.org/10.2139/ssrn.4427548.

Part III
Model

Chapter 4
Equilibrium Conditions and General Assumptions for a Quantitative Geospatial Economic Model

Iman Haqiqi

In this chapter, we introduce the general conditions for mathematical representation of the model theory. This theory draws heavily on the methods underpinning many computable general equilibrium models, including the widely used GTAP (Global Trade Analysis Project) model initially developed by Hertel (1997), one of the authors of this book. The innovation in SIMPLE-G comes in implementing this theory at the grid-cell level and developing a structure that can readily incorporate the biophysical information typically available at this level. The spatial resolution ranges from 30 arcmin (Liu et al. 2017) to 5 arcmin (Haqiqi et al. 2023). We start with the production side of the model, where grid-cell resolution comes into play, then move to consumer demand and trade. The latter elements are unchanged from those of the (nongridded) SIMPLE model (Baldos and Hertel 2013).

1 Introduction

SIMPLE-G is a computational model that links human and environmental systems. It connects changes in global food demand to the dynamics of local land and water resources used in agricultural production. This model is designed for "what if" (counterfactual) analyses to evaluate the impacts of environmental shocks, global socioeconomic changes, and local and global policies on different components of the system. Counterfactual analysis involves asking "what if" questions to explore the possible outcomes of different events or decisions. It is frequently used to evaluate the potential impacts of changes in government policies, technological advancements, and climate change on food production, consumption, and prices. For instance, what

I. Haqiqi (✉)
Center for Global Trade Analysis, Department of Agricultural Economics, Purdue University, West Lafayette, IN, USA
e-mail: ihaqiqi@purdue.edu

© The Author(s) 2025
I. Haqiqi, T. W. Hertel (eds.), *SIMPLE-G*,
https://doi.org/10.1007/978-3-031-68054-0_4

effect would a major grain-producing country choosing to regulate fertilizer applications have on global food prices? Or, what impact would adopting a new agricultural technology have on food production and land use? Alternatively, if a government were to provide subsidies for locally grown fruits and vegetables, how would that affect consumer behavior and diets? These are the kinds of questions being asked in the context of achieving the United Nations Sustainable Development Goals that relate to the food system and associated environmental changes (see Fig. 1.1).

Figure 4.1 illustrates the major components of the SIMPLE-G model using a partial equilibrium modeling framework that characterizes demand for, and supply of, various food commodities and associated production factors. The major elements of this framework are regional food consumption, regional production, global trade, and nutrition.

The model has three main features. First, it considers human systems' responses to environmental changes. This includes changes in decisions about food consumption, agricultural production, land use, and water withdrawals informed by new market prices. Second, it provides a framework for measuring spillover effects transmitted from one location to another location through markets. The main advantage of SIMPLE-G over traditional biophysical models is the inclusion of market-mediated effects and feedback from human systems to natural systems. Including these effects and feedback is critical for designing effective conservation and sustainability policies. By simulating international trade, SIMPLE-G accounts for telecoupled distant regions, enabling analysis of interregional dependencies at a global scale. Third, SIMPLE-G incorporates rich information from the biophysical sciences to inform spatially heterogeneous economic decisions. The ability to address the spatial distribution of activities and explicitly model the spatial heterogeneity associated with biophysical and socioeconomic systems is a significant advance over more aggregated economic models. This feature is critical when analyzing environmental changes or conservation policies that have spatially heterogeneous implications.

1.1   Socioeconomic Determinants of Food Demand

For all the SIMPLE-G specifications that follow, the demand system follows the same SIMPLE (nongridded) regional model. At the regional scale, the consumption of different commodities is a function of population, income, prices, and biofuel demand. Prices are determined endogenously as a function of supply and demand, while population and income changes are exogenous to the model, with increases in per capita income driving diet changes. Within this framework, global food and agricultural markets link changes in population, income, and diet to gridded crop production and associated stresses on land and water resources.

Global food demand is linked to local land use through the derived demand for crops. Total demand for crops in each region comes from four sources. Direct crop demand (i.e., crops consumed directly without processing, including fresh fruits and vegetables and household consumption of grains and oils) is obtained by multiplying

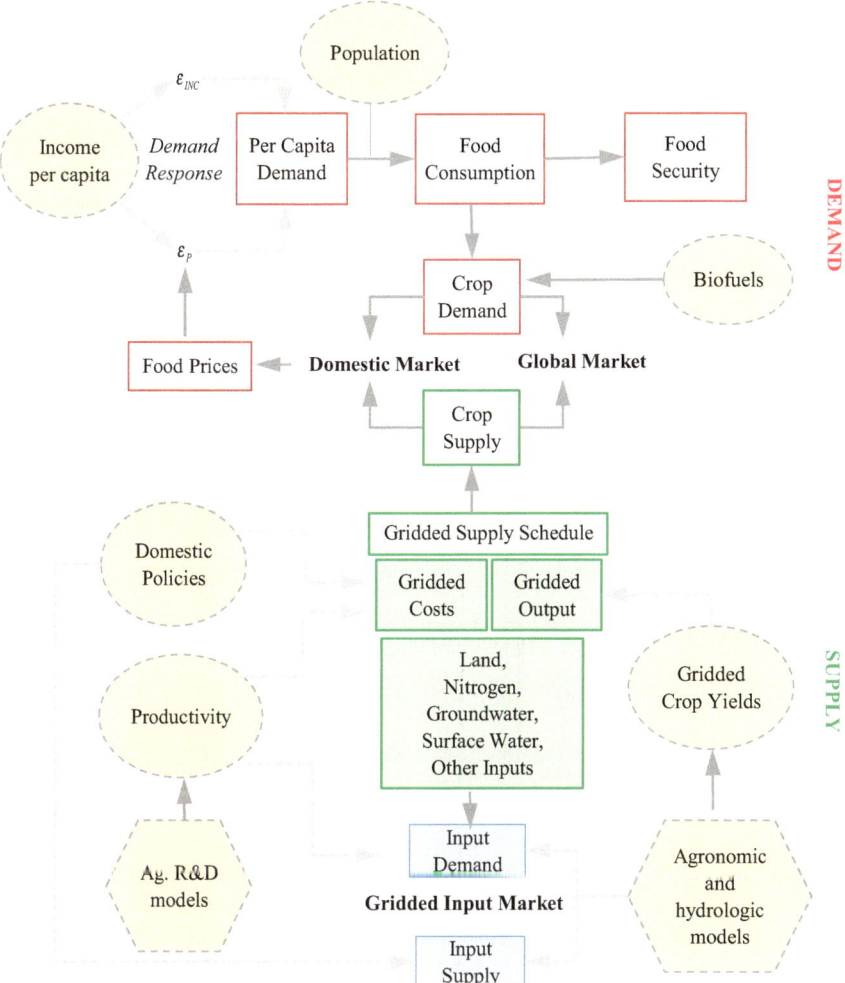

Fig. 4.1 Overview of the SIMPLE-G model
The crop production boxes are in *green*, the gridded input markets are in *blue*, and the demand side boxes are in *red*. The *dashed* shapes are information and models that provide input to SIMPLE-G. The SIMPLE-G model determines the changes in prices and quantities of agricultural inputs (i.e., land, water, fertilizer, labor, and others) and agricultural outputs considering the demand and supply forces

per capita demand for direct crop consumption by population in each region. The three additional components of total demand for crops are indirect demands by the livestock, food processing, and biofuel sectors. Demand for crops in biofuel production is assumed to be exogenously determined by government mandates, but the other demands are a function of income and prices.

The price and income responses of food demand are governed by the regional income elasticities of demand and the regional price elasticities of demand. These factors vary by type of demand (e.g., livestock versus processed foods) and by income level. Food demand generally becomes more inelastic as incomes rise (i.e., wealthier households are less inclined to alter their food consumption in response to higher prices or income changes).

International trade is handled in the same manner as it is in most computable general equilibrium (CGE) models, by differentiating between domestic and foreign goods following Armington's (1969) approach. This imperfect substitution between locally produced food commodities and those imported from global markets is governed by the Armington elasticity of substitution, which permits the model to be calibrated to observed import demand responses. Transformation of local production to exports is also imperfect, reflecting the fact that some producers cannot easily access global markets, thereby allowing the model to mimic observed export supply elasticities.

The nongridded version of SIMPLE (Baldos and Hertel 2013) has just one regional production function, and land use is not disaggregated below the regional level. However, in the SIMPLE-G framework, production is modeled at the grid-cell level; the size of agricultural production units can vary from 50 to 10,000 hectares, depending on the model's resolution. For example, one popular version of SIMPLE-G for the United States (see Chap. 12) disaggregates crop production to the level of 5 arcminutes resulting in more than 75,000 crop production units (grid cells) of about 55–77 km^2 in the continental United States. In SIMPLE-G-US, the other world regions are not disaggregated, allowing for applications focused on the United States, even as the background responses of other regions are accounted for. Of course, such a regional focus is not always desirable, and two applications in Part IV utilize a global-gridded version of SIMPLE-G (Chaps. 16 and 17).

The production functions in SIMPLE-G determine crop outputs as well as local demand for agricultural inputs (i.e., land, water, fertilizer, labor, and other inputs) for any given set of policies and climate regime (i.e., weather). In some versions of SIMPLE-G, we disaggregate rainfed and irrigated crop production, resulting in two distinct production functions in each grid cell, each competing for irrigable land, which allows us to estimate local demand for irrigation water. This approach is important when modeling the impacts of, for example, groundwater sustainability policies as well as adaptation to climate change, which has sharply different impacts on irrigated and nonirrigated crops due to differing responses to heat and water stress.

1.2   Local to Global Supply Linkages

Figure 4.2 illustrates the local-to-global supply of crops. The composition of crop production at the grid-cell level—the most granular level for understanding crop production patterns—is expected to vary by location due to agroecological

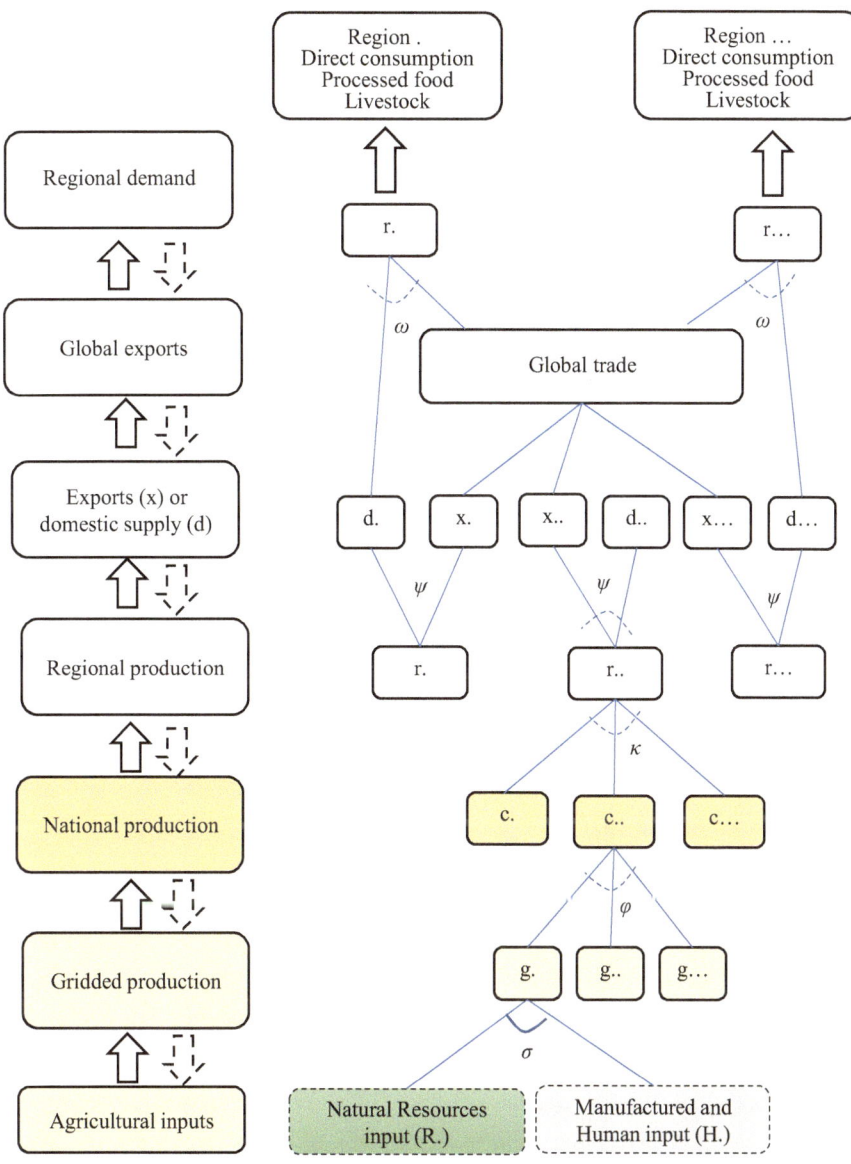

Fig. 4.2 Schematic representation of local-to-global supply linkages in SIMPLE-G

The boxes on the lefthand side describe the activity occurring in the "tree structure" of this figure. At the bottom are the inputs used in crop production at each grid cell (g). The resulting outputs are aggregated to the country level (c) via a CES (constant elasticity of substitution) function. Each region (r) contains multiple countries, and the outputs of these countries are also aggregated using a CES function. The disposition of sales between domestic and export markets is determined at the regional level via a constant elasticity of transformation (CET) function. Exports enter the global market, where supply must equal demand, with global prices adjusting to assure this equilibrium condition. Domestic sales are available for purchase by the four sources of final demand. Domestic and imported goods substitute imperfectly in use according to a CES function conditions. These diverse commodities differentiated by location can be aggregated at the national or regional level based on assumptions about product differentiation. The greater the variety of crops grown within a region is, the smaller the elasticity of substitution between these gridded crop outputs.

Regional crop production can either be exported or supplied domestically (i.e., consumed within the country where they are produced). The decision to export or supply domestically a crop is based on several factors, including the relative price of crops informed by production costs.

1.3   All Grid Cells Are Connected: Leakage and Spillovers

Leakage and spatial spillover effects are two important concepts in the context of conservation, sustainability, and climate change. Leakage refers to the displacement of environmental problems from one location to another as a result of conservation or sustainability policies (Meyfroidt et al. 2020). For example, a policy implemented to reduce deforestation in one area may lead to increased deforestation in other areas, as loggers and farmers move to new locations. Spatial spillover effects refer to the environmental impacts of conservation or sustainability policies that extend beyond the boundaries of the policy area (Johnson et al. 2023). For example, the creation of a protected area may lead to increased economic activity in surrounding areas. These environmental impacts can be both positive and negative, depending on the nature of the economic activity. Leakage and spatial spillover effects can be important considerations for policy makers when designing and implementing conservation, sustainability, and climate change policies. Understanding how these effects may play out is important for designing effective and efficient policies.

This book and the closely related literature contain many examples of leakage and spatial spillover effects in the context of conservation, sustainability, and climate change. Figure 4.3 illustrates how changes in one location can affect close and remote farms. A policy that reduces the use of fertilizer in one area may lead to increased fertilizer use in other areas as farmers and growers seek to maintain their yields (Chap. 14). A policy that reduces deforestation in one country may lead to increased deforestation in other countries as agriculture and industries relocate (Meyfroidt et al. 2010). The development of a renewable energy project (e.g., biofuels or co-firing of coal power plants with biomass) may increase land-use change in surrounding areas as development spreads to new areas (Sun et al. 2020). A groundwater sustainability policy in the Western United States can cause an increase in land use and fertilizer application in Florida (Chap. 16) and affect labor markets, both locally and elsewhere in the country (Ray et al. 2023). By design, SIMPLE-G captures unintended consequences, spillover effects, and leakages stemming from policy interventions.

Leakage and spatial spillover effects can be complex and difficult to predict. There is no one-size-fits-all solution for addressing these effects; the best approach will vary depending on context. However, SIMPLE-G can shed light on the

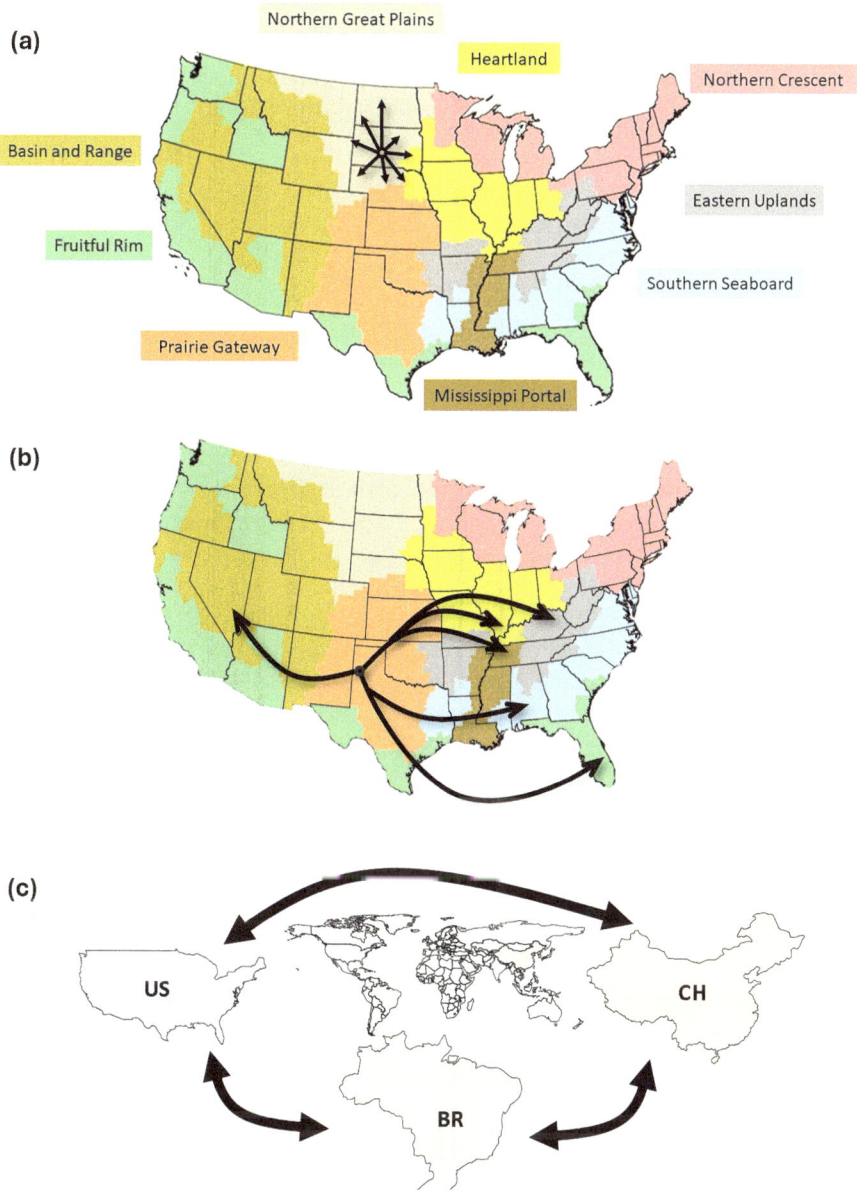

Fig. 4.3 Spillover effects (**a**) to similar crop-producing locations, (**b**) across subregions, and (**c**) across countries

Leakage and spillover effects depend on the degree of mobility of inputs and the degree of product differentiation. Unfavorable conditions in in the US Northern Great Plains can cause an increase in production in neighboring locations. A decline in agricultural activity in the US Prairie Gateway region may lead to changes in agricultural production in other regions of the United States. Here, labor migration can play a critical role. Telecoupled changes in the United States, China, and Brazil may have implications for all three regions. For example, growth in demand from China can cause an increase crop production in the United States and Brazil magnitude and direction of these effects, helping policy makers design policies that are comprehensive and coordinated across a variety of scales (Johnson et al. 2023). It can also serve as a foundation for developing new models and quantitative approaches to investigate leakage and spatial spillover effects.

2 Equilibrium Conditions

At its heart, any SIMPLE-G model is a model of economic equilibrium. Thus, the main equations in the model serve to determine the demand and supply of food commodities and agricultural inputs. Demand is written as a mathematical function that shows the relationships between the quantity demanded and other determinants of demand. It is negatively related to commodity price but positively related to income and scale of production. Similarly, supply is written as a mathematical function that shows the relationship between the quantity supplied and other economic and biophysical determinants of supply. A higher relative price for the commodity being produced motivates greater supply. Price is the equilibrating variable in this system. If an adverse climate shock in one location reduces supply, then a price increase will curb demand and induce additional supplies from other locations, thereby restoring equilibrium. Market equilibrium is defined as the price level at which the quantity of a commodity supplied by sellers (producers) is equal to the quantity demanded by buyers (consumers). In other words, there is no shortage or surplus. As such, this is not a short-run model. Rather, the time frame is long enough such that, if no other shocks were to occur, equilibrium would be restored following the shock or policy being implemented.

2.1 Global Crop Market Equilibrium

Global crop market equilibrium is the state of the market in which the total quantity of crops exported by all countries is equal to the total quantity of crops imported by all countries. This equilibrium is determined by a complex interplay of factors, including the relative prices of crops, the cost of production in each region, and policies. Equation 4.1 represents global crop market equilibrium:

$$\underbrace{\sum_r QGS_{c,r}}_{\text{Total exports}} = \underbrace{\sum_r \sum_b QGD_{c,r,b}}_{\text{Total imports}}, \qquad (4.1)$$

where $QGS_{c,r}$ is the total quantity of global supply of crop c exported by region r and $QGD_{c,r,b}$ is the total quantity of global demand for crop category c imported by buyer type b in region r. The base version of SIMPLE-G employed in this book includes 17 global regions (r), one composite crop category (c), and four types of buyers (b): direct consumers and the livestock, processed food, and biofuels sectors.

Equation 4.1 states that, in long-run equilibrium, the total quantity of crops exported must be equal to the total quantity of crops imported. If there were more crops exported than importers demanded, then the global market would experience a surplus of crops, which would drive prices down until equilibrium was restored. Conversely, if there were more demand for crops to be imported than were actually exported, then the global market would experience a shortage of crops, driving prices up, thereby restoring equilibrium.

2.2 Regional Crop Market Equilibrium

Regional crop market equilibrium is the state of the market in which the total quantity of crops supplied domestically in a region is equal to the total quantity of domestic crops purchased in the region by buyers, including the total quantity of crops used for biofuel production. This equilibrium is determined by an interplay of factors, including the relative prices of crops, the costs of production, and biofuel policies. For each crop category, Eq. 4.2 represents the regional crop market equilibrium:

$$QLS_{c,r} = \sum_b QLD_{c,r,b}, \tag{4.2}$$

where $QLS_{c,\,r}$ is the total quantity of locally supplied crops (domestically produced and supplied to domestic consumers) in the region, and $QLD_{c,\,r,\,b}$ is the quantity of local demand for regionally produced crops that are purchased by each buyer in the region. Total regional crop production, $QTS_{c,\,r}$, is the sum of gridded production in each region, which is equal to exports plus domestic sales:

$$\underbrace{QGS_{c,r} + QLS_{c,r}}_{\text{Total regional supply}} = QTS_{c,r} = \underbrace{\sum_{g \in r} Q_{c,g}}_{\text{Total gridded production}}. \tag{4.3}$$

2.3 Local Input Market Equilibrium (Immobile Inputs)

For agricultural inputs tied to a specific location (e.g., land), the gridded quantity of the factor demanded should be equal to the gridded quantity supplied at the equilibrium price for each location. For example, for each immobile factor i in each grid cell g, we can write

$$QS_{i,g} = QD_{i,g} = Q_{i,g}, \tag{4.4}$$

where $QS_{i,g}$ is the total quantity supply of input i in grid cell g and $QD_{i,g}$ is the total quantity of gridded demand for input i in grid cell g.

2.4 Subregional Input Market Equilibrium (Mobile Inputs)

For mobile factors (e.g., labor, capital, and fertilizer), the quantity of the factor demanded should be equal to the quantity supplied at the equilibrium price of the marketshed shown by index s. Here, the marketshed is defined as a neighborhood of grid cells within which the factor can commute or move. This could be the commuting zone for labor or a regional market for capital or purchased inputs.

$$QS_{j,s} = QD_{j,s} = \sum_{g \in s} Q_{j,g} \tag{4.5}$$

The main difference between the agricultural input market equilibrium conditions for mobile (Eq. 4.4) and immobile (Eq. 4.5) factors of production is that the equilibrium price for mobile factors of production will be equalized across all grid cells, while the equilibrium price for immobile factors of production will vary by grid cell. This is because mobile factors of production can move freely between grid cells in response to differences in prices and wages. As a result, if the price of a mobile factor of production is higher in one location than in another, then mobile factors of production will flow from the lower-price grid cell to the higher-price location until the prices are equalized. On the other hand, immobile factors of production cannot move between grid cells. As a result, the equilibrium price for an immobile factor of production will vary by grid cell depending on local supply and demand for that factor. This is reflected in the evidence that labor wages move together within a commuting zone, but land rents are highly spatially heterogeneous.

3 Behavioral Assumptions

Overall, the SIMPLE-G model accounts for a variety of factors that influence the human and environmental systems. The model's assumptions are based on empirical evidence, but it is important to note that the model is still a significant simplification of the real world (hence its name, SIMPLE!). Here, we review these assumptions and their empirical bases.

On the demand side of food markets, the SIMPLE-G model considers estimated price and income elasticities empirically; that is, the quantity of a product demanded is assumed to be sensitive to changes in its price and consumers' income. Clements and Si (2018) and Gouel and Guimbard (2019) nicely summarize the extensive literature on food demand that has evolved over the past century (Working 1925). The overall conclusion of this literature is that demand for food changes as income increases. At lower income levels, consumers derive the bulk of their calories from cereals and other carbohydrates, with proteins coming from plant-based sources. As incomes rise, consumers add more livestock products (which tend to be more expensive) as well as fresh fruits and vegetables, oils and fats, and sweeteners. Rising incomes translate into more overall consumption as well as more diverse diets. Recent research on the cost of diets across the world suggests that households require an income of roughly US$3/person/day to be able to afford a healthy diet (Bai et al. 2021). However, reaching this threshold does not mean that they will consume a healthy diet. Recent research suggests that the emerging "triple burden" of malnutrition includes households consuming excessive calories but still not attaining a healthy diet (Gómez et al. 2013).

Another important aspect of food demand that has recently received increasing attention is food waste. From a resource point of view, there is little difference between food purchased and consumed and food that is purchased and later discarded. (Food waste generally goes to landfills, where it can contribute to methane gas emissions, Krause et al. 2023.) There is evidence that global food waste is positively correlated with per capita income. Lopez Barrera and Hertel (2020) estimate an S-shaped curve: food waste is very low at the lowest per capita income levels but then increases rapidly in middle income countries (e.g., China over the past two decades) before leveling off at the highest income levels. To date, most international studies of food demand have been unable to distinguish food intake from food waste; therefore, our estimated income elasticities will combine these two sources of demand.

The other important characteristic of food demand is its responsiveness to price. At low-income levels, this price response tends to be quite strong. After all, low-income households have little choice but to curb consumption when food prices increase sharply since food comprises a large share of their total income. However, the price responsiveness of food demand diminishes as incomes rise (Muhammad et al. 2011), Capturing this change is important for any long-run projections as it reflects the fact that consumers' overall willingness to curb consumption in response to scarcity will diminish with economic growth, leading to increased price volatility.

Our assumptions on the producer side of the food system are conventional for economic models. First, we assume that producers choose inputs to minimize their production costs. This hypothesis is testable and has been subjected to statistical testing with mixed results at the level of individual farms (Zereyesus and Featherstone 2017). While not all producers are economically rational in this way, we are not modeling individuals but rather groups of producers (albeit within a given grid cell), and the cost-minimization hypothesis is more likely to hold in groups of producers, where competitive pressures cause irrational producers to downsize or exit the market (Hertel et al. 1996). We assume that, at the margin, producers that are not making an economic profit will eventually cede their land to other, more profitable producers. This is indeed what has been observed over time: The size of individual farms has tended to increase with consolidation (MacDonald 2020). The form of the SIMPLE-G production functions follows the well-established tradition in CGE modeling of nested CES (constant elasticity of substitution) functional forms. This means that the production of crops depends on a combination of different inputs, such as land–water, labor–capital, or a composite of remaining inputs. By carefully specifying the elasticity of substitution between individual inputs or groups of inputs, we are able to mimic observed economic and biophysical responses to changing relative prices. For example, the elasticity of substitution between nitrogen fertilizer and land can be calibrated to reproduce the observed yield response to fertilizer applications (Chap. 14).

On the international trade front, the model assumes that imports are governed by an Armington function. When countries import crops, they consider not only relative prices but also unobserved factors including product variety and other desirable qualities (e.g., taste, transportation restrictions, regulations). Capturing this product differentiation is important for modeling trade in SIMPLE-G, as it determines the degree of connectedness between domestic and international markets in the model. This is handled using Armington (1969) elasticities, which have been the subject of extensive empirical investigation (Hillberry and Hummels 2013); we rely on these estimates to parameterize the model. There is a symmetric Armington elasticity on the export side, where a constant elasticity of transformation (CET) function allocates produced crops between global and local markets. This allows the strength of export supply response from a region to be specified by the model developer.

Finally, for land use, the model assumes grid-cell-specific land supply functions. This means that the model accounts for the various characteristics of land in different locations when determining how much land will be used for agricultural production. Locations will differ in their responses to changes and policies, with implications of the spatial patterns of agricultural production and land use.

Each element of the model (variable) has an assigned value and quantity. The quantity of the variable can be shown as an index (i.e., US$ at constant prices) or as a physical quantity (e.g., tons of crops, hectares of land, cubic meters of water, or kilograms of fertilizer). The model includes equations that describe food demand as well as food production linked to natural resources. There are also parameters that describe behavioral responses or other biophysical dynamics of the system. The model is written as a system of simultaneous equations that determines the decisions on land and water use endogenously along with market prices for food and agricultural inputs. Tables 4.1 and 4.2 summarize the major conditions and functional forms used in SIMPLE-G.

To capture the dynamics necessary to analyze sustainability and food security, the model of the supply side of the economy includes four sets of equations: First, the equations that illustrate the supply of agricultural inputs. Second, the equations that Determine the demand for agricultural inputs. Third, the equations that illustrate the price and expenditure indices to ensure zero profit conditions. Fourth, the equations that determine the spatial allocations of inputs in case of imperfect mobility. The combination of these equations provides the basis for assessing spillover effects. This is a significant feature of SIMPLE-G, and it is important in explicitly modeling

Table 4.1 Market equilibrium conditions in the SIMPLE-G model

Description	Condition	Equation
Global market clearing	$$\underbrace{\sum_r QGS_{c,r}}_{\text{Total exports}} = \underbrace{\sum_r \sum_b QGD_{c,r,b}}_{\text{Total imports}}$$	Eq. (4.1)
Domestic market clearing	$QLS_{c,r} = \sum_b QLD_{c,r,b}$	Eq. (4.2)
Total regional supply	$\underbrace{QGS_{c,r} + QLS_{c,r}}_{\text{Total regional supply}} = QTS_{c,r} = \underbrace{\sum_{g \in r} Q_{c,g}}_{\text{Total gridded production}}$	Eq. (4.3)
Immobile input gridded market clearing	$QS_{i,\,g} = QD_{i,\,g} = Q_{i,\,g}$	Eq. (4.4)
Mobile input subregional or regional market clearing	$QS_{j,s} = QD_{j,s} = \sum_{g \in s} Q_{j,g}$	Eq. (4.5)

Table 4.2 Major assumptions in the SIMPLE-G model

	Assumption and functional forms	Reference
Food demand	Based on empirically estimated price and income elasticities	Muhammad et al. (2011)
Production structure	Composite output based on nested CES (constant elasticity of substitution) inputs	Dixon et al. (1982)
Imports	Using Armington CES: Imperfect substitution between imports and domestic crops	Hertel et al. (2007)
Exports	Using CET (constant elasticity of transformation) for allocation of local production to global and domestic markets	van der Mensbrugghe (2019)
Land use	CET representation of land supply response	Ahmed et al. (2008)
Input supply	Estimated fixed supply elasticities and CET for spatial allocation	Baldos et al. (2020) and Ray et al. (2023)
Relocation	Subregional Armington and product differentiation	Haqiqi et al. (2022)

the spatial reallocation of production in response to socioeconomic and environmental shocks. Overall, the model equations are derived based on multiple assumptions. Table 4.2 summarizes some of these assumptions.

4 Summary

The SIMPLE-G modeling framework requires multiscale market equilibrium for global, regional, and local markets. At each scale, total demand of a commodity or production input should be equal to total supply. Prices will adjust to eliminate any excess supply or demand. Production and consumption are linked through markets and international trade, where prices provide signals about changes in production costs and consumer preferences due to changes in environmental or human systems.

The SIMPLE-G models often employ functional forms that are widely tested and used in many economic studies. Leveraging well-established functional forms with strong empirical grounding, SIMPLE-G is a robust and comprehensive analytical tool for understanding the complex interactions within human and environmental systems.

Acknowledgment and Competing Interests The author acknowledges support from the U.-S. Department of Energy, Office of Science, Biological and Environmental Research Program, Earth and Environmental Systems Modeling, MultiSector Dynamics under Cooperative Agreement DE-SC0022141; the National Science Foundation award #2118329: "NSF Institute for Geospatial Understanding through an Integrative Discovery Environment (I-GUIDE)," the United States Department of Agriculture AFRI grant #2019–67,023-29,679, "Economic Foundations of Long Run Agricultural Sustainability," and the National Science Foundation INFEWS award #1855937, "Identifying Sustainability Solutions Through Global-Local-Global Analysis of a Coupled Water-Agriculture-Bioenergy System."

The findings and conclusions presented in this chapter are those of the authors and should not be construed to represent any official determination or policy of the US Department of Agriculture (USDA), the US government, the Department of Energy (DOE), or the National Science Foundation (NSF). Furthermore, we declare that there is no conflict of interest related to this work.

References

Ahmed, Syud Amer, Thomas W. Hertel, and Ruben Lubowski. 2008. *Calibration of a land cover supply function using transition probabilities*. West Lafayette, IN: Research Memorandum 14. Department of Agricultural Economics, Purdue University, Global Trade Analysis Project (GTAP). https://doi.org/10.21642/GTAP.RM14.

Armington, Paul S. 1969. A theory of demand for products distinguished by place of production. *Staff Papers* 16: 159–178. https://doi.org/10.2307/3866403.

Bai, Yan, Robel Alemu, Steven A. Block, Derek Headey, and William A. Masters. 2021. Cost and affordability of nutritious diets at retail prices: Evidence from 177 countries. *Food Policy* 99: 101983. https://doi.org/10.1016/j.foodpol.2020.101983.

Baldos, Uris Lantz C., and Thomas W. Hertel. 2013. Looking back to move forward on model validation: Insights from a global model of agricultural land use. *Environmental Research Letters* 8: 034024. https://doi.org/10.1088/1748-9326/8/3/034024.

Baldos, Uris Lantz C., Iman Haqiqi, Thomas W. Hertel, Mark Horridge, and Jing Liu. 2020. SIMPLE-G: A multiscale framework for integration of economic and biophysical determinants of sustainability. *Environmental Modelling & Software* 133: 104805. https://doi.org/10.1016/j.envsoft.2020.104805.

Clements, Kenneth W., and Jiawei Si. 2018. Engel's law, diet diversity, and the quality of food consumption. *American Journal of Agricultural Economics* 100: 1–22. https://doi.org/10.1093/ajae/aax053.

Dixon, Peter B. 1982. *ORANI, a multisectoral model of the Australian economy*. New York, NY: Elsevier.

Gómez, Miguel I., Christopher B. Barrett, Terri Raney, Per Pinstrup-Andersen, Janice Meerman, André Croppenstedt, Brian Carisma, and Brian Thompson. 2013. Post-green revolution food systems and the triple burden of malnutrition. *Food Policy* 42: 129–138. https://doi.org/10.1016/j.foodpol.2013.06.009.

Gouel, Christophe, and Houssein Guimbard. 2019. Nutrition transition and the structure of global food demand. *American Journal of Agricultural Economics* 101: 383–403. https://doi.org/10.1093/ajae/aay030.

Haqiqi, Iman, Chris J. Perry, and Thomas W. Hertel. 2022. When the virtual water runs out: Local and global responses to addressing unsustainable groundwater consumption. *Water International* 47: 1060–1084. https://doi.org/10.1080/02508060.2023.2131272.

Haqiqi, Iman, Laura Bowling, Sadia Jame, Uris Baldos, Jing Liu, and Thomas Hertel. 2023. Global drivers of local water stresses and global responses to local water policies in the United States. *Environmental Research Letters* 18: 065007. https://doi.org/10.1088/1748-9326/acd269.

Hertel, Thomas W. 1997. *Global trade analysis: Modeling and applications.* Cambridge University Press.

Hertel, Thomas W., Kyle Stiegert, and Harry Vroomen. 1996. Nitrogen-land substitution in corn production: A reconciliation of aggregate and firm-level evidence. *American Journal of Agricultural Economics* 78: 30–40. https://doi.org/10.2307/1243776.

Hertel, Thomas W., David Hummels, Maros Ivanic, and Roman Keeney. 2007. How confident can we be of CGE-based assessments of free trade agreements? *Economic Modelling* 24: 611–635. https://doi.org/10.1016/j.econmod.2006.12.002.

Hillberry, Russell, and David Hummels. 2013. Chapter 18—Trade elasticity parameters for a computable general equilibrium model. In *Handbook of computable general equilibrium modeling*, ed. Peter B. Dixon and Dale W. Jorgenson, vol. 1, 1213–1269. Handbook of Computable General Equilibrium Modeling SET, Vols. 1A and 1B. Elsevier. https://doi.org/10.1016/B978-0-444-59568-3.00018-3.

Johnson, David R., Stephen Polasky, and Jacob Ricker-Gilbert. 2023. Policy collision: A framework to identify where polycentric, multi-objective sustainability solutions are needed. *Environmental Research Letters* 18: 025004. https://doi.org/10.1088/1748-9326/acb0e4.

Krause, Max, Shannon Kenny, Jenny Stephenson, and Amanda Singleton. 2023. *Quantifying methane emissions from landfilled food waste.* EPA 600-R-23–064. U.S. Environmental Protection Agency Office of Research and Development.

Liu, Jing, Thomas W. Hertel, Richard B. Lammers, Alexander Prusevich, Uris Lantz C. Baldos, Danielle S. Grogan, and Steve Frolking. 2017. Achieving sustainable irrigation water withdrawals: Global impacts on food security and land use. *Environmental Research Letters* 12: 104009. https://doi.org/10.1088/1748-9326/aa88db.

Lopez Barrera, Emiliano, and Thomas W. Hertel. 2020. Global food waste across the income spectrum. Implications for food prices, production and resource use. *Food Policy* 98: 101874. https://doi.org/10.1016/j.foodpol.2020.101874.

MacDonald, James M. 2020. *Consolidation in U.S. agriculture continues.* Amber Waves. https://doi.org/10.22004/ag.econ.302913.

Meyfroidt, Patrick, Thomas K. Rudel, and Eric F. Lambin. 2010. Forest transitions, trade, and the global displacement of land use. *Proceedings of the National Academy of Sciences* 107: 20917–20922. https://doi.org/10.1073/pnas.1014773107.

Meyfroidt, P., J. Börner, R. Garrett, T. Gardner, J. Godar, K. Kis-Katos, B.S. Soares-Filho, and S. Wunder. 2020. Focus on leakage and spillovers: Informing land-use governance in a tele-coupled world. *Environmental Research Letters* 15: 090202. https://doi.org/10.1088/1748-9326/ab7397.

Muhammad, Andrew, James L. Seale Jr., Birgit Meade, and Anita Regmi. 2011. *International evidence on food consumption patterns: An update using 2005 international comparison program data.* Washington, DC: Technical Bulletin 1929. US Department of Agriculture, Economic Research Service.

Ray, Srabashi, Iman Haqiqi, Alexandra E. Hill, J. Edward Taylor, and Thomas W. Hertel. 2023. Labor markets: A critical link between global-local shocks and their impact on agriculture. *Environmental Research Letters* 18: 035007. https://doi.org/10.1088/1748-9326/acb1c9.

Sun, Shanxia, Brayam Valqui Ordonez, Mort D. Webster, Jing Liu, Christopher J. Kucharik, and Thomas W. Hertel. 2020. Fine-scale analysis of the energy–land–water nexus: Nitrate leaching implications of biomass cofiring in the Midwestern United States. *Environmental Science & Technology* 54: 2122–2132. https://doi.org/10.1021/acs.est.9b07458.

van der Mensbrugghe, Dominique. 2019. *The Environmental Impact and Sustainability Applied General Equilibrium (ENVISAGE) model*. https://mygeohub.org/groups/gtap/envisage-docs

Working, Holbrook. 1925. The statistical determination of demand curves. *The Quarterly Journal of Economics* 39: 503. https://doi.org/10.2307/1883264.

Zereyesus, Yacob A., and Allen M. Featherstone. 2017. Empirical analysis of profit maximization and cost minimization behaviour of Kansas farms. *Applied Economics Letters* 24: 1255–1258. https://doi.org/10.1080/13504851.2016.1270407.

Chapter 5
SIMPLE-G Model Specification: Mathematical Equations in a Multiscale Market Equilibrium Model

Iman Haqiqi

Understanding the complex link between land, water, and environmental systems is required for addressing sustainability challenges. Earth observation provides valuable insights into these systems. Traditional earth analysis methods often rely on one-dimensional approaches, focusing on individual components like land cover, hydrology, or biogeochemical processes. However, real-world environmental challenges like climate change, water scarcity, and biodiversity loss necessitate a more holistic understanding that integrates these components with human system. Effective analyses require flexible and adaptable models that incorporate human decisions and responses. This chapter introduces the mathematical representation of SIMPLE-G framework, Simplified International Model of agricultural Prices, Land use, and the Environment- Gridded version. This framework can be applied on different grid cell sizes, from very fine resolution to coarse resolutions, and in multiscale analysis. SIMPLE-G addresses the need for a traceable and comprehensive framework by offering a modular and customizable framework that combines various geospatial data sources and modeling techniques. This allows researchers to tailor the model to specific research questions and environmental contexts, with a deeper understanding of the complex dynamics shaping our planet.

This chapter introduces five successively more complex model specifications of SIMPLE-G. Each model is designed for specific research purposes but can be applied to a wide range of questions. Each successive model has new features and new data requirements. Table 5.1 summarizes the diverse models discussed in this chapter. We start with the simplest model, then explain how to add new dimensions, such as new assumptions, new agricultural inputs, or new production technologies. However, new inputs may differ in the degree of spatial mobility or their implications for the spatial scope of their markets. Therefore, new market-clearing

I. Haqiqi (✉)
Center for Global Trade Analysis, Department of Agricultural Economics, Purdue University, West Lafayette, IN, USA
e-mail: ihaqiqi@purdue.edu

© The Author(s) 2025
I. Haqiqi, T. W. Hertel (eds.), *SIMPLE-G*,
https://doi.org/10.1007/978-3-031-68054-0_5

Table 5.1 SIMPLE-G models in Part III

	Agricultural inputs	New information	New features	Potential application
1	Land, other	Intensive and extensive margins	Regional and local markets	Deforestation, biodiversity
2	Land, water, other	Yield response to water	Subregional markets	Nonland changes
3	Land, water, fertilizer, other	Yield response to fertilizer	Market rigidity	Nitrogen policies
4	Land, water, fertilizer, other	Land conversion, yield response to fertilizer by technology	Multiple activities, land allocation	Water stress, yield, water quality, fertilizer policies
5	Land, surface water, groundwater, equipment, fertilizer, other	Water resources substitutions, irrigation technology	Product differentiation	Water scarcity, water management

conditions and new marketsheds must also be introduced. For example, fertilizer markets are typically spatially connected, while labor markets may be less well connected at the regional level. New parameters will also be added for each new feature, which may require new information to inform the model. For example, introducing fertilizer to a model may require information about yield response to fertilizer, which must be used to inform the new behavioral parameters.

1 Model 1: The Basic Two-Input Model

This section describes the simplest of the gridded models, SIMPLE-G1; each grid cell has one production practice, two agricultural inputs (R for natural resources and H for manufactured inputs and human services), and one composite crop commodity. Table 5.2 summarizes the equations of the production part of this model. For the demand side, we follow nongridded SIMPLE equations, which are common across all the models described in this chapter (see Table 5.2).

1.1 Production Structure

In SIMPLE-G, we adopt a nested constant elasticity of substitution (CES) structure, which can accommodate varied types of technological information about production processes. Figure 5.1 represents the production structure of the basic model, the local market for the natural resource input, and the regional market for the augmented human input. Here, R is a composite of agricultural inputs provided by natural and environmental system (e.g., land and water), and H is a composite of agricultural inputs provided by human systems (e.g., labor, capital, and fertilizer). The natural

Table 5.2 Summary of gridded equations for SIMPLE-G-1.2.1

Variable[a]	Equation	Description
Natural resources input market		
q_g^R	$q_g^R + a_g^R = q_g - a_g - \sigma_g\left(p_g^R - a_g^R - p_g - a_g\right)$	Agricultural demand for natural resources
p_g^R	$q_g^R = \eta_g^R p_g^R + s_g^R$	Economic supply of natural resources
Human services and manufactured input markets		
q_g^H	$q_g^H + a_g^H = q_g - a_g - \sigma_g\left(p_g^H - a_g^H - p_g - a_g\right)$	Demand for human and manufactured input
p_g^H	$p_g^H = p_r^H$	Price of human and manufactured input
p_r^H	$q_r^H = \eta_r^H p_r^H + s_r^H$	Supply of human and manufactured input
q_r^H	$Q_r^H = \sum_{g \in r} Q_g^{H}$ [b]	Total subregional/regional input supply
Crop markets		
q_g	$q_g - a_g = \theta_g^H\left(q_g^H + a_g^H\right) + \theta_g^R\left(q_g^R + a_g^R\right)$	Gridded zero profits
p_g	$p_g = p_r$	Gridded price of crops
$q_r = f(p_r)$	$Q_r = \sum_{g \in r} Q_g$ [b]	Total regional crop supply

[a]This column shows matching variables that are determined by each equation (usually price or quantity) with the matching dimension
[b]These equations are written in levels, not as percentage change

input markets are geographically separate, assuming that they are not directly linked; therefore, the natural input is not mobile. That is, land does not move from one grid cell to another grid cell. However, the human input market is aggregated, reflecting the fact that the grid cells compete for human input. Additional, human input can move among locations, which causes prices for the human input to move together across grid cells.

The single-nest Constant Elasticity of Substitution (CES) gridded production function in Fig. 5.1 comprises three key behavioral equations that are obtained from microeconomic theory based on assumptions of cost minimization, coupled with free entry and exit from these activities.

1.2   Input Demand Equations

In keeping with the analytical approach, model condensation and nonlinear solution strategy described in Chap. 8, we write these equations in linearized (percentage change) form (Dixon 1982). The cost shares are updated over the course of the nonlinear model solution. The following three equations pertain to the top-level nest,

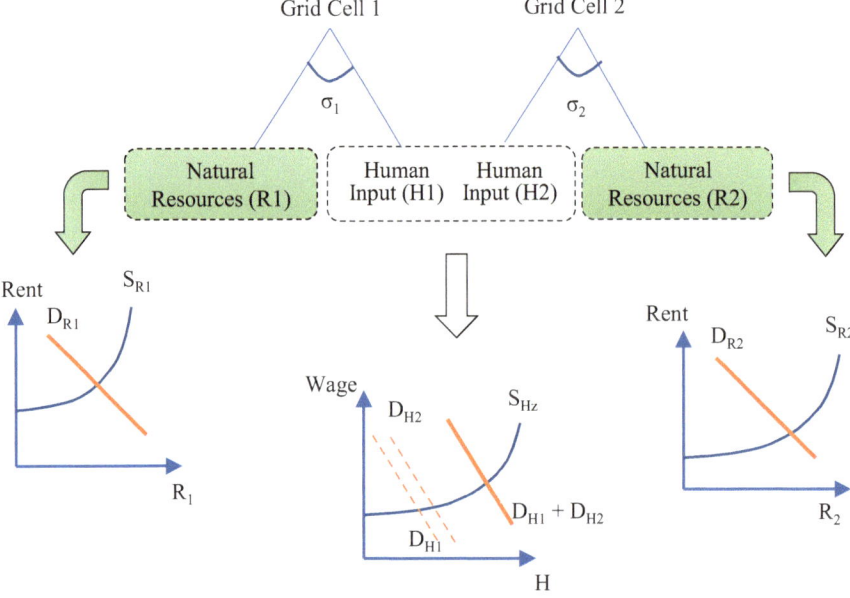

Fig. 5.1 Basic gridded production structure of a simple two-input model and associated markets

in which natural resources (R) and other inputs (H) are combined, in variable proportions, to produce aggregate crop output:

$$q_g^R + a_g^R = q_g - a_g - \sigma_g\left(p_g^R - a_g^R - p_g - a_g\right) : \text{demand for natural input,} \quad (5.1)$$

$$q_g^H + a_g^H = q_g - a_g - \sigma_g\left(p_g^H - a_g^H - p_g - a_g\right) : \text{demand for human input,} \quad (5.2)$$

$$p_g + a_g = \theta_g^H\left(p_g^H - a_g^H\right) + \theta_g^R\left(p_g^R - a_g^R\right) : \text{zero profits,} \quad (5.3)$$

where lowercase variables denote percentage changes in the corresponding upper-case, "levels" variables (i.e., $p = 100(dP/P)$ is the percentage change in crop price and $a = 100(dA/A)$ is the percentage change in total factor productivity). The variables p_g, q_g, and a_g denote the percentage changes in gridded crop price, gridded crop production, and overall crop productivity, respectively. Similarly, p_g^j, q_g^j, and a_g^j denote the percentage changes in input price, input demand, and factor-augmenting productivity, respectively. Finally, θ_g is the share of each input in total crop production costs.

The zero-profit equation is the consequence of our assumption of unrestricted entry and exit from the crop sector. If the output price rises, given unchanged technology and input prices, then the sector will experience excess profits. This will attract new entrants or encourage the expansion of existing producers, which will drive up input prices and drive down output prices until zero pure economic profits are restored. Manipulation of these equations yields the following, equivalent, quantity-based expression of this condition, which we use in the model to facilitate our condensation strategy:

$$q_g - a_g = \theta_g^H \left(q_g^H + a_g^H \right) + \theta_g^R \left(q_g^R + a_g^R \right). \tag{5.4}$$

Thus, the percentage change in demand for natural resources depends on changes in technology (a, a^R), total crop output (q), and the price of natural resources (p^R) relative to an index of all input costs (p). Equation 5.1 shows that the demand for natural resources, q_g^R, will increase with increasing crop production, q_g. As the substitution elasticity is a positive parameter, the demand for natural resources will also increase with a reduction in input price (or cost) of natural resources, p_g^R. Finally, the demand for natural resources will increase with an increase in crop prices. The impact of productivity improvement on factor prices depends on the magnitude of the substitution elasticity and price responses. Holding the input prices constant, an increase in productivity will generally increase both output level and demand for both inputs.

In Chap. 7, we will discuss how the elasticity of substitution, σ, can be calibrated to reproduce grid-cell- and practice-specific agronomic characteristics of crop production. It is evident that a large substitution elasticity will result in a much greater response to, for example, a tax on natural resource use in crop production. Therefore, σ is a key parameter in sustainability analysis.

1.3   Natural Resource Input Supply Equations

Input supply functions are written as mathematical equations that show the relationship between the price of agricultural inputs and the quantity of the input that is supplied to the market by the seller. The supply function for the natural resource inputs is typically written for each grid cell in the following linearized form:

$$q_g^R = \eta_g^R p_g^R + s_g^R, \tag{5.5}$$

where q is the percentage change in quantity supplied, p is the percentage change in price, η is the supply elasticity in the linearized supply function, and s is the shifter of the supply curve. The supply elasticity is a fixed parameter and defined as the percentage change in the quantity supplied divided by the percentage change in the price, holding everything else constant. The parameter of supply elasticity is positive, which means that the quantity supplied increases as the price increases.

1.4 Human and Produced Input Supply Equations

The supply function for human-produced inputs (e.g., labor, capital, fertilizer) is typically written in the following linearized form:

$$q_r^H = \eta_r^H p_r^H + s_r^H, \tag{5.6}$$

where r is an index for regions. Equation 5.6 shows that the total supply of human input in each region depends on regional prices (e.g., wages) and regional supply elasticities. Additionally, a shift in supply at the regional level (s_r) can affect the input market. Note that Eq. 5.6 can be specified for subregions depending on the application, indicating less mobility among human inputs.

1.5 Market Equilibrium Equations

For a natural resource input, the market-clearing condition implies that the gridded economic supply of the natural resource input should be equal to the gridded agricultural economic demand for natural resource input at the equilibrium price. For a human input, the market-clearing condition holds at the regional level. Table 5.2 summarizes the production side equations for Model 1 (SIMPLE-G-1.2.1).

Box 5.1 describes an application of Model 1 in SIMPLE-G-Mini that is designed for understanding the determinants of local responses and decisions about resources.

Box 5.1 SIMPLE-G Mini and AnalyseGE: The Minimodel for Understanding Gridded Production and Gridded Markets

Before moving on to more complex models, it is useful to pause and reflect on what can be learned from working solely with the supply side of the model. (The theory behind this gridded analysis was introduced in Part II.) If the market prices of outputs and variable inputs are known, then the outcome in each grid cell can be treated as an independent solution to a problem in which the market prices and grid-cell-specific policies (e.g., groundwater sustainability policies) are treated as exogenous shocks and we can solve the model for the local prices and quantities. We dub this the "minimodel" since it can be readily solved and analyzed in depth on a grid cell by grid cell basis. Of course, to know what the market price changes are for any comprehensive policy, one must first solve the "maxi-model" (i.e., the full-blown SIMPLE-G model with consumption, trade, and production in the rest of the world). This is therefore a sequential exercise.

The great advantage of working with the minimodel is that we have access to theoretical and numerical results that can facilitate insights and help us

(continued)

Box 5.1 (continued)

identify key parameters in the model and behavioral assumptions governing the outcomes (recall Part II.2). Thus, for example, referring to Eq. 3.9 of Box 3.1 (reproduced in a simplified form that assumes $a = 0$), we can see that the change in production in a given grid depends on three components: the extensive margin, the intensive margin, and the conservation policy (see also Ray et al. 2023). Each of these components hinges on grid cell parameters and data. % change in production (Q): $q = p^* \frac{\nu_R}{\theta_R} \Gamma_R + p^* \frac{\sigma \theta_H}{\theta_R} \Gamma_\sigma - \phi_R \Gamma_R$.

Therefore, any surprising outcome at the grid-cell level can be traced back to the underlying parameter and data combinations, drawing our attention back to the foundational work of estimating these parameter values and data: Are these values sound? If so, we may have uncovered an important finding. If not, then we need to revisit our data and/or parameters, which, in turn, will alter the grid-cell outcome.

To facilitate analysis with the minimodel, we use the GEMPACK software "AnalyseGE," which allows the user to load the solution, database, and parameter values into the algebraic representation of the model in order to quickly evaluate every component of an equation, such as that shown above, so that the analyst can quickly identify which term and which parameter is driving the unusual outcome. The concept of a minimodel is original to SIMPLE-G modeling, and this novel invention is important step forward in global-to-local sustainability analysis. It is developed in more detail in the application in Chap. 11.

2   Model 2: Introducing Subregional Input Markets

Model 2 introduces subregional markets and multiple inputs, which requires slightly different production functions and a slightly different supply structure. The production function is written in a nested CES form, a CES function of multiple composite inputs. In a nested CES function, the inputs to a production process are grouped into different nests that reflect the multiple stages of a cost-minimizing producer's decision making. For example, in Fig. 5.2, the farmer first decides how much irrigation water to apply to a field, after which they then decide how much of the human inputs to apply. The land–water composite input will have a price index and a quantity index (usually weighted based on values). Then, the composite land–water input is included in the top nest, where it is combined with human inputs.

In a nested CES function, demand for each individual input is written as a function of the demand for the composite input, which is linked to the scale of production. More output requires additional composite input. Here, the demand for the composite land–water input can be expressed as

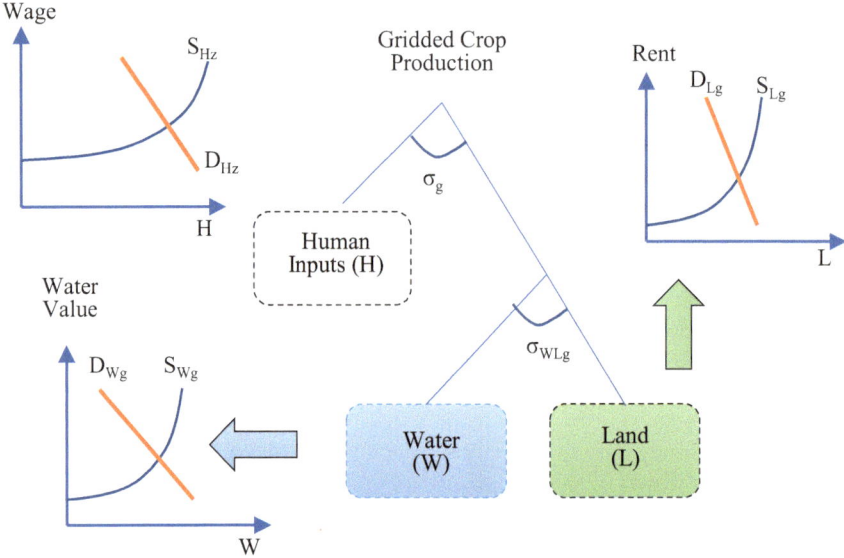

Fig. 5.2 Basic production structure of a SIMPLE-G model with three-input nested CES
The presence of two natural resource inputs, land and water, allows for different availability of these
two critical inputs

$$q_g^{LW} = q_g - a_g - \sigma_g\left(p_g^{LW} - p_g - a_g\right), \tag{5.7}$$

where LW is the land–water composite, q is percentage change in quantity, and p is
the percentage change in price. The structure of Eq. 5.7 is similar to the demand for
natural resources in Model 1: The demand for land–water composite depends on the
production scale effect, $q_g - a_g$, and the substitution effect, $-\sigma_g\left(p_g^{LW} - p_g - a_g\right)$.

The demand for individual land and water inputs is derived under the assumption
of cost minimization, subject to the CES functional form:

$$q_g^L + a_g^L = q_g^{LW} - \sigma_g^{LW}\left(p_g^L - a_g^L - p_g^{LW}\right) \tag{5.8}$$

$$q_g^W + a_g^W = q_g^{LW} - \sigma_g^{LW}\left(p_g^W - a_g^W - p_g^{LW}\right) \tag{5.9}$$

$$p_g^{LW} = \beta_g^L p_g^L + \beta_g^W p_g^W \tag{5.10}$$

Equation 5.10 introduces a price index for the land–water composite, which is a
weighted average of percentage changes in land rent and value of water. The cost
shares sum to 1 and are shown by β, which represents the value shares in the land–
water composite. This is different from θ, which was introduced in Model 1 to
represent the input cost shares in total costs.

In a subregional market for inputs, the constant elasticity supply function for the human inputs (e.g., labor, capital, and fertilizer) is typically written in the following linearized form:

$$q_z^H = \eta_z p_z^H + s_z^H,$$ (5.11)

where z is an index that denotes the scope of the human-produced input markets, ranging from the grid-cell to the regional level. Depending on factor mobility, there are three potential relationships between gridded prices and regional prices for the human inputs:

$$\text{Equilibrium} : \begin{cases} p_g^H = p_z^H = p_r^H & \text{if } z = r, \text{and perfect mobility} \\ p_g^H = p_z^H \neq p_r^H & \text{if } z \subset r, \text{and perfect mobility} \\ p_g^H \neq p_z^H \neq p_r^H & \text{if } z \subset r, \text{and imperfect mobility} \end{cases}$$ (5.12)

Depending on the size of grid cells and assumptions about mobility, index z can show one of the following:

- All the grid cells in a region are paid a uniform input price (or move at the same rate) as a result of perfect mobility of inputs. There is no mobility across regions.
- A subset of grid cells in a region faces a uniform input price for the subregion and perfect mobility of inputs within the subregion, but there is no mobility across subregions. Input prices are equal for all grid cells in a subregion but not for all grid cells in a region.
- An individual grid cell: Assuming no mobility and heterogeneous unrelated markets for inputs. There is no mobility across grid cells. Input prices can vary for grid cells in a subregion and in a region.

Table 5.3 summarizes the equations for Model 2 (SIMPLE-G-1.3.1).

2.1   Factor Mobility

The economic production factors (e.g., land, water, and human inputs) can be broadly categorized into mobile and immobile factors. In SIMPLE-G, this corresponds to spatial mobility (i.e., the ability of production factors to move freely among geographical locations). This mobility is essential for efficient resource allocation. For an efficient outcome, the marginal productivity of a given input must be equalized across locations and activities. Since cost-minimizing producers equate the value of marginal productivity for each input to its price, an efficient allocation of the input is achieved when the changes in input prices are equal in all grid cells. This also reduces spatial income inequality and promotes social mobility. Capital in the form of machinery can be mobile because it can be moved or used in

Table 5.3 Gridded equations for SIMPLE-G-1.3.1 with three inputs

Variable[a]	Equation	Description
p_z^H	$q_z^H = \eta_z p_z^H + s_z^H$	Subregional supply of human input
p_g^W	$q_g^W = \eta_g^W p_g^W + s_g^W$	Gridded supply of water input
p_g^L	$q_g^L = \eta_g^L p_g^L + s_g^L$	Gridded supply of land input
q_g^H	$q_g^H + a_g^H = q_g - a_g - \sigma_g\left(p_g^H - a_g^H - p_g - a_g\right)$	Gridded demand for human input
q_g^W	$q_g^W + a_g^W = q_g^{LW} - \sigma_g^{LW}\left(p_g^W - a_g^W - p_g^{LW}\right)$	Gridded demand for water input
q_g^L	$q_g^L + a_g^L = q_g^{LW} - \sigma_g^{LW}\left(p_g^L - a_g^L - p_g^{LW}\right)$	Gridded demand for land input
q_g^{LW}	$q_g^{LW} = q_g - a_g - \sigma_g\left(p_g^{LW} - p_g - a_g\right)$	Gridded demand for land–water composite (*LW*) input
p_g^{LW}	$p_g^{LW} = \beta_g^L p_g^L + \beta_g^W p_g^W$	Gridded price for *LW*
q_g	$q_g - a_g = \theta_g^L\left(q_g^L + a_g^L\right) + \theta_g^W\left(q_g^W + a_g^W\right) + \theta_g^H\left(q_g^H + a_g^H\right)$	Gridded zero profits
$q_r = f(p_r)$	$Q_r = \sum_{g \in r} Q_g$ [b]	Total regional crop supply
q_z^H	$Q_z^H = \sum_{g \in z} Q_g^H$ [b]	Total subregional input supply
p_g^H	$p_g^H = p_z^H$	Gridded price of mobile input
p_g	$p_g = p_r$	Gridded price of crops

[a]This column shows matching variables that is determined by each equation (usually price or quantity) with matching dimension
[b]These equations are written in levels not the percentage change

different locations. Technology is typically also mobile within a region because it can be transferred from one location to another. Examples of immobile factors of production include land, natural resources, and infrastructure. Land, natural resources, and infrastructure are considered immobile due to their nature and high cost of moving them. Of course, there can be interactions. For example, the construction of canals can facilitate the movement of water from one location to another.

For immobile factors, demand and supply must be in equilibrium in each individual grid cell. The prices of immobile factors can vary significantly by location, and they are not directly related. For mobile factors, farmers compete across a wide geographical range and synchronize mobile input prices across space. For example, fertilizer prices are usually very similar across locations; but land rents are usually different.

2.2 Marketsheds

A marketshed is a geographical area that represents the catchment area for a particular market. It is the area from which a market draws its supplies and demands. The size and shape of a marketshed can vary depending on the type of market, the market's location, and transportation infrastructure in the area.

For some research questions, it is necessary to determine the scope of factor mobility. For example, while the labor can typically move freely from one grid cell to neighbor grid cells or other locations within a certain distance, it is not easy to move across national borders. The concept of a labor-market commuting zone is related to the concept of a marketshed in that it defines a geographical area from which workers commute to a particular location. This information can be used to estimate the size of the potential labor force for a particular location. This topic comes up in the application documented in Chap. 13.

3 Model 3: Introducing Factor Market Rigidity

In Model 3, we introduce fertilizer as a distinct agricultural input for the purposes of analyzing sustainability related to water quality and food security. We also introduce factor market rigidity. Figure 5.3 illustrates the production tree structure and markets.

In this system, nitrogen (N) fertilizer use is determined endogenously in the model by considering relative prices, technology, substitution possibilities, and overall output level. The potential for nitrogen–land substitution is grid-cell- and activity-specific and is obtained from agronomic yield functions. The price of N fertilizer is determined at the regional level through a market-clearing condition wherein regional supply equals demand, which is in turn determined by aggregating N use across all grid cells and practices.

The four-input model has four input supply functions, each of which depends on the elasticities and shifter variables at the relevant scale. These functions can be expressed as

$$q_g^L = \eta_g^L p_g^L + s_g^L, \tag{5.13}$$

$$q_g^W = \eta_g^W p_g^W + s_g^W, \tag{5.14}$$

$$q_r^N = \eta_r^N p_r^N + s_r^N, \tag{5.15}$$

$$q_z^H = \eta_z p_z^H + s_z^H, \tag{5.16}$$

where L, W, N, H are land, water, N fertilizer, and other human-produced inputs, respectively; g, z, and r are indices for the grid cells, marketsheds, and regional scopes; q is the percentage in quantity supplied; η is the supply elasticity; p is the

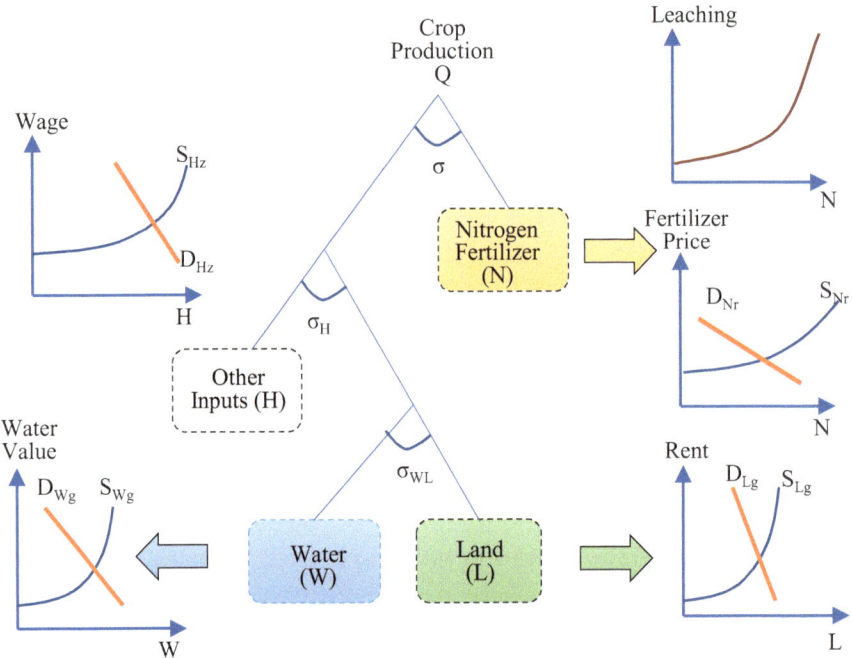

Fig. 5.3 Basic production structure with four inputs

The addition of fertilizer at the top of the production tree allows for the incorporation of information about the responsiveness of crop output to additional fertilizer applications. The leaching function varies by grid cell

percentage change in market price; and s is the percentage change due to exogenous shifters.

The demand for each composite and individual input is derived following the nested CES structure summarized in Table 5.4.

Explicitly modeling market rigidity in gridded economic models is important for several reasons. First, it allows researchers to better understand the geographical impact of conservation and sustainability policies, which often lead to leakage and other unanticipated effects. Second, it provides a framework for better understanding the market-mediated effects. Third, it allows policy makers to design more effective policies to address market rigidity and its implications for conservation policies.

To introduce rigidity in markets, we use a constant elasticity of transformation (CET) function, which is analogous to the CES function; rather than determining demands, it governs supply response to changing relative prices. In linearized form, it is expressed as

$$q_g^H = q_z^H - \tau_z^H \left(p_g^H - p_z^H \right), \tag{5.17}$$

Table 5.4 Gridded equations for SIMPLE-G-1.4.1 multiscale markets

Variable[a]	Equation	Description
p_g^L	$q_g^L = \eta_g^L p_g^L + s_g^L$	Supply of land input
p_g^W	$q_g^W = \eta_g^W p_g^W + s_g^W$	Supply of water input
p_r^N	$q_r^N = \eta_r^N p_r^N + s_r^N$	Supply of fertilizer input
p_z^H	$q_z^H = \eta_z p_z^H + s_z^H$	Supply of other inputs
q_g^L	$q_g^L + a_g^L = q_g^{LW} - \sigma_g^{LW}\left(p_g^L - a_g^L - p_g^{LW}\right)$	Demand for land input
q_g^W	$q_g^W + a_g^W = q_g^{LW} - \sigma_g^{LW}\left(p_g^W - a_g^W - p_g^{LW}\right)$	Demand for water input
q_g^N	$q_g^N + a_g^N = q_g - a_g - \sigma_g\left(p_g^N - a_g^N - p_g - a_g\right)$	Demand for fertilizer input
q_g^H	$q_g^H + a_g^H = q_g^{HLW} - \sigma_g^{HLW}\left(p_g^H - a_g^H - p_g^{HLW}\right)$	Demand for other inputs
q_g^{LW}	$q_g^{LW} + a_g^{LW} = q_g^{HLW} - \sigma_g^{HLW}\left(p_g^{LW} - a_g^{LW} - p_g^{HLW}\right)$	Demand for land–water composite (LW)
q_g^{HLW}	$q_g^{HLW} = q_g - a_g - \sigma_g\left(p_g^{HLW} - p_g - a_g\right)$	Demand for composite HLW
p_g^{LW}	$p_g^{LW} = \beta_g^L p_g^L + \beta_g^W p_g^W$	Price for LW
p_g^{HLW}	$p_g^{HLW} = \beta_g^H p_g^H + \beta_g^{LW} p_g^{LW}$	Price for composite HLW
q_g	$q_g - a_g = \theta_g^L\left(q_g^L + a_g^L\right) + \theta_g^W\left(q_g^W + a_g^W\right)$ $+\theta_g^N\left(q_g^N + a_g^N\right) + \theta_g^H\left(q_g^H + a_g^H\right)$	Zero profits
q_r	$Q_r = \sum_{g \in r} Q_g$ [b]	Total crop supply
q_z^H	$Q_z^H = \sum_{g \in z} Q_g^H$ [b]	Total input supply
p_g^H	$q_g^H = q_z^H - \tau_z^H\left(p_g^H - p_z^H\right)$	Supply of other inputs
p_g^N	$p_g^N = p_z^N$	Price of nitrogen fertilizer
p_g	$p_g = p_r$	Price of crops

[a]This column shows matching variables that are determined by each equation (usually price or quantity) with matching dimension
[b]These equations are written in levels, not in percentage change

where q_g^H is the percentage change in supply of input to grid cell g; q_z^H is the percentage change in total supply in marketshed z; $\tau_z^H \leq 0$ is the transformation elasticity parameter determining the responsiveness of transformation in the supply function; p_g^H is the factor price (wage) in location g; and p_z^H is the marketshed average factor price. Equation 5.17 shows that the supply of an input to one location depends on the factor price (wage) in other locations in addition to the local price. However, as long as the transformation elasticity is not infinite, factor prices can differ across grid cells within the marketshed.

3.1 Market Rigidity

Market rigidity refers to the inability of market quantities to adjust to changes in spatially disaggregated supply and demand conditions. Spatial market rigidity is important in global-gridded economic modeling because it can have a significant impact on a variety of economic outcomes, including wages, employment, land use, and agricultural production patterns. In the case of farmworkers, labor market rigidity refers to the fact that workers are not perfectly mobile across jobs and locations. In the case of capital, market rigidity can arise from the inability to move agricultural capital to another location. This typically depends on the share of immobile installed capital and infrastructure. Farm capital in the form of machinery is easier to move to another location than buildings and irrigation canals are. Factor immobility gives rise to spatially heterogeneous prices for the same input, with prices being (weakly) related.

Labor market immobility within a marketshed is an important rigidity explored in Chap. 13. This immobility can stem from market forces, institutions, or individual preferences. For example, labor market rigidity can arise due to geographical barriers, employment contracts, local attachments, social/cultural differences, language barriers, or simply lack of information on job opportunities and wage differentials. This can make it difficult for employers and employees to adjust to changes in demand or supply and can lead to farmworker wage disparities across locations, as those in high-yield areas can ask for higher compensations.

The CET function outlined above is an adaptable tool used in SIMPLE-G to model a variety of types of factor market rigidity. By including the difference in degree of mobility between different locations, economists can gain insights into the impacts of global and local changes on the spatial reallocation of resources and agricultural production.

4 Model 4: Multiple Activities and Land Allocation

Thus far, we have defined only one production function for each grid cell. Here, we introduce two activities in each grid cell, with different production functions, cost structures, input usages, yields, and revenue streams in each. Most of the applications in Part IV distinguish between irrigated and rainfed production, although this approach could also be used to explore other differences (e.g., traditional and commercial production). Figure 5.4 divides cropland area into rainfed and irrigated practices, and the different land rents for these two activities are endogenously determined in the model. Land rents are also grid-cell-specific and depend on local biophysical characteristics, irrigation yield gaps, prices, and the technologies available to each production unit. The allocation of land to rainfed and irrigated production in response to a policy shock is determined by their relative returns (i.e., land rental rates), which are determined endogenously for each grid cell on the basis of

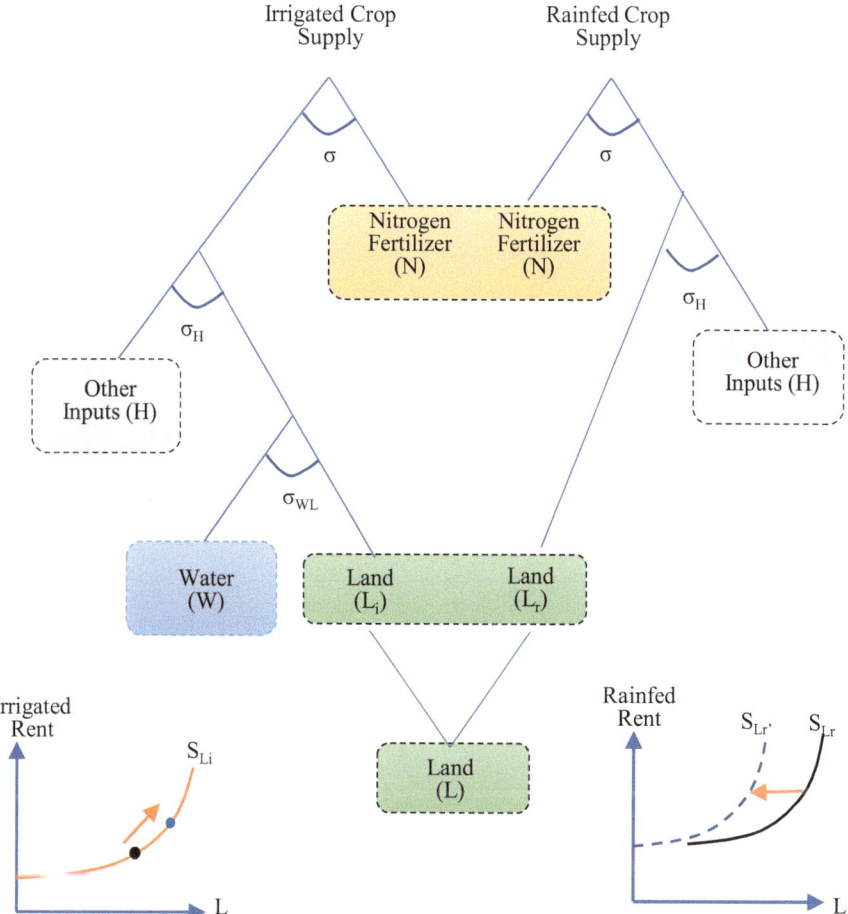

Fig. 5.4 Basic production structure with land allocation
The presence of two different practices in each grid cell allows for cropland for to be converted from one use to another based on the CET (constant elasticity of transformation) parameter

movement between land uses determined by a constant elasticity of transformation function (Ahmed et al. 2008). The key parameter in this function is the elasticity of transformation between irrigated and rainfed cropland. This elasticity measures the responsiveness of the rainfed–irrigated crop mix ratio to changes in relative returns to the two activities. A larger absolute value of this elasticity indicates an easier transformation of cropland between irrigated and rainfed categories. The conversion of land from rainfed to irrigated cropping is heavily influenced by institutional and biophysical constraints. In the United States, these constraints depend critically on state and local water laws, which vary greatly by region.

We use a quantity-preserving CET function for this particular application (van der Mensbrugghe and Peters 2020). A QCET function is a function that can be used to allocate land to different land cover types (e.g., agriculture, forestry, urban development) or agricultural practices (e.g., irrigated and rainfed land). This function allows us to account for the different characteristics of land, such as yield. In the context of land allocation, the CET function is used to model the conversion of various land uses. The QCET function is important because it preserves the total quantity of land. The traditional CET does not have this feature. Instead, it imposes a constraint on the overall economic value of the land. This quantity-preserving feature is important when tracking total physical units of land area is of central importance.

The multiactivity model introduces a new index for each variable. Here, we introduce l for land types, and the demand for land type l is written as follows:

$$q_{g,l}^L + a_{g,l}^L = q_{g,l}^{LW} - \sigma_{g,l}^{LW}\left(p_{g,l}^L - a_{g,l}^L - p_{g,l}^{LW}\right). \tag{5.18}$$

Equation 5.18 is similar to the land demand in Model 3, with the addition of the land type index. We also introduce new composite inputs: $q_{g,l}^{LW}$, which is a land and water composite, and $q_{g,l}^{HLW}$, which a composite of human input, land, and water. The default model has two land types: irrigated and rainfed. To allocate land to different land types, we need to introduce three new equations:

$$q_{g,l}^L = q_g^L - \tau_g^L\left(p_{g,l}^L - p_g^{\dagger L}\right), \tag{5.19}$$

$$p_g^{\dagger L} = \sum_l \theta_{g,l}^{\dagger L} p_{g,l}^L, \tag{5.20}$$

$$p_g^L + q_g^L = \sum_l \theta_{g,l}^L\left(p_{g,l}^L + q_{g,l}^L\right), \tag{5.21}$$

where $q_{g,l}^L$ is the percentage change in supply of land to different land types; τ_g^L is the transformation elasticity representing the flexibility in land conversion; $p_g^{\dagger L}$ is the weighted average change in land rents, where the weights are defined as quantity shares; $\theta_{g,l}^{\dagger L}$ is the quantity share; and $\theta_{g,l}^L$ is the value share in the land allocation structure. Together, Eqs. 5.19–5.21 comprise the QCET supply system.

4.1 The Advantage of Splitting Rainfed and Irrigated Production

Irrigation water is another focal point of SIMPLE-G. Irrigation water supply and demand are endogenously determined for each grid cell, but they are linked to exogenous environmental factors. For example, heat stress may increase the water requirements of crops grown in a grid cell, or a drought may reduce the environmental supply of water. Hydrological dynamics are not directly modeled and are treated exogenously. However, SIMPLE-G can be readily paired with a hydrological model to shed light on, for example, the economic consequences of changing basin-level water scarcity or interbasin transfers of water (Liu et al. 2017; Woo et al. 2022).

Water withdrawals are endogenously determined through the interaction of supply constraints and irrigation demand for crop production. Demand for water depends on the irrigation area, production levels, technology, and relative prices. This includes likely adaptation channels and adjustment mechanisms. We consider change in irrigation extent (Haqiqi and Hertel 2019), crop production location, change in irrigation technology, change in water intensity, and trade (Haqiqi and Hertel 2016).

4.2 Using the QCET to Govern Cropland Supply

Just as cropland is allocated across irrigated and rainfed production, the allocation of total land area between cropland and other uses (e.g., pastureland, forests, and other land cover types) can also be governed by a quantity-preserving CET (QCET) function. Cropland supply is a function of the CET parameter and the share of cropland in the grid cell and varies with the cropland rental rate relative to the returns to other uses. In SIMPLE-G, the other land uses are not explicitly modeled, so those land rental rates are deemed exogenous. However, they may be shocked to simulate the impact of intensified competition for land. An important feature of the QCET function is that as cropland expands and covers more of the available land in the grid cell, the cropland supply response diminishes and eventually falls to zero when the entire grid cell is devoted to crop production. The CET parameter can be calibrated to estimated elasticities of current cropland supply (see Chap. 7). In some of the applications featured in Part IV, the QCET function is replaced by a simple cropland supply elasticity, which does not explicitly account for the origin of this cropland or the use to which unused cropland might revert. Table 5.5 summarizes the production-side equations for Model 4 (SIMPLE-G-2.4.1).

5 Model 5: Introducing Product Differentiation

In economics, product differentiation refers to the existence of different varieties of a product that are perceived by consumers as being imperfect substitutes, due to a variety of factors including physical attributes, brand reputation, or quality. Product differentiation opens the possibility of commodity prices varying by product origin and composition, with significant implications for gridded modeling in the output market and input markets. This approach is widely used in international trade (Armington 1969), and it has proven effective for modeling the flows of goods

Table 5.5 Gridded equations for SIMPLE-G-2.4.1

Variable[a]	Equation	Description
p_g^L	$q_g^L = \eta_g^L p_g^L + s_g^L$	Gridded supply of land input
p_g^W	$q_g^W = \eta_g^W p_g^W + s_g^W$	Gridded supply of water input
p_r^N	$q_r^N = \eta_r^N p_r^N + s_r^N$	Regional supply of fertilizer input
p_z^H	$q_z^H = \eta_z p_z^H + s_z^H$	Supply of human input
$q_{g,l}^L$	$q_{g,l}^L + a_{g,l}^L = q_{g,l}^{LW} - \sigma_{g,l}^{LW}\left(p_{g,l}^L - a_{g,l}^L - p_{g,l}^{LW}\right)$	Gridded demand for land input
$q_{g,l}^W$	$q_{g,l}^W + a_{g,l}^W = q_{g,l}^{LW} - \sigma_{g,l}^{LW}\left(p_{g,l}^W - a_{g,l}^W - p_{g,l}^{LW}\right)$	Gridded demand for water input
$q_{g,l}^N$	$q_{g,l}^N + a_{g,l}^N = q_{g,l} - a_{g,l} - \sigma_{g,l}\left(p_{g,l}^N - a_{g,l}^N - p_{g,l} - a_{g,l}\right)$	Gridded demand for fertilizer input
$q_{g,l}^H$	$q_{g,l}^H + a_{g,l}^H = q_{g,l}^{HLW} - \sigma_{g,l}^{HLW}\left(p_{g,l}^H - a_{g,l}^H - p_{g,l}^{HLW}\right)$	Gridded demand for other inputs
q_g^{LW}	$q_g^{LW} + a_{g,l}^{LW} = q_{g,l}^{HLW} - \sigma_{g,l}^{HLW}\left(p_{g,l}^{LW} - a_{g,l}^{LW} - p_{g,l}^{HLW}\right)$	Gridded demand for land–water composite (LW)
$q_{g,l}^{HLW}$	$q_{g,l}^{HLW} = q_{g,l} - a_{g,l} - \sigma_{g,l}\left(p_{g,l}^{HLW} - p_{g,l} - a_{g,l}\right)$	Gridded demand for composite HLW
$p_{g,l}^{LW}$	$p_{g,l}^{LW} = \beta_{g,l}^L p_{g,l}^L + \beta_{g,l}^W p_g^W$	Gridded price for LW
$p_{g,l}^{HLW}$	$p_{g,l}^{HLW} = \beta_{g,l}^H p_{g,l}^H + \beta_{g,l}^{LW} p_{g,l}^{LW}$	Gridded price for composite HLW
$q_{g,l}$	$q_{g,l} - a_{g,l} = \theta_{g,l}^L\left(q_{g,l}^L + a_{g,l}^L\right) + \theta_{g,l}^W\left(q_{g,l}^W + a_{g,l}^W\right)$ $+ \theta_{g,l}^N\left(q_{g,l}^N + a_{g,l}^N\right) + \theta_{g,l}^H\left(q_{g,l}^H + a_{g,l}^H\right)$	Gridded zero profits
$p_{g,l}^L$	$q_{g,l}^L = q_g^L - \tau_g^L\left(p_{g,l}^L - p_g^{\dagger L}\right)$	QCET land allocation
$p_g^{\dagger L}$	$p_g^{\dagger L} = \sum_l \theta_{g,l}^{\dagger L} p_{g,l}^L$	QCET price index
q_g^L	$p_g^L + q_g^L = \sum_l \theta_{g,l}^L\left(p_{g,l}^L + q_{g,l}^L\right)$	Ensure revenue fully exhausted
q_z^H	$Q_z^H = \sum_{g \in z}\sum_l Q_{g,l}^{H~b}$	Total subregional input supply
q_r^N	$Q_r^N = \sum_{g \in r}\sum_l Q_{g,l}^{N~b}$	Total subregional input supply
q_r	$Q_r = \sum_{g \in r}\sum_l Q_{g,l}^{~b}$	Total regional crop supply
$p_{g,l}^H$	$p_{g,l}^H = p_z^H$	Gridded price of mobile input
$p_{g,l}^N$	$p_{g,l}^N = p_r^N$	Gridded price of fertilizer
$p_{g,l}$	$p_{g,l} = p_r$	Gridded price of crops

[a]This column shows matching variables that are determined by each equation (usually price or quantity) with matching dimension
[b]These equations are written in levels, not in percentage change

between countries and regions (Tinbergen 1962; Anderson 1979; Hillberry and Hummels 2003, 2008; Chaney 2008, 2018; Hillberry and Hummels 2013).

In SIMPLE-G, the composite crop output from any given grid cell is likely unique in its product composition. While we call them all "crops" and assume that their returns move in tandem over the long run, they do in fact differ. Indeed, the Food and Agriculture Organization of the United Nations (FAO) identifies 175 different crop categories; even at this fine level of detail, there are still some composite crop types. In short, it is not possible—particularly in a SIMPLE model—to distinguish all the different crop types. However, many of these differences are likely to matter for consumers. For example, within the product category of wheat, hard red winter wheat is used for baking purposes and durum wheat is used to make pasta. Both are listed as wheat, but they are, in fact, different products in the eyes of food manufacturers and consumers, generally grown in different locations. For this reason, the SIMPLE model allows the prices of crops from different locations to differ. In some versions (e.g., Models 1–4), this differentiation is applied only at the regional level and therefore governs international trade. However, Model 5 also applies this differentiation regionally to capture differences in product composition across grid cells and individual countries (Haqiqi et al. 2022, 2023).

5.1 Modeling Product Differentiation

The equilibration of supply and demand for crops occurs at the level of market regions. Within the market regions in SIMPLE, crop demands are an aggregate of four end uses: direct consumption, food processing, livestock demand, and biofuels. Demands may be satisfied from either domestic or global markets, depending on relative prices, following the method suggested by Armington (1969), which results in imperfect substitution between domestic and foreign products. Symmetrically, on the supply side, producers "transform" their products imperfectly between domestic and global markets. They do not shift completely from the domestic to the foreign market when a small price difference emerges. This permits us to calibrate the model to observed data in which similar products are both imported and exported from the same country.

Accommodating product differentiation in the crop market requires four new equations. The main subregional Armington equations are expressed as

$$q_g = q_s - \varphi_s (p_g - p_s), \tag{5.22}$$

$$q_s = q_r - \kappa_r (p_s - p_r), \tag{5.23}$$

where q_g is the demand for crops produced in grid cell g; q_s is the demand for all the crops produced in subregion s; φ_s is the substitution elasticity parameter that determines the substitutability based on preferences; q_r is total demand for crops produced in region r; and κ_r is a parameter that determines the substitutability between subregional crop composites. For example, in Fig. 4.2, the parameter φ determines the degree of similarity of crops within each subregion. In Fig. 4.2, the parameter κ illustrates the imperfect substitutability across subregions and between products coming out of subregions.

5.2 Modeling Differentiated Inputs

Model 5 also introduces differentiated inputs, particularly for irrigation water, reflecting the fact that groundwater and surface water have different physical and economic characteristics. Differences in physical characteristics (e.g., salinity and mineral content) can affect their suitability for different uses. For example, groundwater is often preferred for drinking water because it is typically cleaner than surface water. However, surface water is often preferred for irrigation because it may be more readily available. In addition to their physical characteristics, groundwater and surface water also have different economic characteristics. Groundwater extraction is typically more expensive than surface water extraction, but it is also more reliable. Surface water is typically less expensive to extract than groundwater, but it is also more susceptible to contamination and drought. The different physical and economic characteristics of groundwater and surface water lead to product differentiation in the market for water. Model 5 captures this differentiation by introducing the two distinct sources of irrigation water and treating them as imperfect substitutes (Fig. 5.5).

5.3 Nested CES Production Function for Model 5

Figure 5.5 presents the two nested CES production functions for each grid cell in Model 5. As in Model 4, we distinguish between rainfed and irrigated activities, and the top nest determines the yield response to fertilizer applications. The "other inputs" are a composite of water, land, and the remaining variable inputs. This is followed by a CES nest combining land and irrigation water. If crop output is strictly proportional to the amount of irrigation water delivered, then the elasticity of substitution between land and irrigation water is zero (i.e., no deficit irrigation). On the other hand, if a reduction in water delivered to the crop is not accompanied by a proportionate reduction in output, then this elasticity is greater than zero and it captures the potential for deficit irrigation (i.e., achieving the same output level with less water but more land). The next CES nest in Fig. 5.5 combines irrigation water with equipment. The associated elasticity of substitution at the bottom of this production tree reflects the potential for conserving irrigation water through investments in, for example, drip irrigation to replace sprinkler or canal-based irrigation capital. This is a key sustainability parameter discussed in Chap. 7.

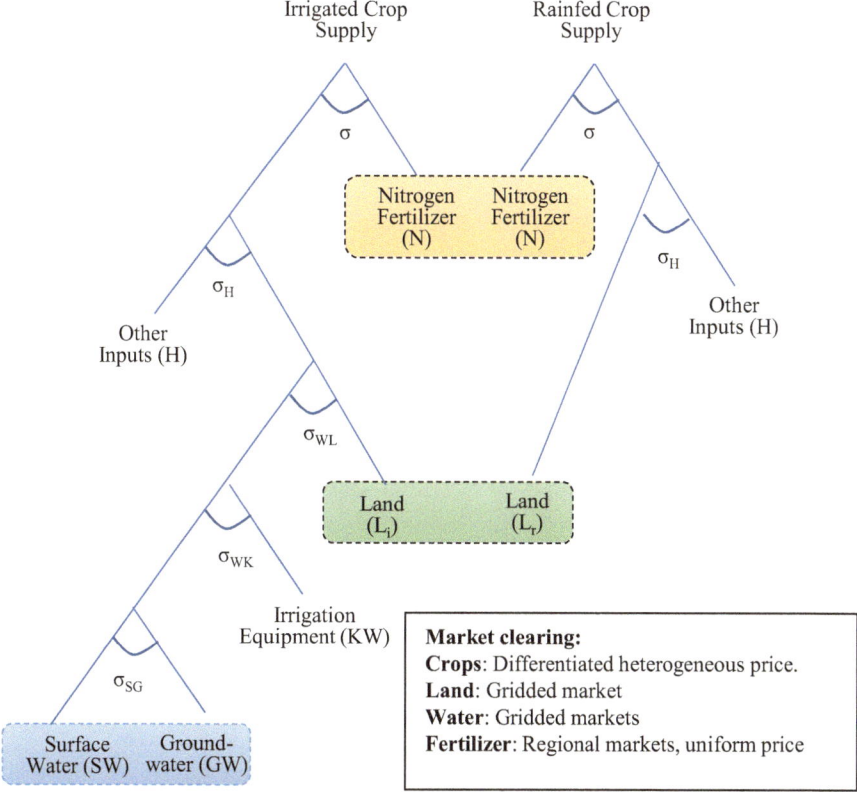

Fig. 5.5 Structure of crop production at each grid cell
Elasticities of substitution are denoted by σ. The equilibrium quantity and price of land and water are determined at the grid-cell level. Irrigated and rainfed practices compete for land. Land supply depends on total cropland supply and the elasticity of transformation between irrigated and rainfed land

The final CES nest in Fig. 5.5 combines surface and groundwater to create an irrigation water composite. The rationale for this nesting is that surface and groundwater extractions often coexist in a given grid cell, despite differences in cost. As previously noted, the two sources of water offer farmers different characteristics. Groundwater, for example, is available on demand and is largely independent of current weather conditions. Surface water must be delivered to the farm, often by infrastructure that is out of the farmer's control. The equations in Table 5.6 describe the substitution possibilities at each level in the production "tree."

Crop production is the result of representative producers' profits maximization, subject to technology, prices, policies, and resource constraints. The crop production technologies (both rainfed and irrigated production) in each grid cell allow for substitution between N fertilizer, water, land, and other inputs (an aggregate of capital, labor, other chemicals, and energy). The particular mix of inputs employed

Table 5.6 Gridded equations for SIMPLE-G-2.6.1

	Equation	Description
Old equations		
p_g^W	$q_g^W = \eta_g^W p_g^W + s_g^W$	Gridded supply of water input
$p_{g,l}$	$Q_r = \sum_{g\in r}\sum_l Q_{g,l}$ [b]	Regional crop markets and price index
	$p_{g,l} = p_r$	
New water		
p_g^{GW}	$q_g^{GW} = \eta_g^{GW} p_g^{GW} + s_g^{GW}$	Supply of groundwater
p_g^{SW}	$q_g^{SW} = \eta_g^{SW} p_g^{SW} + s_g^{SW}$	Supply of surface water
p_z^{KW}	$q_z^{KW} = \eta_z^{KW} p_z^{KW} + s_z^{KW}$	Supply of irrigation equipment and inputs
p_g^W	$p_g^W = \beta_g^{SW} p_g^{SW} + \beta_g^{GW} p_g^{GW} + \beta_g^{KW} p_g^{KW}$	Irrigation price index
q_g^{SW}	$q_g^{SW} = q_g^{SG} - \sigma_g^{SG}\left(p_g^{SW} - p_g^{SG}\right)$	Demand for surface water
q_g^{GW}	$q_g^{GW} = q_g^{SG} - \sigma_g^{SG}\left(p_g^{GW} - p_g^{SG}\right)$	Demand for groundwater
q_g^{KW}	$q_g^{KW} = q_g^W - \sigma_g^W\left(p_g^{KW} - p_g^W\right)$	Demand for irrigation inputs
q_g^{SG}	$q_g^{SG} = q_g^W - \sigma_g^W\left(p_g^{SG} - p_g^W\right)$	Demand for total surface water and groundwater
q_z^{KW}	$Q_z^{KW} = \sum_{g\in z} Q_g^{KW}$	Market clearing for irrigation inputs
p_g^{KW}	$p_g^{KW} = p_r^{KW}$	Price index for irrigation inputs
New product differentiation		
q_g	$q_g = q_s - \varphi_s(p_g - p_s)$	Subregional Armington (bottom layer)
q_s	$q_s = q_r - \kappa_r(p_s - p_r)$	Subregional Armington (top layer)
p_s	$p_s = \sum_{g\in s}\sum_l \theta_{g,l} p_{g,l}$	Subregional crop price index
p_r	$p_r = \sum_{s\in r}\theta_s p_s$	Regional crop price index

in a grid cell depends on relative prices, government policies, and production possibilities. Output levels expand or contract to ensure zero pure economic profits over the long run. Thus, unlike downscaling approaches to the gridded analysis of land use, the spatial pattern of production is endogenously determined. Crop producers within a given grid cell are price takers, as they are assumed to have no market power.

Table 5.6 summarizes the equations that introduce product differentiation and differentiated water sources. The equations for supply and demand for irrigation inputs follow an approach similar to that introduced earlier. The substitution elasticities are key parameters in demand and are derived from the nested CES functions; supply elasticities are also key parameters in supply functions. Here, *GW*, *SW*, and *KW* represent groundwater, surface water, and irrigation equipment, respectively. While supply of *GW* and *SW* are grid-cell-specific, the supply of *KW* occurs at the regional level, assuming that irrigation capital is a mobile factor of production.

6   Consumer Demand and Trade

The one common module across all of the SIMPLE-G specifications in Models 1–5 is the demand system, which follows the (nongridded) SIMPLE model (Baldos and Hertel. 2013; Hertel and Baldos 2016). This section describes the relevant equations taken from the SIMPLE regional model. We have modified the notation slightly but kept the demand structure intact. Table 5.7 summarizes these equations for quick reference.

6.1   Equations for Socioeconomic Determinants of Food Demand

At the regional scale, consumer demand for different commodities is a function of population, income, and prices. Equilibrium prices are determined endogenously as a function of supply and demand, while population and income changes are exogenous to the model, with increases in per capita income driving diet changes. Total demand for crops in a given region comes from four sources. The direct crop demand is calculated by multiplying per capita demand for crop consumption by population in each region. Then, we sum the total direct demand for crops in final consumption with the indirect (derived) demands in the livestock, food processing, and biofuel sectors. The demand for crops in biofuel production is assumed to be exogenously determined by government mandates. The livestock and food processing sectors' demands for crops are endogenous and modeled using CES production functions

Table 5.7 Major equations for the demand side of SIMPLE-G

Equation	Description
$q_{r,f} = \varepsilon^p_f p_{r,f} + \varepsilon^y_f y_{r,f}$ $\varepsilon^y_{r,f} = \alpha^y_{r,f} + \beta^y_{r,f} \ln Y_r **$ $\varepsilon^p_{r,f} = \alpha^p_{r,f} + \beta^p_{r,f} \ln Y_r **$	Consumer demand for food and its elasticities
$qc^V_r + ac^V_r = qo^V_r - ao^V_r - \sigma^V_r\left(pc^V_r - ac^V_r - p^V_r - ao^V_r\right)$ $p^V_r = \theta^V_r pc^V_r + \left(1 - \theta^V_r\right)pn^V_r$	Livestock (V) demand for crops as feed
$qc^F_r + ac^F_r = qo^F_r - ao^F_r - \sigma^F_r\left(pc^F_r - ac^F_r - p^F_r - ao^F_r\right)$ $p^F_r = \theta^F_r pc^F_r + \left(1 - \theta^F_r\right)pn^F_r$	Processed food (F) demand for crops
$qm_{r,b} = qc_{r,b} - \omega_r(pm_r - pc_r)$ $qd_{r,b} = qc_{r,b} - \omega_r(pd_r - pc_r)$ $pc_r = \theta^m_r pm_r + \theta^d_r pd_r$ $pm_r = pw$	Demand for imported and domestic crops
$qx_r = q_r - \psi_r(px_r - p_r)$ $qd_r = q_r - \psi_r(pd_r - p_r)$ $p_r = \theta^x_r px_r + \theta^s_r pd_r$ $px_r = pw$	Exports and domestic supply

that combine the raw crop input with other inputs used in livestock or processed food production.

One of the best-understood patterns of economic development is Engel's Law, which states that as per capita income rises, the share of income devoted to food will decrease (Clements and Chen 1996). The SIMPLE framework captures this relationship by allowing the income elasticity of demand for food (ε_i^y, the propensity to spend incremental income on food) to evolve with per capita income (Y) based on the estimated parameters α_i^y and β_i^y. The price elasticity of food demand (ε_i^p) is derived similarly. In a general form,

$$\varepsilon_i^y = \alpha_i^y + \beta_i^y \ln Y; \tag{5.24}$$

$$\varepsilon_i^p = \alpha_i^p + \beta_i^p \ln Y. \tag{5.25}$$

Equations 5.24 and 5.25 are indexed by food type (i). SIMPLE distinguishes between direct consumption of crops and indirect consumption through either livestock product consumption or processed food consumption. This results in the following equations describing the evolution of per capita demand for each type of food product:

$$q_i = \varepsilon_i^p p_i + \varepsilon_i^y y. \tag{5.26}$$

Specifically, for each region r and commodity i, we can write consumption demand as

$$qcons_{r,i} = \varepsilon_{r,i}^p p_{r,i} + \varepsilon_{r,i}^y y_r, \tag{5.27}$$

where *qcons* shows the demand for commodities that include crops, livestock, processed food, and nonfood. While the income elasticity of demand for food is typically positive (i.e., the demand for food increases as income increases), it can also be negative, showing a dietary shift with income growth. The price elasticity of demand for food is negative, which means that the demand for food decreases as the price of food increases.

For each food commodity, we introduce a production function to show different uses of crops in the food sectors. For each use (activity), the production function combines crops as one input with other inputs and supplies a food commodity to the final consumer. In each food sector, demand for crops is determined assuming a CES function. The linearized form of demand can be written as

$$qc_r^V + ac_r^V = qo_r^V - ao_r^V - \sigma_r^V \left(pc_r^V - ac_r^V - p_r^V - ao_r^V \right)$$

$$p_r^V = \theta_r^V pc_r^V + \left(1 - \theta_r^V \right) pn_r^V \tag{5.28}$$

$$qc_r^F + ac_r^F = qo_r^F - ao_r^F - \sigma_r^F \left(pc_r^F - ac_r^F - p_r^F - ao_r^F \right)$$

$$p_r^F = \theta_r^F pc_r^F + \left(1 - \theta_r^F \right) pn_r^F$$

where qc is the percentage change in crop demand for use in processed food (F) or livestock (V) production region r; ac is related to partial factor productivity; qo is the percentage change in activity output; ao is the total factor productivity in activity i; σ is the substitution elasticity, which differ by region and use; pc is the regional crop price index; and p is the regional food commodity price index. The equilibrium condition is that the commodity supplied from each activity should be the quantity demanded, given the absence of trade for these processed products in this model:

$$qo_r^f = qcons_{r,f}. \tag{5.29}$$

6.2   Equations for Global Trade Flows

As with consumer demand, SIMPLE-G follows the SIMPLE model in its specification of international trade. It relies on product differentiation by origin (Armington 1969), identifying domestic and international goods separately. Heterogeneous consumers are assumed to choose between domestic and imported goods, based on relative prices and preferences; when aggregated, the demand for imports and domestic goods follows a CES function (Anderson et al. 1989). The more homogeneous consumers are (both in preferences and market access), the greater the elasticity of substitution and the more readily that aggregate demand shifts between imported and domestic crop products. Producers are also assumed to be heterogeneous, with different access to international markets. Analogous to what occurs among consumers, this results in a CET market supply function for domestic and export markets. In addition to the elasticities of substitution and transformation between domestic and international goods, the initial penetration of exports and imports into any given market will play a key role in determining the extent of price transmission from global to domestic markets. This can differ between consumers and producers. Finally, it is important to note that SIMPLE-G does not model international trade in livestock or processed food products. These commodities are more lightly traded than are crops, with much of the food processing occurring locally, often facilitated by foreign direct investment. Nonetheless, this remains an important limitation of SIMPLE-G. A proposal for relaxing this is explored in Part V of this book, where SIMPLE-G is nested with a general equilibrium model in which all commodities are traded.

The demand for imports depends on the prices of domestically produced crops and import prices, as well as the Armington elasticity of substitution:

$$qm_{r,b} = qc_{r,b} - \omega_r(pm_r - pc_r), \tag{5.30}$$

$$qd_{r,b} = qc_{r,b} - \omega_r(pd_r - pc_r), \tag{5.31}$$

$$pc_r = \theta_r^m pm_r + \theta_r^d pd_r, \tag{5.32}$$

$$pm_r = pw, \tag{5.33}$$

where m is imports and d is domestic markets; qm is the percentage change in demand for imported crops; qd is percentage change in demand for domestic crops; qc is total demand for crops by buyer; pc is average buyers' price index; pm is the imports price index; pw is the global average price index of all crops; θ is the value share in price index; and ω is the Armington substitution elasticity parameter.

Imports are obtained from the global market for crops, which is fulfilled through exports. Producers in each region face a choice between selling their goods within their country or exporting them to the global market. This decision involves a trade-off, which is captured by the (negative-valued) CET parameter. This parameter indicates how easily producers can switch between domestic and global markets. A higher absolute value for this elasticity shows that producers are more responsive to price changes, (i.e., they are more likely to focus on exporting their goods when foreign prices are comparatively attractive). In contrast, a lower absolute value for this elasticity suggests that producers have less access to global markets and are therefore less affected by the price differentials that may emerge between domestic and global markets:

$$qx_r = q_r - \psi_r(px_r - p_r), \tag{5.34}$$

$$qd_r = q_r - \psi_r(pd_r - p_r), \tag{5.35}$$

$$p_r = \theta_r^x px_r + \theta_r^s pd_r, \tag{5.36}$$

$$px_r = pw, \tag{5.37}$$

where x is exports and d is domestic markets; qx is the percentage change in supply for exports to global markets; qd is percentage change in supply to domestic markets; q is total regional supply of crops; p is average producer price index; px is the exports price index; pw is the global average price index of all crops; θ is the value share in price index; and ψ is the (negative) CET transformation elasticity parameter.

7   Summary

This chapter presented mathematical equations for five specifications of the SIMPLE-G gridded production and markets. In Model 1, agricultural production is a function of human and environmental inputs with different degrees of mobility across space. This framework is useful for understanding the tradeoffs and synergies between food security and environmental sustainability as well as evaluating spill-over effects. In Model 2, land and water inputs are explicitly introduced, enabling detailed modeling of land and water use at each grid cell. Model 3 adds fertilizer inputs and their markets to enable water quality assessments. In Model 4, multiple local activities are introduced at each grid cell, showing competition over local land and water resources. Finally, in Model 5, groundwater and surface water resources are separated, enabling analysis of specific groundwater sustainability policies. These models are examples of the SIMPLE-G framework. While they provide a solid foundation for sustainability analysis, it is important to keep in mind that additional dimensions may need to be introduced to ensure a comprehensive analysis, depending on the research question. By doing so, we can enhance our understanding and arrive at more accurate conclusions. The models introduced in this chapter are both flexible and scalable. With appropriate data preparation, they can be applied to various grid cell sizes, ranging from very fine to coarse resolutions, as needed for multiscale analysis. As you learn more about data preparation and parameter selection in the upcoming chapters, we invite you to consider how these tools can be harnessed to uncover new insights and drive meaningful discoveries in your research.

Acknowledgment and Competing Interests The author acknowledges support from the U.-S. Department of Energy, Office of Science, Biological and Environmental Research Program, Earth and Environmental Systems Modeling, MultiSector Dynamics under Cooperative Agreement DE-SC0022141; the National Science Foundation award #2118329: "NSF Institute for Geospatial Understanding through an Integrative Discovery Environment (I-GUIDE)," the United States Department of Agriculture AFRI grant #2019-67023-29679, "Economic Foundations of Long Run Agricultural Sustainability," and the National Science Foundation INFEWS award #1855937, "Identifying Sustainability Solutions Through Global-Local-Global Analysis of a Coupled Water-Agriculture-Bioenergy System."

The findings and conclusions presented in this chapter are those of the authors and should not be construed to represent any official determination or policy of the US Department of Agriculture (USDA), the US government, the Department of Energy (DOE), or the National Science Foundation (NSF). Furthermore, we declare that there is no conflict of interest related to this work.

References

Ahmed, Syud Amer, Thomas W. Hertel, and Ruben Lubowski. 2008. *Calibration of a land cover supply function using transition probabilities*. West Lafayette: Research Memorandum 14. Department of Agricultural Economics, Purdue University, Global Trade Analysis Project (GTAP). https://doi.org/10.21642/GTAP.RM14.

Anderson, James E. 1979. A theoretical foundation for the gravity equation. *The American Economic Review* 69: 106–116.

Anderson, Simon P., André De Palma, and Jacques-François Thisse. 1989. Demand for differentiated products, discrete choice models, and the characteristics approach. *The Review of Economic Studies* 56: 21–35. https://doi.org/10.2307/2297747.

Armington, Paul S. 1969. A theory of demand for products distinguished by place of production. *Staff Papers* 16: 159–178. https://doi.org/10.2307/3866403.

Baldos, Uris Lantz C., and Thomas W. Hertel. 2013. Looking back to move forward on model validation: Insights from a global model of agricultural land use. *Environmental Research Letters* 8: 034024. https://doi.org/10.1088/1748-9326/8/3/034024.

Chaney, Thomas. 2008. Distorted gravity: The intensive and extensive margins of international trade. *American Economic Review* 98: 1707–1721. https://doi.org/10.1257/aer.98.4.1707.

———. 2018. The gravity equation in international trade: An explanation. *Journal of Political Economy* 126: 150–177. https://doi.org/10.1086/694292.

Clements, Kenneth W., and Dongling Chen. 1996. Fundamental similarities in consumer behaviour. *Applied Economics* 28: 747–757. https://doi.org/10.1080/000368496328498.

Dixon, Peter B. 1982. *ORANI, a multisectoral model of the Australian economy*. 1st ed. New York: Elsevier.

Haqiqi, Iman, Laura Bowling, Sadia Jame, Uris Baldos, Jing Liu, and Thomas Hertel. 2023. Global drivers of local water stresses and global responses to local water policies in the United States. *Environmental Research Letters* 18: 065007. https://doi.org/10.1088/1748-9326/acd269.

Haqiqi, Iman, and Thomas W. Hertel. 2016. Decomposing irrigation water use changes in equilibrium models. In *Annual meeting*. Boston: Agricultural and Applied Economics Association.

———. 2019. Estimating water withdrawal response to environmental stresses. In *Annual meeting*. Atlanta, GA: Agricultural and Applied Economics Association. https://doi.org/10.22004/ag.econ.291097.

Haqiqi, Iman, Chris J. Perry, and Thomas W. Hertel. 2022. When the virtual water runs out: Local and global responses to addressing unsustainable groundwater consumption. *Water International* 47: 1060–1084. https://doi.org/10.1080/02508060.2023.2131272.

Hertel, Thomas W., Lantz C. Uris, and Baldos. 2016. *Global change and the challenges of sustainably feeding a growing planet*. New York: Springer.

Hillberry, Russell, and David Hummels. 2003. Intranational home bias: Some explanations. *The Review of Economics and Statistics* 85: 1089–1092. https://doi.org/10.1162/003465303772815970.

———. 2008. Trade responses to geographic frictions: A decomposition using micro-data. *European Economic Review* 52. Elsevier: 527–550. https://doi.org/10.1016/j.euroecorev.2007.03.003.

———. 2013. Trade elasticity parameters for a computable general equilibrium model. In *Handbook of computable general equilibrium modeling*, vol. 1, 1213–1269. Elsevier. https://doi.org/10.1016/B978-0-444-59568-3.00018-3.

Liu, Jing, Thomas W. Hertel, Richard B. Lammers, Alexander Prusevich, Uris Lantz C. Baldos, Danielle S. Grogan, and Steve Frolking. 2017. Achieving sustainable irrigation water withdrawals: Global impacts on food security and land use. *Environmental Research Letters* 12: 104009. https://doi.org/10.1088/1748-9326/aa88db.

van der Mensbrugghe, Dominique, and Jeffrey C. Peters. 2020. *Volume preserving CES and CET formulations*. Working Paper 87. Department of Agricultural Economics, Purdue University, West Lafayette, IN: Global Trade Analysis Project (GTAP). https://doi.org/10.21642/GTAP.WP87.

Ray, Srabashi, Iman Haqiqi, Alexandra E. Hill, J. Edward Taylor, and Thomas W. Hertel. 2023. Labor markets: A critical link between global-local shocks and their impact on agriculture. *Environmental Research Letters* 18: 035007. https://doi.org/10.1088/1748-9326/acb1c9.

Tinbergen. 1962. An analysis of world trade flows. In *Shaping the world economy: Suggestions for an international economic policy*, ed. Jan Tinbergen, 262–293. New York: Twentieth Century Fund.

Woo, Jungha, Lan Zhao, Danielle S. Grogan, Iman Haqiqi, Richard Lammers, and Carol X. Song. 2022. C3F: Collaborative container-based model coupling framework. In *Practice and experience in advanced research computing*, 1–8. Boston, MA: ACM. https://doi.org/10.1145/3491418.3530298.

Chapter 6
Benchmark Data: Integrating Biophysical and Economic Information in a Consistent Geospatial Dataset

Iman Haqiqi and Uris Lantz C. Baldos

Existing environmental databases often suffer from fragmentation, inconsistencies, and limited spatial representation. In addition, economic representation at the geospatial level is often limited. However, the SIMPLE-G dataset aims to overcome these limitations by providing a unified platform for exploring the interactions between land, water, and the environment. Each data point within the dataset is geographically referenced with details about economic values and biophysical variables, enabling researchers to conduct spatial analyses and uncover crucial insights into localized resource use, economic patterns, and sustainability challenges. Furthermore, the dataset's comprehensive scope, which encompasses both production and consumption aspects of the agri-food system, fosters a holistic understanding of resource flows and economic dependencies. This data resource empowers researchers to tackle complex environmental questions and propose data-driven solutions for promoting sustainable land and water management practices. While the chapter concentrates on the SIMPLE-G-US dataset (Baldos et al. 2020; Haqiqi et al. 2023), a similar approach is taken for other global and regional models (Liu et al. 2017; Haqiqi et al. 2022).

The SIMPLE-G model uses two types of information: benchmark data and behavioral parameters. Benchmark data include information describing the equilibrium conditions of input use, agricultural production, trade, and food consumption in the reference year. The values of the variables are expected to change and are updated to reflect the new economic conditions following a SIMPLE-G simulation. The benchmark data are collected from various sources, including international databases, satellite imagery, climate and hydrological data, and national censuses. These data must be processed and made consistent to produce gridded data that can be used in the model.

I. Haqiqi (✉) · U. L. C. Baldos
Center for Global Trade Analysis, Department of Agricultural Economics, Purdue University, West Lafayette, IN, USA
e-mail: ihaqiqi@purdue.edu

© The Author(s) 2025
I. Haqiqi, T. W. Hertel (eds.), *SIMPLE-G*,
https://doi.org/10.1007/978-3-031-68054-0_6

Behavioral parameters refer to information that describes how the model responds to changes in various factors (e.g., prices and income). These parameters are based on independent estimates of consumer and producer behavior. They govern consumers' and producers' responses to changes in the economic, climatic, and environmental conditions of agricultural production. By definition, these parameters are fixed over the course of the simulation.

Before describing the gridded data, we explain how the regional (non-gridded) SIMPLE database is constructed.

1 Non-gridded Data

The non-gridded SIMPLE database includes information on macroeconomic variables, commodity consumption, regional trade for crops, and regional production of crops, livestock, and processed food. Additional data on input and output carbon emissions for each commodity (Aguiar et al. 2019), land-use change emissions (West et al. 2010), and food security metrics (FAO 2020) are also available for model versions with environmental and food security modules. A publicly available version of the non-gridded database and compatible model version has detailed information for 150+ countries for the year 2017 (Baldos 2023).

1.1 Regional Production and Trade

The regional crop production database in SIMPLE-G is constructed using country-level economic and agricultural data taken from several sources. Crop production, producer crop price, and cropland area data are sourced from the United Nations Food and Agricultural Organization's FAOSTAT database (FAO 2020). Global prices are computed for each crop and are used to convert quantity of crop production into corn-equivalent production using the ratio of each crop's global price and the global price of corn. The value of crop production is also calculated using these data. Since we assume that the benchmark dataset represents a long-run equilibrium, we impose zero profit conditions in the crop sector. Therefore, the annual flow of land and nonland input costs in each region can be calculated using the value of crop production and the input cost shares from the GTAP database (Aguiar et al. 2019). The crop sale shares across the domestic and international markets are taken from the GTAP database (Aguiar et al. 2019).

1.2 Regional Consumption

On the demand side, the benchmark data include the value and quantity of crops used in direct food consumption, for feed in the livestock sectors, as raw inputs in the processed food industries, and as feedstocks in the biofuel sector. The amount of crop feedstock used by the biofuel sector in each region is calculated using the sales shares of the crop sector in the bioenergy sectors from the GTAP-BIO V.9 database (Taheripour et al. 2017). Remaining crop quantities are allocated to food, feed, and processed food input use using crop purchase shares in the local and global markets using share data from the GTAP database (Aguiar et al. 2019).

Consumer demand data in the model are mainly driven by population and per capita income. Population and real gross domestic product (GDP) data are taken from FAOSTAT (FAO 2020), and these are used to calculate the per capita income in each region. The food security module in SIMPLE includes food security metrics based on average food consumption in each region (Baldos and Hertel 2014). Following Neiken (2003), each region has a unique distribution of average caloric consumption and a minimum daily dietary energy intake which delineates the fraction of the distribution below this minimum threshold. The nutrition data include the prevalence of caloric undernutrition and the undernutrition headcount for each region. This distribution shifts depending on changes in average dietary energy consumed. The module also captures shifts in the caloric composition of food by linking food caloric content to per capita income levels (Baldos and Hertel 2013). Food security data from FAOSTAT (FAO 2020) are used to initialize the nutritional information for each region.

1.3   Food Processing and Livestock Production

The value of consumption for livestock and processed food commodities is equal to the total value of output for these sectors, given the assumption that total revenue is equal to total consumption expenditure for each commodity. Under zero-profit conditions, the total value of output is equal to the total production costs. The total cost of crop inputs in the livestock and processed food sectors is calculated using the crop prices and crop use quantities. The values of non-crop inputs are then computed from the total cost of crop inputs and input cost shares. The input cost shares for livestock and processed food commodities are taken from the GTAP database (Aguiar et al. 2019).

1.4   Agricultural Greenhouse Gas Emissions

Some model versions (e.g., Chap. 10) report changes in agricultural greenhouse gas (GHG) emissions, specifically emissions from crop, livestock, and processed food production as well as land-use change emissions due to cropland expansion. Data on GHG emissions from agricultural production are based on the GTAP database (Aguiar et al. 2019), which reports CO_2 emissions from fossil fuel combustion using detailed energy volume data from the International Energy Agency and combustion factors from the Revised 1996 *IPCC Guidelines for National Greenhouse Gas Inventories* (IPCC/OECD/IEA 1997). GHG emissions from methane, nitrous oxide and fluorinated, gases are based on the non-CO_2 GTAP database (Chepeliev 2023), which uses FAOSTAT data (FAO 2020) for agricultural emissions. Land-use change emissions from converting natural land into cropland rely on the global carbon stocks calculated by West et al. (2010). These carbon stocks are constructed using spatially explicit datasets on potential vegetation and soil carbon.

2   Benchmark Gridded Data: US Example

SIMPLE-G requires benchmark gridded data for key economic and biophysical variables describing the initial equilibrium of the crop economy. The preferred gridded data sources vary by region. Table 6.1 lists the variables and underlying data sources for the United States. Comparable data sources have been compiled for several other regions (see the applications in Part IV). Each model includes a gridded dataset that includes cropland use, crop production, nitrate leaching, and water use.

Given the centrality of the gridded database to the use and credibility of SIMPLE-G, this section provides a brief overview of how to construct a SIMPLE-G dataset. Figure 6.1 describes the workflow for constructing the benchmark data and behavioral parameters for a US-focused version of SIMPLE-G, utilizing gridded

Table 6.1 Gridded data for SIMPLE-G-US

Data	Source	Description of process
Cropland area	USDA CDL, USGS MIrAD	The share of cropland area in each grid cell is calculated from 30 m resolution. The share of irrigated area from cropland is calculated from 25 m resolution
Value of land	USDA Census of Agriculture	Land area is multiplied by rent in USD per ha assuming uniform rent within a county
Value of crops produced	USDA Census of Agriculture, GCWM by crop	The value of crops sold by county is distributed to grid cells within a county based on gridded pattern from GCWM and irrigated yield gap
Quantity index of crops	Estimated	The quantity index follows the value of crops sold divided by reference crop price to create a price adjusted corn-equivalent index
Value of irrigation	USGS and USDA	Irrigation cost shares are estimated based on county-specific data on share of sprinkler, drip irrigation, surface water, and groundwater
Volume of water	USGS	Water volume follows land multiplied by water intensity in m^3/ha
Quantity of N fertilizer	Cao et al. (2018), Agro-IBIS	Total fertilizer application follows land multiplied by kg/ha
Value of N fertilizer	Estimated	Value of fertilizer follows quantities multiplied by price USD per kg

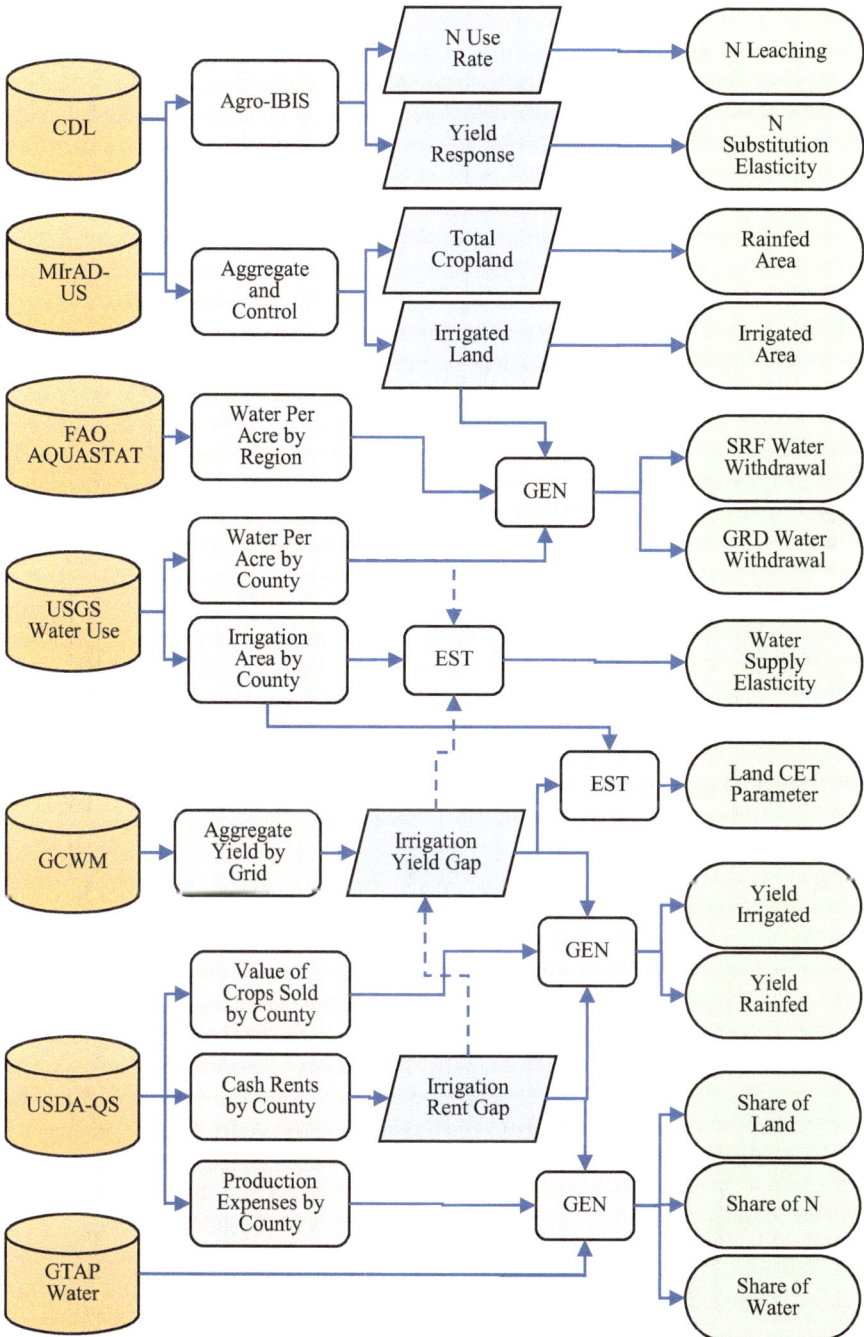

Fig. 6.1 Overview of main data and parameter processing method for SIMPLE-G-US. "GEN" represents a computation that does not involve statistical regressions; "EST" is a process that includes statistical estimation. Other versions, (gridded World, gridded China, and gridded Brazil) follow similar flows but employ rich global and national data sources.

data for the United States and employing regional information for other parts of the world. Other versions (e.g., SIMPLE-G-China, SIMPLE-G-Brazil) follow similar flows but employ the best available national data sources from those countries. In addition to the primary source data, there are two types of methodologies for generating the gridded data: "GEN" represents a computation that does not involve statistical regressions, and "EST" is a process that includes statistical estimation.

Although resolution is not a constraint for SIMPLE-G (see computational section for examples with millions of grid cells), current versions work with crop production at the level of georeferenced grid-cell units at 5 arcmin resolution (squares with sides of 9.26 km at the equator). Most of our collaborators in the biophysical sciences use this resolution. We have added gridded information for US crop production covering both irrigated and nonirrigated practices and including the value and quantity of crop output, land use, N fertilizer input, water, and other aggregated inputs.

2.1 Cropland Area: Rainfed and Irrigated

Cropland area data were obtained from the USDA Cropland Data Layer (Han et al. 2012) at a resolution of 30 meters and aggregated to 5 arcmin. Irrigated cropland data were obtained from the Moderate Resolution Imaging Spectroradiometer (MODIS) Irrigated Agriculture Dataset for the United States (MIrAD-US) provided by the US Geologic Survey (USGS) at a resolution of 250 m (Brown and Pervez 2014), aggregated to 5 arcmin. These data determine the distribution of irrigated and nonirrigated cropland in the United States.

2.2 Crop Production: Price-Adjusted Quantity Index

Aggregated output for each grid cell is recorded as the corn-equivalent total crop output, calculated as the sum of the value the crops sold divided by the price of corn in the base year, yielding "corn-equivalent tons of output." We take the value of crops sold per acre from USDA National Agricultural Statistics Service by county (USDA-NASS 2019) and use simulated yields from the Global Crop Water Model (GCWM) (Siebert and Döll 2010) to generate gridded yield for the grid cells in each county. The GCWM data are aggregated over all crops using USDA/FAO actual prices to calculate the corn-equivalent crop output for each grid cell.

We split the base data into irrigated and rainfed crop production using various satellite datasets and county-level information from the USDA and USGS, as well as the simulated yield of irrigated and nonirrigated crops. For yield estimation, we assume that grid cell aggregated yield per hectare is equal to the county average in which the grid cell is located. We split the gridded total production into irrigated and nonirrigated components using total, irrigated, and nonirrigated land as obtained from MIrAD-US and the cropland data layer (CDL) and the ratio of rainfed to irrigated yields in a given grid cell as estimated by Siebert and Döll (2010) for 29 crop categories and aggregated to all crops according to production value weights. Total cropland area from the CDL is matched with USDA county-level cropland to ensure consistency of yield and area at the county level.

2.3   Nitrogen Fertilizer Applications

Nitrogen (N) fertilizer application rates per hectare per year for each grid cell were obtained from Cao et al. (2018) for all crops at 5 km and from Agro-IBIS (the agricultural version of the Integrated Biosphere Simulator) for rainfed and irrigated production of major crops (Lark et al. 2022). This product provided high-resolution (8 arcsecond) land cover and nutrient application maps across the continental United States for the period of 1750 to 2017, accounting for the nutrient legacies of historical land use/cover. The Agro-IBIS land cover categories are based on vegetation type simulations. This product is also consistent with our landcover data, which is based on several gridded land cover datasets and historical county-level USDA Census of Agriculture data. Irrigation maps from Agro-IBIS were created using the MIrAD-US and historical data from the USDA Census of Agriculture. With its rich set of information on gridded economic activity, the agroecological Agro-IBIS model helped us estimate irrigated versus nonirrigated fertilizer application rates.

2.4   Volume of Water Withdrawals

Irrigation water withdrawal rates are estimated using county-level USGS water use data (Maupin et al. 2014). We calculate total water withdrawal per irrigated hectare and allocate it to groundwater and surface water using county-level USGS water use data by source. For some version of SIMPLE-G a quantity index is introduced for water which requires conversion to m3 to show the level of variables.

2.5   Value of Crop Production

Calculating the value of crops produced at the grid-cell level requires several steps. First, satellite data with sufficient resolution are acquired to identify cropland within each grid cell. These data are combined with gridded maps of irrigated areas to separate irrigated and rainfed areas in each grid cell. To avoid the computational issues that arise when extremely small pieces of cropland are modeled, we assign a minimum area to each grid cell in the model if it is cultivated. Next, utilizing the agronomic GCWM data, crop yields and areas are estimated by crop at the grid-cell level. Agronomic models incorporate factors such as weather, soil conditions, and crop management practices to simulate crop growth and predict yields by irrigation technology. The combination of land area and yield provides information on production by crop. Next, information on the average prices of crops at the national level is collected from FAOSTAT. For each grid cell, crop-specific values are calculated from the agronomic model by multiplying production by the corresponding crop price, considering the crop area data layer derived from the satellite data. This will provide an estimate of the total value of crops produced within that grid cell by irrigation technology. Since our goal is to match the USDA census data by county, we apply the gridded patterns of production value to the USDA county-level data on the value of crops produced in each county to obtain census-consistent, gridded estimates of the value of crop production.

This approach provides a comprehensive and spatially explicit assessment of crop production values, providing granular insights into agricultural productivity and economic contributions.

2.6   Value of Land

The annual payment to land, also known as land rent, is a crucial component of the economic model. It represents the compensation that landowners receive for the use of their land in agricultural production. We assume that these same landowners consider alternative uses, where feasible; therefore, each grid cell has an upward-sloping land supply curve, reflecting the opportunity cost of crop land. We consider explicit payments from rental lands to landowners as well as implicit self-payment of landowner farmers. The value of water is calculated based on irrigated and rainfed cash rents by county in the United States.

In SIMPLE-G, total annual payments to land play a key role in constructing the database. At the grid-cell level, the calculation of land rent typically involves two key factors: land area and rental rates by irrigation type. While land area information can be obtained from various sources (e.g., satellite imagery), information about land rent requires surveys or census data. Land rents represent the per unit payment, typically expressed in terms of dollars per hectare (or per acre) per year, and vary depending on several factors, including soil quality, proximity to markets, access to irrigation, crop mix, and overall demand for land in the county. Understanding these factors and their influence on land rents is crucial for understanding spatial heterogeneity in cost structures across a region.

To calculate the total payment to land in a grid cell, the gridded land area is multiplied by the average county-level rental rate, as reported by the USDA. This calculation provides an estimate of the overall compensation that landowners receive, implicitly and explicitly, for the use of their land for agricultural purposes. It is important to note that this calculation represents an average and may not accurately reflect the specific rental payments in a given grid cell.

2.7 Value of Irrigation

The central challenge with regard to valuing water is that there is usually no market for water. Only a small portion of irrigation water is traded in markets. Even where they exist, market prices for water rarely reflect the true value added of irrigation water in agriculture. Thus, we developed an alternative approach to estimate the value of water. We start from the land rent. The cost share (θ) for cropland (L) for each grid-cell (g) is defined as

$$\theta^L_{g,l} = QL_{g,l}PL_{g,l}/QY_{g,l}PY_{g,l} \tag{6.1}$$

where QY and QL are production level and land area, and PY and PL are price index and land rents, respectively. As noted, much of the cropland is owned by the enterprise or individual farming the land. Thus, the true value added of land is not observed. Therefore, we infer rental rates from average cash rents reported by other farms in the neighborhood. In combination of CDL cropland data, we then calculate the total rental costs land. We have constructed an empirical model to estimate the cost shares of water inputs for groundwater, surface water, and irrigation input. We assume that the distribution of cost shares is driven by the volume share of water extracted from groundwater (XGW) and surface water (XSW), extent of area irrigated by each source (LGW and LSW, respectively), and shares of area equipped with higher efficiency irrigation technology such as microdrip irrigation (XMIC) and sprinklers (XSPR). The cost shares for groundwater, surface water, and irrigation equipment inputs are estimated using the following equations from Haqiqi (2023):

$$\begin{aligned}
\theta^{GW}_{l,g} &= \alpha^G_0 + \alpha^G_1 XGW_{l,g} + \alpha^G_1 LGW_{l,g} \\
\theta^{SW}_{l,g} &= \alpha^S_0 + \alpha^S_1 XSW_{l,g} + \alpha^S_1 LSW_{l,g} \\
\theta^{KW}_{l,g} &= \alpha^K_0 + \alpha^K_1 XMIC_g + \alpha^K_2 XSPR_g
\end{aligned} \tag{6.2}$$

Equation (6.2) considers the associated costs of irrigation, including the labor, fuel, and other inputs required for water withdrawal, transport, and application. Therefore, we can empirically estimate the irrigation costs using the observed variables at the county level. In a statistical framework, we partially attribute the spatial differences in fuel, labor, and other nonland material to irrigation. Specifically, county-level nonland costs are estimated as a function of irrigated areas in acres, surface water withdrawals in gallons per acre, groundwater withdrawals in gallons per acre, share of area with sprinkler irrigation, and share of area with microdrip irrigation. We differentiate the marginal costs of groundwater and surface water irrigation. County-level expense data are obtained from USDA Census of Agriculture for total costs and input costs, including fuel, labor, fertilizer, seeds, chemicals, and other inputs for 2002, 2007, 2012, and 2017. Information on physical area and volume of water (IR-WGWFr, IR-WSWFr, IR-IrSpr, and IR-IrMic) is obtained from USGS Estimated Use of Water in the United States County-Level Data for 2005, 2010, and 2015.

3   Summary

The SIMPLE-G benchmark data include a consistent geospatial dataset that provides information on both economic values and biophysical variables for a reference year. At the regional level, this dataset includes exports, imports, income, and consumer expenditures on different food and non-food categories. At the grid-cell level, this includes the value of crop sales and production costs such as land, water, and labor expenses. The biophysical information includes cropland area, volume of water withdrawals, fertilizer applications, crop yields, and different land uses. These rich data enable researchers to conduct in-depth geospatial economic and sustainability analyses related to agriculture, land, water, and environment.

Acknowledgment and Competing Interests   The authors acknowledge support from the National Science Foundation award #2118329: "NSF Institute for Geospatial Understanding through an Integrative Discovery Environment (I-GUIDE)," the U.S. Department of Energy, Office of Science, Biological and Environmental Research Program, Earth and Environmental Systems Modeling, MultiSector Dynamics under Cooperative Agreement DE-SC0022141; the United States Department of Agriculture AFRI grant #2019-67023-29679, "Economic Foundations of Long Run Agricultural Sustainability," and the National Science Foundation INFEWS award #1855937, "Identifying Sustainability Solutions Through Global-Local-Global Analysis of a Coupled Water-Agriculture-Bioenergy System."

The findings and conclusions presented in this chapter are those of the authors and should not be construed to represent any official determination or policy of the US Department of Agriculture (USDA), the US government, the Department of Energy (DOE), or the National Science Foundation (NSF). Furthermore, we declare that there is no conflict of interest related to this work.

References

Aguiar, Angel, Maksym Chepeliev, Erwin L. Corong, Robert McDougall, and Dominique van der Mensbrugghe. 2019. The GTAP data base: Version 10. Journal of global. *Economic Analysis* 4: 1–27. https://doi.org/10.21642/JGEA.040101AF.

Baldos, Uris Lantz C. 2023. SIMPLE database and model for base year 2017. https://doi.org/10.13019/RPZW-BX12.

Baldos, Uris Lantz C., and Thomas W. Hertel. 2013. Looking back to move forward on model validation: Insights from a global model of agricultural land use. *Environmental Research Letters* 8: 034024. https://doi.org/10.1088/1748-9326/8/3/034024.

———. 2014. Global food security in 2050: The role of agricultural productivity and climate change. *Australian Journal of Agricultural and Resource Economics* 58: 554–570. https://doi.org/10.1111/1467-8489.12048.

Baldos, Uris Lantz C., Iman Haqiqi, Thomas W. Hertel, Mark Horridge, and J. Liu. 2020. SIMPLE-G: A multiscale framework for integration of economic and biophysical determinants of sustainability. *Environmental Modelling & Software* 133. Elsevier: 104805. https://doi.org/10.1016/j.envsoft.2020.104805.

Brown, Jesslyn F., and Md Shahriar Pervez. 2014. Merging remote sensing data and national agricultural statistics to model change in irrigated agriculture. *Agricultural Systems* 127: 28–40. https://doi.org/10.1016/j.agsy.2014.01.004.

Cao, Peiyu, Lu Chaoqun, and Yu. Zhen. 2018. Historical nitrogen fertilizer use in agricultural ecosystems of the contiguous United States during 1850–2015: Application rate, timing, and fertilizer types. *Earth System Science Data* 10: 969–984. https://doi.org/10.5194/essd-10-969-2018.

Chepeliev, Maksym. 2023. GTAP-power data base: Version 11. *Journal of Global Economic Analysis* 8: 101–135. https://doi.org/10.21642/JGEA.080203AF.

FAO. 2020. *FAOSTAT: Food and agriculture data*. Rome: Food and Agriculture Organization of the United Nations. http://faostat.fao.org/. Accessed 1 Dec 2020.

Han, Weiguo, Zhengwei Yang, Liping Di, and Richard Mueller. 2012. CropScape: A web service based application for exploring and disseminating US conterminous geospatial cropland data products for decision support. *Computers and Electronics in Agriculture* 84: 111–123. https://doi.org/10.1016/j.compag.2012.03.005.

Haqiqi, Iman. 2023. The value of water in US agriculture: Integrating spatially and temporally heterogeneous hydroclimatic and economic data (version 1.0). MyGeoHUB. https://doi.org/10.13019/9MXE-T280.

Haqiqi, Iman, Chris J. Perry, and Thomas W. Hertel. 2022. When the virtual water runs out: Local and global responses to addressing unsustainable groundwater consumption. *Water International* 47: 1060–1084. https://doi.org/10.1080/02508060.2023.2131272.

Haqiqi, Iman, Laura Bowling, Sadia Jame, Uris Baldos, Jing Liu, and Thomas Hertel. 2023. Global drivers of local water stresses and global responses to local water policies in the United States. *Environmental Research Letters* 18: 065007. https://doi.org/10.1088/1748-9326/acd269.

IPCC/OECD/IEA. 1997. *1996 IPCC guidelines for national greenhouse gas inventories*. Bracknell: IPCC/OECD/IEA UK Meteorological Office.

Lark, Tyler J., Nathan P. Hendricks, Aaron Smith, Nicholas Pates, Seth A. Spawn-Lee, Matthew Bougie, Eric G. Booth, Christopher J. Kucharik, and Holly K. Gibbs. 2022. Environmental outcomes of the US Renewable Fuel Standard. *Proceedings of the National Academy of Sciences* 119. Proceedings of the National Academy of Sciences: e2101084119. https://doi.org/10.1073/pnas.2101084119.

Liu, Jing, Thomas W. Hertel, Richard B. Lammers, Alexander Prusevich, Lantz C. Uris, Danielle S. Baldos, and Grogan, and Steve Frolking. 2017. Achieving sustainable irrigation water withdrawals: Global impacts on food security and land use. *Environmental Research Letters* 12: 104009. https://doi.org/10.1088/1748-9326/aa88db.

Maupin, Molly A., Joan F. Kenny, Susan S. Hutson, John K. Lovelace, Nancy L. Barber, and Kristin S. Linsey. 2014. *Estimated use of water in the United States in 2010*. Reston, VA: US Geological Survey. https://doi.org/10.3133/cir1405.

Neiken, Loganaden. 2003. FAO methodology for estimating the prevalence of undernourishment. In *International Scientific Symposium on Measurement and Assessment of Food Deprivation and Undernutrition*. Rome: Food and Agriculture Organization of the United Nations. http://www.fao.org/3/Y4249E/y4249e06.htm.

Siebert, Stefan, and Petra Döll. 2010. Quantifying blue and green virtual water contents in global crop production as well as potential production losses without irrigation. *Journal of Hydrology* 384: 198–217. https://doi.org/10.1016/j.jhydrol.2009.07.031.

Taheripour, Farzad, Luis Pena-Levano, and Wally Tyner. 2017. *Introducing first and second generation biofuels into GTAP 9 Data Base*. Research Memorandum 29. Department of Agricultural Economics, Purdue University, West Lafayette, IN: Global Trade Analysis Project (GTAP). 10.21642/GTAP.RM29.

USDA-NASS. 2019. *Value of crop sold per acre*. Washington, DC: US Department of Agriculture, National Agricultural Statistics Service. https://data.nal.usda.gov/dataset/nass-quick-stats. Accessed 15 Dec 2019.

West, Paul C., Holly K. Gibbs, Chad Monfreda, John Wagner, Carol C. Barford, Stephen R. Carpenter, and Jonathan A. Foley. 2010. Trading carbon for food: Global comparison of carbon stocks vs crop yields on agricultural land. *Proceedings of the National Academy of Sciences* 107: 19645–19648. https://doi.org/10.1073/pnas.1011078107.

Chapter 7
Behavioral Parameters: Capturing Geospatial Heterogeneity in Economic Decisions and Responses

Iman Haqiqi

Tackling fundamental environmental and societal challenges requires a comprehensive geospatial understanding of the interactions among food, agriculture, land, water, and environmental systems. However, this understanding must be achieved with a keen eye on spatial heterogeneity, as environmental responses to economic and policy interventions, resource availability, and climate change are rarely uniform across landscapes. Failure to consider spatial variation can lead to flawed models, misguided policies, and unintended consequences.

To address this challenge, this chapter introduces a geospatial dataset that includes economic behavioral parameters. This dataset is specifically designed for spatially heterogeneous analyses within gridded economic modeling focused on land, water, and environmental sustainability. The dataset incorporates key parameters such as land supply elasticity, water supply elasticity, land conversion possibilities, and substitution elasticity, offering detail and resolution for capturing the spatial responses of production systems to economic and environmental drivers.

Economic parameters are rarely available at gridded scales, yet they are essential for modeling how different regions within a landscape will respond to changes. By integrating these parameters with readily available environmental and economic data, researchers can unlock new avenues for analyzing and predicting spatially differentiated land-use change, water resource allocation, and agricultural production under various sustainability scenarios.

This chapter describes the construction of the SIMPLE-G-US dataset with all crops composite, highlighting its specific advantages and unique aspects in spatially heterogeneous analysis. Through examples, the application of the dataset is showcased in Part IV, demonstrating its potential to contribute to gridded economic modeling. Other internally consistent SIMPLE-G datasets include the global dataset

I. Haqiqi (✉)
Center for Global Trade Analysis, Department of Agricultural Economics, Purdue University, West Lafayette, IN, USA
e-mail: ihaqiqi@purdue.edu

© The Author(s) 2025
I. Haqiqi, T. W. Hertel (eds.), *SIMPLE-G*,
https://doi.org/10.1007/978-3-031-68054-0_7

(Haqiqi et al. 2022), the corn–soy dataset (Sun et al. 2020, Liu et al. 2023), and the Brazil dataset (Wang et al. 2022, 2024).

Parameters in the model represent behavior leading to the dynamics of the SIMPLE-G system. The model parameters in each simulation are fixed and are usually obtained from other studies or empirical estimations. Table 7.1 summarizes the main parameters used in the model. We describe some of these parameters in the following sections.

1 Regional Parameters

This section introduces the data sources for regional parameters that determine how price changes and income growth influence food selection, how consumers choose between domestic and imported goods, and how livestock and processed food sectors adjust their inputs in response to changing prices.

1.1 *Consumer Demand*

Consumer demand elasticities, which govern how consumers react to food price changes and income growth, are based on the work of Muhammad et al. (2011), who use cross-section international data to estimate food demand systems spanning the full range of national per capita incomes. To capture the impacts of dietary upgrading in the model, equations relating the price and income elasticities of food demand to the log of per capita incomes are embedded in the model. The estimation of these linkages is described in Hertel and Baldos (2016).

1.2 *Spatial Reallocation and Trade Elasticities*

Armington-type elasticity parameters are important components of the model. Armington substitution elasticities measure the degree of substitutability between imported and domestic goods or between goods from different sources of origin. Hertel et al. (2007) estimate the Armington substitution elasticities for GTAP sectors based on bilateral transport costs among a subset of countries. The average elasticity for crops is around 3. However, these estimates are subject to uncertainty and sensitivity. It is advisable to use a range of values or conduct sensitivity analyses when applying these elasticities in trade models. We also assume a symmetric value of 3 on the export side.

Table 7.1 Parameters in the gridded model

Parameter	Description	Source
Supply elasticities		
η^L	Cropland supply elasticity	Empirical estimations (Villoria et al. 2023)
η^{GW}	Groundwater supply elasticity	Empirical estimations (Haqiqi et al. 2023)
η^{SW}	Surface water supply elasticity	Empirical estimations (Haqiqi et al. 2023)
η^N	N fertilizer supply elasticity	Uniform value (Baldos et al. 2020)
η^H	Other agricultural inputs supply elasticity	Uniform value (Baldos et al. 2020)
QCET land allocation parameters		
τ	QCET transformation elasticity parameter for land allocation	Empirical estimations for each class of water rights*
Substitution elasticities		
σ	Substitution elasticity between fertilizer and other inputs composite	Based on yield response to fertilizer from Agro-IBIS (Liu et al. 2023)
σ^{HLW}	Substitution elasticity between water-land composite and other inputs composite	Empirical estimations of intensive margin*
σ^{LW}	Substitution elasticity between water and land	Based on yield response to deficit irrigation*
σ^W	Substitution elasticity between water and other irrigation inputs	Based on irrigation technology and crop mix*
σ^{SG}	Substitution elasticity between groundwater and surface water	Uniform value (Baldos et al. 2020)
Subregional Armington parameters		
φ	Substitution possibility between varieties of aggregated crop outputs of grid cells	International trade literature (Hertel et al. 2007)
κ	Substitution possibility between varieties of aggregated crop outputs of subregions	International trade literature (Hertel et al. 2007)
Consumer and trade elasticities		
ω	Imports substitution elasticity	International trade literature (Hertel et al. 2007)
ψ	Export transformation elasticity	Symmetric to imports
α	Demand elasticity parameter	Muhammad et al. (2011)
β	Demand elasticity parameter	Muhammad et al. (2011)
σ^V	Substitution elasticity in livestock production sector	Calibrated to target historical changes in feed use (see Baldos and Hertel 2012)
σ^F	Substitution elasticity in processed food sector	Zero value (i.e., fixed proportion in production)
Other parameters		
Multiple parameters	Water quality and nitrogen leaching	Based on leaching response to fertilizer from Agra-IBIS (Liu et al. 2023)
Multiple parameters	Greenhouse gas emissions factors	Based on GTAP and other sources (Aguiar et al. 2019)
Multiple parameters	Nutrition and food security parameters	Following Neiken (2003) (Baldos and Hertel 2013)

(Note: * shows the work in progress)

1.3 Input Substitution Elasticities for Livestock and Processed Food Sectors

The input substitution elasticity for the livestock sector is based on the historical calibration documented in Baldos and Hertel (2012) over the period from 2001 to 2006. Feed use consumption, taken from FAOSTAT, for high income regions is targeted by adjusting the input substitution elasticity. Processed food sectors assume the Leontief production function given zero values for these substitution elasticities.

2 Gridded Parameters

This section describes the data sources for parameters that determine the spatially heterogeneous agricultural decisions. They include parameters related to supply of agricultural inputs (land and water) and demand of inputs derived from agricultural production from Constant Elasticity of Substitution (CES) functions.

2.1 Cropland Supply Parameters

The cropland supply function determines the willingness to supply cropland at a given level of land rents. The main parameter in this function is the supply elasticity, which represents the percentage change in cropland supply as a response to a 1% change in average cropland rent. Regional cropland supply elasticities are taken from the SIMPLE model, and gridded land supply elasticities for the United States are based on the statistical model developed by Villoria and Liu (2018) using gridded data from the Americas. This elasticity was estimated based on the propensity of land to be converted to crop cultivation from other uses (including pasture and forestry) conditional on the profitability of land while controlling for agroecological conditions. In the model, this parameter ranges from 0 to 0.755. A larger value indicates a greater likelihood of switching to cropland from other uses for a given increase in cropland rents.

2.2 Cropland Conversion (Transformation) Parameters

A quantity-preserving constant elasticity of transformation (QCET) function determines the allocation of land to irrigated and rainfed activities. The transformation elasticity parameter determines the percentage change in the ratio of each land type in response to a 1% change in relative rents to the two activities. This parameter shows the flexibility of converting cropland to irrigated or rainfed production. In the United States, this parameter is estimated following Jame et al. (2017) based on local water rights law. Different water rights can restrict the extension or intensity of irrigation at one location. For this estimation, we employ county level information from the USDA, including cash rents for irrigated and nonirrigated cropland as well as total irrigated and nonirrigated cropland area by county. We assume that all the grid cells in a county follow the estimated parameter for that county.

2.3   Water Supply Parameters

The water supply elasticity determines how the economic supply of water will change in response to changes in the price or value of water. Without observing market price behavior, it is difficult to estimate this parameter. Thus, we use hydrological information to create a database of water supply elasticities based on availability and current withdrawals. We assume that water supply at each grid cell is limited by hydrological constraints. Figure 7.1 illustrates two examples. This form of water supply function is slowly increasing at the beginning (up to A) and then rapidly increasing (after B) when approaching the asymptote (C). With adverse changes in hydroclimatic conditions, the cost schedule may shift to S_2 (depending on the natural supply of water).

Withdrawal of water is constrained by the maximum amount of water available in each grid cell after subtracting nonagricultural water use. We assume that the supply elasticity of water varies by grid cell and depends on the ratio of water extracted relative to the sustainable extraction level (R). We assume a three-parameter Fréchet function for water supply with calibrated parameters ω_0, ω_1, ω_2, ω_3:

$$\varepsilon_g = \omega_0 + \frac{\omega_1}{\left(\omega_2 + R_g\right)^{\omega_3}}, \tag{7.1}$$

where R is calculated as the ratio of annual withdrawal to annual groundwater recharge or as the ratio of annual withdrawal to annual available surface water. We calibrate this supply function separately for surface water and groundwater at each grid cell based on economic and hydrologic information, including annual water withdrawal for crop irrigation, the sustainable extraction level of water by source, and the estimated value of water, as described previously.

Haqiqi et al. (2023) have calibrated the gridded water supply schedules for the continental United States to the benchmark year 2010 based on the ratio of ground-water withdrawal to groundwater recharge (Gleeson et al. 2016; Reitz et al. 2017). For the US grid cells, the water supply elasticity is calculated using the withdrawal-to-recharge ratio and the empirically estimated parameters ω_0, ω_1, ω_2, and ω_3. These values are estimated using water withdrawal data from the USGS for 2010 and the

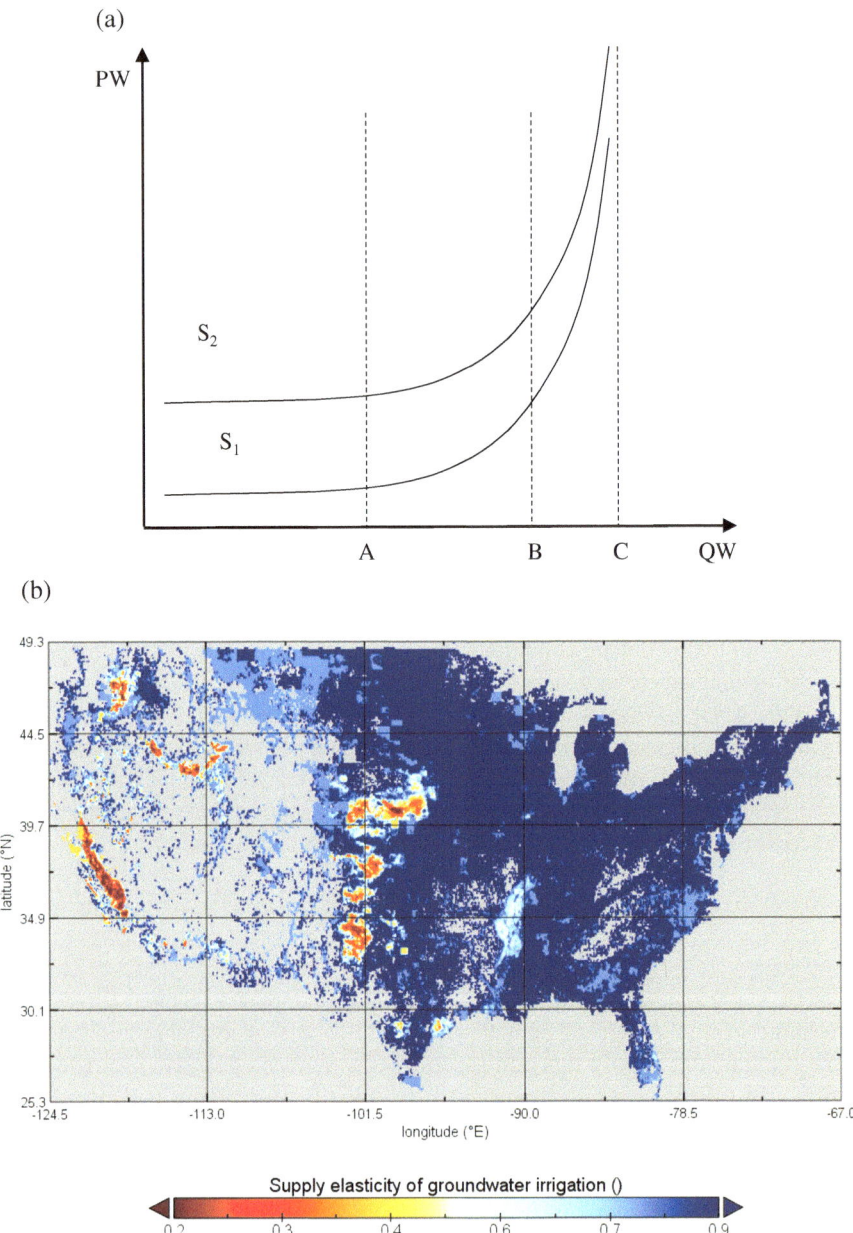

Fig. 7.1 (**a**) Economic supply for water with maximum availability determined by asymptote C. (**b**) Estimated groundwater elasticities in the continental United States

estimated value of water (Haqiqi et al. 2016). In this version, ω_1 is 0.50, ω_2 is 0.30, ω_3 is 0.45, and ω_0 is 0 for groundwater and -0.05 for surface water. Then, we apply the estimated function to all the grid cells to determine the unique water supply elasticity for each grid cell. The information about groundwater recharge is taken from the Annual Estimate of Recharge (Reitz et al. 2017). Figure 7.1b shows the supply elasticity based on the ratio of groundwater withdrawal to local recharge around the year 2010. The red color shows the locations with a very rapid depletion of groundwater which are mostly in California and the Ogallala Aquifer. For some locations, more groundwater is withdrawn in 1 year than flows in as over 10 years. The High Plains Aquifer, the Central Valley of California, the Snake River Basin, and western Washington show dramatic levels of unsustainability based on this index.

Using previous versions of SIMPLE-G, researchers have taken other approaches for estimating these elasticities. Haqiqi et al. (2018) consider variable water supply elasticity (i.e., the elasticity would adjust depending on the distance to the asymptote). These authors calculate the maximum surface water available at each grid cell after subtracting nonagricultural water use from locally generated runoff (Wolock 2003) while maximum available groundwater available was determined with groundwater stock (Gleeson et al. 2016; Befus et al. 2017).

2.4 Elasticities of Substitution in Crop Production

The substitution elasticity between N fertilizer and other inputs is an important parameter in the model. This elasticity determines the likely changes in N application rate in response to change in relative price of N fertilizer. We estimate this parameter for each grid cell, following Liu et al. (2023) to establish a framework for estimating this parameter. We begin by obtaining the yield response functions from Agro-IBIS, then combine this response function with estimated yields of irrigated and nonirrigated crops to find the substitution elasticity for irrigated and rainfed crop production (Liu et al. 2023).

2.5 Other Parameters

Liu et al. (2018) have estimated leaching parameters from Agro-IBIS to construct a nutrient-leaching module for all the grid cells in SIMPLE-G-US. This nonlinear leaching function shows that nutrient leaching will increase quadratically when N application rate increases. The parameters are specific to the unique biophysical characteristics of each grid cell, including soil type, irrigation, and land cover (Liu et al. 2023).

3 Summary

Considering geospatially heterogeneous behavioral parameters is important for understanding the interactions among agriculture, land, water, and environmental systems. We introduced a geospatial dataset of economic parameters specifically designed for analysis within gridded economic modeling. This dataset includes key parameters—such as land supply elasticity, water supply elasticity, land conversion possibilities, and substitution elasticity—that allow researchers to capture the spatial responses of production systems to economic and environmental drivers.

The chapter described the construction and sources of the SIMPLE-G-US dataset, which encompasses all crops in a composite commodity. The regional parameters determine how price changes and income growth influence the production and consumption of goods and services in different regions. The gridded parameters determine local decisions and economic responses to policies and shocks. This dataset is an example of how to employ biophysical information to calibrate and estimate parameters required in analyzing and predicting spatially differentiated decisions. We encourage researchers to consider the spatial variation in economic parameters and to incorporate this knowledge into their analyses to address the complex challenges associated with food, agriculture, land, water, and environmental systems.

Acknowledgment and Competing Interests The authors acknowledge support from the National Science Foundation award #2118329: "NSF Institute for Geospatial Understanding through an Integrative Discovery Environment (I-GUIDE)," the U.S. Department of Energy, Office of Science, Biological and Environmental Research Program, Earth and Environmental Systems Modeling, MultiSector Dynamics under Cooperative Agreement DE-SC0022141; the United States Department of Agriculture AFRI grant #2019-67023-29679, "Economic Foundations of Long Run Agricultural Sustainability," and the National Science Foundation INFEWS award #1855937, "Identifying Sustainability Solutions Through Global-Local-Global Analysis of a Coupled Water-Agriculture-Bioenergy System."

The findings and conclusions presented in this chapter are those of the authors and should not be construed to represent any official determination or policy of the US Department of Agriculture (USDA), the US government, the Department of Energy (DOE), or the National Science Foundation (NSF). Furthermore, we declare that there is no conflict of interest related to this work.

References

Aguiar, Angel, Maksym Chepeliev, Erwin L. Corong, Robert McDougall, and Dominique van der Mensbrugghe. 2019. The GTAP data base: Version 10. *Journal of Global Economic Analysis* 4: 1–27. https://doi.org/10.21642/JGEA.040101AF.

Baldos, Uris Lantz C., Iman Haqiqi, Thomas W. Hertel, Mark Horridge, and J. Liu. 2020. SIMPLE-G: A multiscale framework for integration of economic and biophysical determinants of sustainability. *Environmental Modelling & Software* 133. Elsevier: 104805. https://doi.org/10.1016/j.envsoft.2020.104805.

Baldos, Uris Lantz C., and Thomas W. Hertel. 2013. Looking back to move forward on model validation: Insights from a global model of agricultural land use. *Environmental Research Letters* 8: 034024. https://doi.org/10.1088/1748-9326/8/3/034024.

———. 2012. *SIMPLE: A simplified international model of agricultural prices, land use and the environment*. West Lafayette, IN: Working Paper 70. Department of Agricultural Economics, Purdue University, Global Trade Analysis Project (GTAP). https://doi.org/10.21642/GTAP. WP70.

Befus, Kevin M., Scott Jasechko, Elco Luijendijk, Tom Gleeson, and M. Bayani Cardenas. 2017. The rapid yet uneven turnover of Earth's groundwater. *Geophysical Research Letters* 44: 5511–5520. https://doi.org/10.1002/2017GL073322.

Gleeson, Tom, Kevin M. Befus, Scott Jasechko, Elco Luijendijk, and M. Bayani Cardenas. 2016. The global volume and distribution of modern groundwater. *Nature Geoscience* 9: 161. https://doi.org/10.1038/ngeo2590.

Haqiqi, Iman, Farzad Taheripour, Jing Liu, and Dominique van der Mensbrugghe. 2016. Introducing irrigation water into GTAP data base version 9. *Journal of Global Economic Analysis* 1: 116–155. https://doi.org/10.21642/JGEA.010203AF.

Haqiqi, Iman, Laura C. Bowling, Sadia A. Jame, Thomas W. Hertel, Uris Baldos, and Jing Liu. 2018. Global drivers of land and water sustainability stresses at mid-century. *Purdue Policy Research Institute (PPRI) Policy Briefs* 4: 7.

Haqiqi, Iman, Chris J. Perry, and Thomas W. Hertel. 2022. When the virtual water runs out: local and global responses to addressing unsustainable groundwater consumption. *Water International 47* (7): 1060–1084. https://doi.org/10.1080/02508060.2023.2131272.

Haqiqi, Iman, Laura Bowling, Sadia Jame, Uris Baldos, Jing Liu, and Thomas Hertel. 2023. Global drivers of local water stresses and global responses to local water policies in the United States. *Environmental Research Letters* 18: 065007. https://doi.org/10.1088/1748-9326/acd269.

Hertel, Thomas W., David Hummels, Maros Ivanic, and Roman Keeney. 2007. How confident can we be of CGE-based assessments of free trade agreements? *Economic Modelling* 24: 611–635. https://doi.org/10.1016/j.econmod.2006.12.002.

Hertel, Thomas W., Uris Lantz C. Baldos. 2016. *Global change and the challenges of sustainably feeding a growing planet*. New York: Springer.

Jame, Sadia A., Laura C. Bowling, Thomas W. Hertel, Liu Jing, and Iman Haqiqi. 2017. The influence of US water law on irrigation expansion. In *Global Economic Analysis in the 21st Century: Challenges and Opportunities: 20th Annual Conference on Global Economic Analysis*. West Lafayette: Center for Global Trade Analysis, Purdue University.

Liu, Jing, Thomas W. Hertel, Laura Bowling, Sadia Jame, Christopher Kucharik, and Navin Ramankutty. 2018. Evaluating alternative options for managing nitrogen losses from corn production. *Purdue Policy Research Institute (PPRI) Policy Briefs* 4: 9.

Liu, Jing, Laura Bowling, Christopher Kucharik, Sadia Jame, Lantz C. Uris, Larissa Jarvis Baldos, Navin Ramankutty, and Thomas W. Hertel. 2023. Tackling policy leakage and targeting hotspots could be key to addressing the "wicked" challenge of nutrient pollution from corn production in the U.S. *Environmental Research Letters* 18: 105002. https://doi.org/10.1088/1748-9326/acf727.

Muhammad, Andrew, James L. Seale Jr, Birgit Meade, and Anita Regmi. 2011. *International evidence on food consumption patterns: An update using 2005 international comparison program data*. Washington, DC: Technical Bulletin 1929. US Department of Agriculture, Economic Research Service.

Neiken, Loganaden. 2003. *FAO methodology for estimating the prevalence of undernourishment*. Rome: FAO.

Reitz, Meredith, Ward E. Sanford, Gabriel Senay, and J. Cazenas. 2017. Annual estimates of recharge, quick-flow runoff, and ET for the contiguous U.S. using empirical regression equations. *Journal of the American Water Resources Association* 53: 961983. https://doi.org/10.1111/1752-1688.12546.

Sun, Shanxia, Brayam Valqui Ordonez, Mort D. Webster, Jing Liu, Christopher J. Kucharik, and Thomas Hertel. 2020. Fine-scale analysis of the energy–land–water nexus: nitrate leaching implications of biomass cofiring in the Midwestern United States. *Environmental science & technology, 54* (4): 2122–2132.

Villoria, Nelson B., and Jing Liu. 2018. Using continental grids to improve understanding of global land supply responses: Implications for policy-driven land use changes in the Americas. *Land Use Policy* 75: 411–419. https://doi.org/10.1016/j.landusepol.2018.04.010.

Villoria, Nelson, Jing Liu, Iman Haqiqi, Shourish Chakravarty, Michael Delgado, Alfredo Cisneros-Pineda, and Tom Hertel. 2023. *Gridded cropland supply elasticity for the continental United States.* (5 arc-min spatial resolution) (version 1.0). MyGeoHUB. https://doi.org/10.13019/37T9-0E88.

Wolock, David M. 2003. *Base-flow index grid for the conterminous United States.* Reston: Open-File Report 2003–263. US Geological Survey. https://doi.org/10.3133/ofr03263.

Wang, Z., G.B. Martha, J. Liu, C.Z. Lima, and T.W. Hertel. 2024. Planned expansion of transportation infrastructure in Brazil has implications for the pattern of agricultural production and carbon emissions. *Science of The Total Environment 928*: 172434.

Wang, Zhan, Geraldo B. Martha Jr, Jing Liu, and Cicero Zanetti de Lima. 2022. Climate Change, Irrigation Expansion and Impacts on Agriculture Production: An Integrated Multi-Scale Analysis of Brazil by 2050. Agricultural and Applied Economics Association (AAEA) Conferences, 2022 Annual Meeting, July 31-August 2, Anaheim, California. https://doi.org/10.22004/ag.econ.322560

Chapter 8
Computation and Baseline: Efficient Methods for Solving a Large System of Equations for Projection and Scenario Analysis

Iman Haqiqi and Uris Lantz C. Baldos

The use of quantitative models is critical for sustainability studies because they provide geospatial insights and enable the analysis of the impact of climate extremes and conservation policies on food security and environmental sustainability. Traditional regional models may provide national-level insights, but they miss out on the crucial details that can be identified by using a geospatial approach. Explicit modeling of spatial outcomes enables targeted interventions and fosters more equitable solutions for creating a sustainable future. Additionally, these models can offer valuable insights into resource management and potential trade-offs between economic growth and environmental sustainability by explicitly incorporating land and water constraints.

However, the development and implementation of such models comes with significant computational hurdles. The large amount of data involved—coupled with the complex interactions between diverse economic, land, and water systems—creates a demanding challenge. These models may consist of a system of simultaneous equations with millions of unknowns with limited parallel computation possibilities. Traditional computational approaches often fail to efficiently handle the calculations required to solve these simulations.

To overcome these difficulties, the SIMPLE-G model offers a scalable and effective approach. This chapter introduces the key features and advantages of the SIMPLE-G computational approach and demonstrates how it can significantly contribute to gridded quantitative economic analysis. The chapter introduces and explains the key features and advantages of the SIMPLE-G computation and showcases how SIMPLE-G can contribute to advancing sustainability research and policy

I. Haqiqi (✉) · U. L. C. Baldos
Center for Global Trade Analysis, Department of Agricultural Economics, Purdue University, West Lafayette, IN, USA
e-mail: ihaqiqi@purdue.edu

I. Haqiqi, T. W. Hertel (eds.), *SIMPLE-G*,
https://doi.org/10.1007/978-3-031-68054-0_8

103

analysis, then discusses how the baseline and scenarios can be defined and implemented.

1 Computer Implementation

The SIMPLE-G database and model are prepared and solved with the GEMPACK modeling suite (Horridge et al. 2018). This software package is specifically designed to solve large-scale economic equilibrium models with numerous markets and agents and is used by economists to simulate the impacts of a wide range of policy issues, including trade, taxation, environmental regulation, and development. It is also used for dynamic projections of the economy. GEMPACK offers several advantages, including a user-friendly algebraic notation that is easy to learn and use, which makes it accessible even to users without a strong programming background. The database files can also readily store multiscale and multidimensional variables. Other attractive features of this software are discussed below, but the unique advantage of GEMPACK in the context of multiscale modeling is its ability to condense the model and later backsolve for key endogenous variables. This makes it a computationally efficient software package for solving SIMPLE-G models.

1.1 Condensation

Solution times can be substantial for an equilibrium model with many equations and with complex interconnections among the unknown variables (e.g., the market responds to farmer decisions even as the farmers respond to market outcomes).[1] A typical global SIMPLE-G application might include two million grid cells and 17 regions. A system of about 20 equations (recall Chap. 5) determines crop output for each of those grid cells, given grid-level exogenous settings and the endogenous price of output (which is the same for all cells in a given region), for a total of approximately 40 million grid-level equations. For each region, up to 100 other equations add up the grid-cell outputs to obtain total crop supply or interrelate region-level prices and quantities. The overwhelming majority of equations are clearly at the grid-cell level. A nonlinear system of 20 million equations would be impossibly slow to solve and might require enormous amounts of computer memory. We can greatly reduce the number of equations by substitution (i.e.,

---

[1]Researchers have designed multiple algorithms to reduce solution time. Most algorithms iterate between two phases: a linear algebra phase that solves a first-order approximation to the nonlinear equation system and a "formula" phase that updates variable values and recomputes coefficients of the linear system. In GEMPACK, solution time for the linear phase rises with the square or cube of the number of equations, while time for the formula phase tends to increase only linearly.

condensation). For example, we can write the fertilizer demand equation as follows (the grid index is omitted here):

$$q_N = q - a - a_N - \sigma(p_N - a_N - p - a). \qquad (8.1)$$

Then, we could replace each occurrence of q_N in *other* equations by the righthand side of Eq. 8.1,

$$q - a - a_N - \sigma(p_N - a_N - p - a), \qquad (8.2)$$

thereby dropping the first equation (8.1) from the system. By condensing out the grid-cell-specific variables, we can dramatically reduce the size of the model to be solved. After the system is solved, we can use the above equation to recover (or backsolve for) values of q_N.

Economic modelers often use such techniques, manually performing such substitutions in their model specification files. The drawback is that the necessary algebra is difficult, and the remaining equations become extremely complicated and nontransparent, especially when a large number of substitutions are performed. The GEMPACK software is able to solve the algebra to perform such substitutions (and their backsolves) automatically, reaping a performance gain while leaving the model specification (TABLO) file in its original, simpler, uncondensed form.

For SIMPLE-G, *all* grid-level equations are substituted out, leaving a regional-level system (for 17 regions) of modest size: 1700 equations. Such a system takes very little time to solve. However, the coefficients of the system involve grid-level calculations; the time taken to solve these is proportional to the number of grid cells. Hence, the solution time increases only linearly as a function of the number of grid cells (see Fig. 8.1).

1.2   Linearization

GEMPACK can automatically translate the original equation system into a linearized system (reformulated as a system of first-order partial differential equations). Alternatively, modelers can themselves specify conveniently interpretable linearized forms of the underlying behavioral equations. In this case, the modeler must also supply the requisite update formulas to ensure that all relevant variables are properly updated over the course of the nonlinear solution of this linear system. Clever representation of the model can facilitate condensation as well as a more rapid solution of the model. In SIMPLE-G, all of the cross-grid-cell interactions are transmitted through regional market prices, allowing us to substitute out all of the variables with a grid-cell index. Once we know the regional crop, fertilizer, irrigation capital, and other input prices, we can backsolve for crop output, input use, land prices, and the shadow price of irrigation water in each grid cell independently.

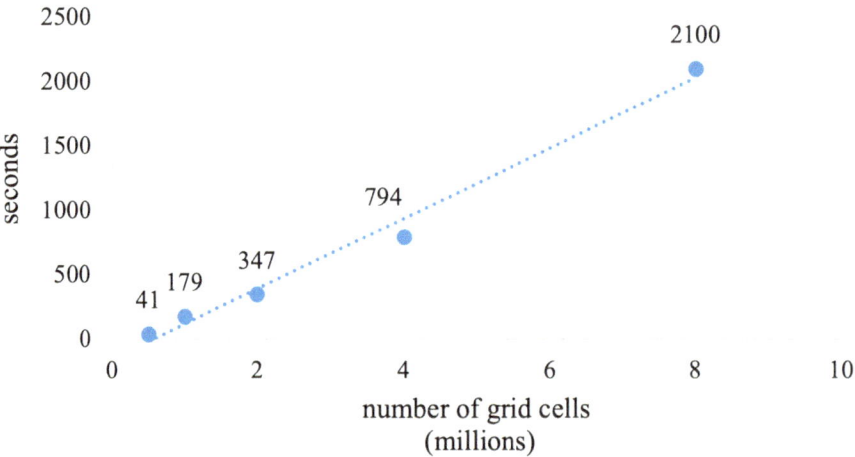

Fig. 8.1 Linear relationship between solution time for SIMPLE-G-US and number of grid cells
Condensation allows users to solve a SIMPLE-G system of equations with 1 million grid cells (~10
million endogenous variables) on a laptop in a few minutes. CPU = core i5-7500, RAM = 16 GB

Since the model is nonlinear, the cost shares, price, and quantity variables must be
updated at each step in the solution process. Consequently, the model is solved by
multistep methods such as the Euler method or Gragg's modified midpoint method
(Pearson 1991). The large system of linear equations is solved using sparse matrix
techniques (Schiffmann and Jerie 2019). Richardson extrapolation is used to
improve accuracy (Pearson 1991). This linearized approach has proven capable of
accurately solving very large, nonlinear models (e.g., model in Fig. 8.1 contains
eight million grid cells).

1.3   Decomposition

In addition to these features, GEMPACK has other extensions that prove invaluable
in SIMPLE-G applications (see Part IV). It provides a way to formulate inequality
constraints, or nondifferentiable equations as complementarities (Bach and Pearson
1996), which can be important in sustainability analyses. It also offers a technique to
decompose changes in model variables due to several shocks into components due to
each individual shock (Harrison et al. 2000). We will illustrate this technique in
Chap. 12.

1.4 Computation "In the Cloud"

The web application version of SIMPLE-G permits users to simulate, explore, and visualize the results of SIMPLE-G without installing the GEMPACK program or any visualization software. It works based on "containers" and Linux versions of GEMPACK programs run on the GeoHub server (https://mygeohub.org/groups/glassnet/res/tools). The web application also includes presolved experiments and demonstrations on how to run the model and analyze results based on the policy briefs presented at the 2018 Conference on Long Run Sustainability of US Agriculture (Haqiqi et al. 2018; Liu et al. 2018).

The latest version of the web application allows users to run their own experiments, which could range from global projections of food production and food demand to grid-level analysis. Users need to provide a GEMPACK header array (HAR) data file that includes the values for gridded shocks or growth rates and a GEMPACK command (CMF) text file that includes the configuration details. The tool can be used to explore the impacts of different scenarios of climate change, water scarcity, population growth, income growth, and biofuel mandates on food production, water resources, and land use. Ongoing improvements to the tool will make it more user friendly and easy to use. Figure 8.2 illustrates an example of SIMPLE-G results visualized on MyGeoHub.

It is also possible to link SIMPLE-G to models and data from different domains (e.g., hydrology, crop science, climate science, and ecology). This enables researchers to address complex, real-world grand challenge problems that require

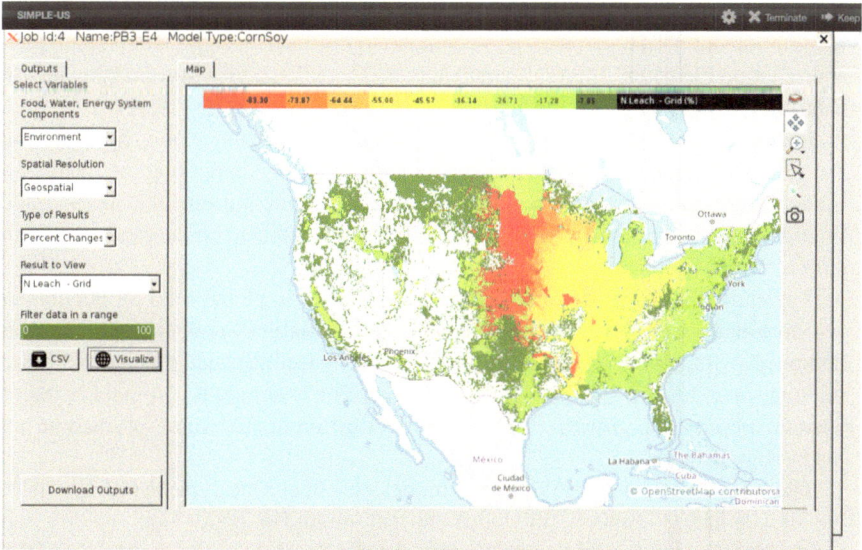

Fig. 8.2 Sample window from the SIMPLE-G web application at https://mygeohub.org/tools/simpleus

in-depth collaboration among researchers from multiple disciplines. Such linkages often involve harnessing multiscale and multidimensional data and models with unidirectional connections. For example, Agro-IBIS provides information to SIMPLE-G through the simulated yield response to N fertilizer. SIMPLE-G can also provide information on economically motivated changes in N application rates as inputs to Agro-IBIS to estimate N leaching under different scenarios.

It is also possible to couple SIMPLE-G with other models, which—in general— faces two major problems. First, it is difficult for collaborators to understand each other's data/models, a precondition for effective collaboration. Second, the infrastructure is usually unavailable to run the coupled systems. To address these challenges, Woo et al. (2022) designed C3F, a Collaborative Container-Based Model Coupling Framework, which can accelerate model integration and linking efforts by leveraging advanced cyberinfrastructures (e.g., high-performance computing and virtual containers). This framework is used to couple the Water Balance Model (WBM) and SIMPLE-G (Baldos et al. 2020; Grogan et al. 2022; Woo et al. 2022). While land-use decisions are exogenous in WBM and hydrologic water supply is exogenous in SIMPLE-G, they are used as bridges between the models. To address the differences in temporal and spatial resolutions, additional postprocessing tools are developed to transfer the information at consistent scales.

2   Building a Baseline

Developing a baseline is critical for evaluating model outcomes and it serves as a point of reference for assessing the impact of policy interventions and/or different assumptions in model parameters. A comparison of baseline simulation results with a simulation based on different parameters or shocks and provides insights into the robustness of model predictions.

For counterfactual (what-if) analysis, a baseline is a reference point against which the simulation results of a scenario can be compared. By comparing the results of a model simulation to a baseline, it is possible to assess the impacts of different policy interventions, technological innovations, and climate change on the food system and agricultural water and land use.

Typically, the SIMPLE baseline captures changes in global food production and resource use given a trend in macroeconomic variables such as population and income growth as well as technological change. A baseline usually includes different regional rates for the drivers of global change. For example, it can include region-specific population growth rates as well as differentiated rates of income and technology growth.

There are many ways to define a model baseline. One typical baseline is the business-as-usual scenario, in which current trends in population, income, and technology are assumed to continue into the future. A baseline scenario could be based on historical data, projections of future trends, or a combination of both. The specific approach used will depend on the research question being addressed.

2.1 Historical Baseline

In the context of model validation, a historical baseline replicates the observed state of the system over a historical period. This baseline serves as a reference point for comparing model predictions with observations to assess the accuracy and reliability of the model. For SIMPLE, the historical baseline involves simulating the observed changes in crop production, cropland, and prices, given observed changes in the key drivers of food supply and demand.

2.2 Future Baseline

In SIMPLE and SIMPLE-G, the future baseline scenario typically requires inputs on population and per capita incomes, feedstock demand for biofuel use, and productivity growth trends for the crops, livestock, and processed food sectors. Other drivers can include future impacts of climate change on crop productivity as well as cropland supply shifters due to urbanization or regional demands for ecosystem services.

Constructing a future baseline involves gathering data from various sources to project future trends and dynamics within the system. Population, income, and biofuel production can be specified to follow long-run growth scenarios such as the Shared Socioeconomic Pathways (O'Neill et al. 2014) or other global economic projections.

2.3 Policies and Scenarios

The baseline serves as a foundation for analyzing the potential impacts of future technological advancements and climate change on the food system. Another use of the baseline is to assess the impacts of future sustainability policies, trade policies, or agricultural protection policies.

Constructing future baselines and alternative scenarios requires convergence science and multidisciplinary collaborations. For example, hydrological model simulations may be used to inform SIMPLE-G about the future availability of water or possible water sustainability policies. Agronomic crop model simulations may be used to inform SIMPLE-G about future yield trends for irrigated and rainfed crops. Additionally, population dynamic models can inform about future changes in food demand and the labor force. For some examples of baseline construction, see Chap. 10; Chap. 12 develops a groundwater sustainability scenario.

3 Summary

The process of going from complex ideas and theories to practical and useful conclusions is filled with difficulties and obstacles. This requires highly non-linear computable models of linked human and environmental systems. The high dimensionality of these models, large data requirements, and computational hurdles necessitate careful trade-offs between realism, tractability, and policy relevance. A theoretically simplified yet spatially detailed model like SIMPLE-G offers promising computational prospects for understanding relationships within environmental and human systems. In addition, SIMPLE-G computational method demonstrates the potential to overcome computational barriers of solving large-scale multi-system geospatial models.

The feasibility and efficiency of SIMPLE-G computation relies on condensation and linearization, two important techniques used in GEMPACK to reduce the complexity and size of the model equations and to solve them efficiently. Condensation involves simplifying the model by eliminating some equations and variables through substitution, which reduces the number of equations and variables that need to be solved. This makes the model more manageable and faster to run. On the other hand, linearization involves reformulating the model equations as a system of first-order partial differential equations that are linear in percentage changes or changes in variables. This allows for the use of a solution method which calculates the movements in the endogenous variables away from their initial values in response to movements in exogenous variables away from their initial values. These two techniques improve the computational efficiency and accuracy of the model solution.

Moving forward, continued research efforts focused on developing efficient solution algorithms, harnessing new data sources, and enhancing model interpretability are key to bridging the gap between the inherent complexity of these models and their practical utility in tackling sustainability challenges. Ultimately, we can harness the power of multi-scale multi-system models to better understand the interactions between economy, environment, and society, enabling informed and sustainable decision-making at all levels.

Acknowledgment and Competing Interests The authors acknowledge support from the U.S. Department of Energy, Office of Science, Biological and Environmental Research Program, Earth and Environmental Systems Modeling, MultiSector Dynamics under Cooperative Agreement DE-SC0022141; the National Science Foundation award #2118329: "NSF Institute for Geospatial Understanding through an Integrative Discovery Environment (I-GUIDE)," the United States Department of Agriculture AFRI grant #2019-67023-29679, "Economic Foundations of Long Run Agricultural Sustainability," and the National Science Foundation INFEWS award #1855937, "Identifying Sustainability Solutions Through Global-Local-Global Analysis of a Coupled Water-Agriculture-Bioenergy System."

The findings and conclusions presented in this chapter are those of the authors and should not be construed to represent any official determination or policy of the US Department of Agriculture (USDA), the US government, the Department of Energy (DOE), or the National Science Foundation (NSF). Furthermore, we declare that there is no conflict of interest related to this work.

References

Bach, Christian Friis, and Ken Pearson. 1996. *Implementing quotas in GTAP using GEMPACK or how to linearize an inequality*. Technical Paper 4. Department of Agricultural Economics, Purdue University, West Lafayette: Global Trade Analysis Project (GTAP). https://doi.org/10.21642/GTAP.TP04.

Baldos, Uris Lantz C., Iman Haqiqi, Thomas W. Hertel, Mark Horridge, and J. Liu. 2020. SIMPLE-G: A multiscale framework for integration of economic and biophysical determinants of sustainability. *Environmental Modelling & Software* 133: 104805. https://doi.org/10.1016/j.envsoft.2020.104805.

Grogan, Danielle S., Shan Zuidema, Alex Prusevich, Wilfred M. Wollheim, Stanley Glidden, and Richard B. Lammers. 2022. Water balance model (WBM) v.1.0.0: A scalable gridded global hydrologic model with water-tracking functionality. *Geoscientific Model Development* 15: 7287–7323. https://doi.org/10.5194/gmd-15-7287-2022.

Haqiqi, Iman, Laura C. Bowling, Sadia A. Jame, Thomas W. Hertel, Uris Baldos, and Jing Liu. 2018. Global drivers of land and water sustainability stresses at mid-century. *Purdue Policy Research Institute (PPRI) Policy Briefs* 4: 7.

Harrison, W. Jill, J. Mark Horridge, and K.R. Pearson. 2000. Decomposing simulation results with respect to exogenous shocks. *Computational Economics* 15: 227–249. https://doi.org/10.1023/A:1008739609685.

Horridge, J.M., Michael Jerie, Dean Mustakinov, and Florian Schiffmann. 2018. *GEMPACK manual*. Victoria University, Centre of Policy Studies/IMPACT Centre.

Liu, Jing, Thomas W. Hertel, Laura Bowling, Sadia Jame, Christopher Kucharik, and Navin Ramankutty. 2018. Evaluating alternative options for managing nitrogen losses from corn production. *Purdue Policy Research Institute (PPRI) Policy Briefs* 4: 9.

O'Neill, Brian C., Elmar Kriegler, Keywan Riahi, Kristie L. Ebi, Stephane Hallegatte, Timothy R. Carter, Ritu Mathur, and Detlef P. van Vuuren. 2014. A new scenario framework for climate change research: The concept of shared socioeconomic pathways. *Climatic Change* 122: 387–400. https://doi.org/10.1007/s10584-013-0905-2.

Pearson, K.R. 1991. *Solving nonlinear economic models accurately via a linear representation. Preliminary Impact Paper IP-55*. Parkville: University of Melbourne Impact Research Centre. https://doi.org/10.22004/ag.econ.295068.

Schiffmann, Florian, and Michael Jerie. 2019. Improving the performance of sparse LU decomposition in GEMPACK. In Warsaw, Poland.

Woo, Jungha, Lan Zhao, Danielle S. Grogan, Iman Haqiqi, Richard Lammers, and Carol X. Song. 2022. C3F: Collaborative container-based model coupling framework. In *PEARC '22: Practice and experience in advanced research computing*. Boston: Association for Computing Machinery. https://doi.org/10.1145/3491418.3530298.

Chapter 9
Model Validation: Comparing Gridded and Regional Simulations to Observations

Iman Haqiqi, Zhan Wang, and Uris Lantz C. Baldos

Model validation is important to ensure that a model is accurate and reliable. The general goal of validation is to compare a model's predictions or projections to actual data to check whether the model can simulate the conditions that are observed independently of the model and its input data. Since it is generally not possible to validate all aspects of a high-dimensional, multi-scale model, such validation exercises generally have a focus area that depends on the purpose for which it will be used. This varies by application.

Each discipline has a different approach to validation depending on the research purpose and complexity of the processes being modeled (Hansen and Heckman 1996; Rykiel 1996; Oreskes 1998; Biondi et al. 2012; Ngo and See 2012; Kersebaum et al. 2015; van Vliet et al. 2016). For example, climate models seek to explain underlying physical processes and make predictions and projections about a changing climate. In agronomic models, validation often involves estimating out-of-sample yields and biophysical processes. In hydrological models, validation may focus on replicating observed streamflow and river discharge at another location or at another point in time. Economic models aim to explain economic decisions and provide policy insights and evaluate the impacts of future changes. They typically exhibit greater uncertainty due to the confounding effects of human behavior, which does not follow the laws of physics, biology, or chemistry. These economic models generally include many parameters and lean heavily on theoretical structures obtained from microeconomic theoretical foundations.

I. Haqiqi (✉) · Z. Wang · U. L. C. Baldos
Center for Global Trade Analysis, Department of Agricultural Economics, Purdue University, West Lafayette, IN, USA
e-mail: ihaqiqi@purdue.edu

© The Author(s) 2025
I. Haqiqi, T. W. Hertel (eds.), *SIMPLE-G*,
https://doi.org/10.1007/978-3-031-68054-0_9

1 Validation Challenge When Multiple Drivers Interact Across Multiple Scales

The SIMPLE-G model involves economic decisions about land use and water withdrawals at the grid-cell level. Here, the focus is on geospatial validation of economic decisions about cropland and water withdrawals. However, land-use changes are the result of many mutually influential local, regional, and global drivers that together shape the global cropland patterns. Thus, the model must be able to simulate complex processes in order to represent the richness in observed changes in land-use patterns. Unfortunately, few of these models are validated: Advances in model validation techniques have been much slower than new model developments (van Vliet et al. 2016). Accordingly, there is no agreed-upon set of methods for assessing the results of integrated agricultural, economic, water, and land-use models (Razavi and Gupta 2015; Razavi et al. 2021). Here, we draw on validation methods from economics, sociohydrology, and land-use modeling.

There are several possible ways to validate a geospatial economic model like SIMPLE-G. However, the initial step is benchmark replication or calibration. Comparing the results of a model to a set of observed data will ensure that the model can replicate a base reference condition. To perform benchmark replication, the model is first calibrated to the benchmark data (e.g., 2010). The model's predictions should match the benchmark data. Backcasting. another validation method, employs the model to make predictions about observed past events and then compares the model's predictions to actual observed outcomes from independent data sources not used in the calibration process. If the predictions are accurate according to predefined criteria, then it is likely that the model will be accurate in the future or for unobserved conditions. This ensures that the structural processes that can explain changes in the system are modeled correctly. Please note that we use the terms 'prediction' and 'projection' interchangeably.

A complementary approach to economic model validation is to use a sensitivity analysis, which involves varying the model's assumptions, parameters, and drivers to investigate how the predictions change. This helps identify the assumptions and parameters of the model that are most important in determining outcomes. Sensitivity analyses are also important for ensuring the robustness of the model findings (Haqiqi et al. 2023).

While economists try to explain the observations, there are inevitably prediction errors and unobservable variables. This challenge is even more daunting for a geospatial economic model with numerous unobservable local variables. Uncertainty quantification and characterization can help uncover sources of uncertainty in the results and reveal their implications for the major conclusions of each study.

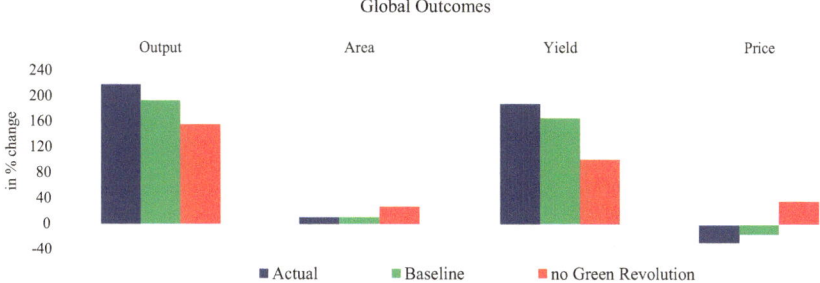

Fig. 9.1 SIMPLE model global validation. SIMPLE is capable of partitioning the actual tripling of crop output (*first blue bar*) into area and yield changes (*second and third blue bars*) while also reproducing the historical change in crop price over the period 1961–2006 (Hertel et al. 2014) The *green bars* are model predictions, and the *red bars* represent a counterfactual experiment exploring the impact of eliminating gains from the Green Revolution, which was a focal point of this chapter (Hertel et al. 2014, p. 13800)

2   Validation at the Global Level

SIMPLE-G is nested within the broader SIMPLE model, which is a global model of crop production, land use, and the environment. In general, the process of model validation requires four steps. First step involves selecting a period wherein the model outcomes will be compared to actual data. In this case, previous work has validated the SIMPLE model against observations from 1961—the start of the FAOSTAT data series for global agriculture—to 2006, just prior to the global food and energy crisis that began in 2007. The second step in general model validation is to project the model backward in time to generate a historical database using actual changes in key model drivers (e.g., 2006–1961). At this point it is possible to compare the model's performance to observed changes. However, forward-looking simulations are often easier to apply; therefore, it is common to follow with the third and the fourth steps, projecting the model forward to the current period using the historical database and then comparing projected with observed changes over the historical period. The key model drivers include population, per capita income growth, and productivity growth for the crop, livestock, and processed food sectors. Figures 9.1 and 9.2 are taken from Hertel et al. (2014) and Hertel and Baldos (2016) to compare the performance of the SIMPLE model in predicting changes in global production, prices, and yields as well as regional crop production over this 45-year period.

At the global level, the SIMPLE model is capable of capturing the direction of historical changes in crop production, cropland use, and crop prices. However, regional historical changes in crop output and cropland use are more difficult to replicate due to region-specific drivers, which are omitted from this historical simulation. Explicit government policies could shape regional patterns of agricultural production, as was in the case for Brazil, where agriculture was initially heavily

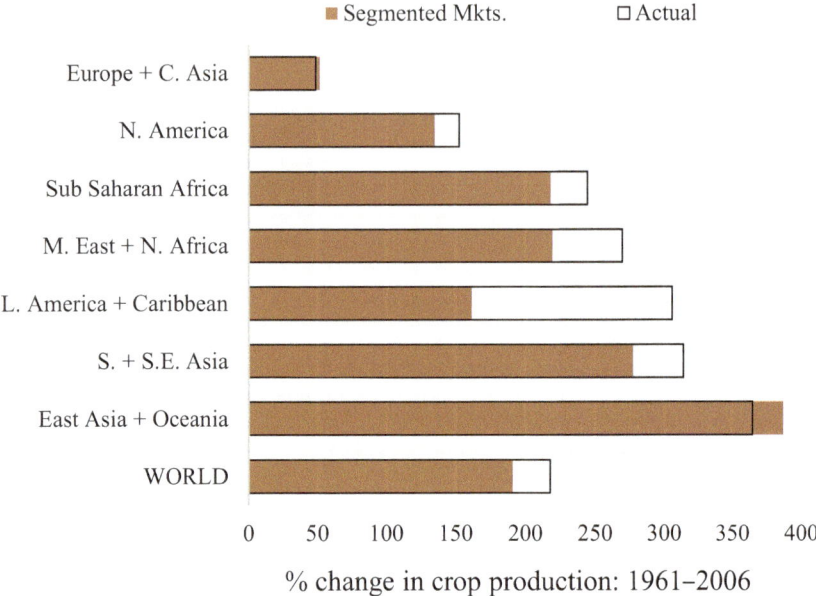

Fig. 9.2 SIMPLE model regional validation. The model is largely capable of reproducing the regional pattern of output growth over the period 1961–2006 (Hertel and Baldos 2016, p. 198) Market segmentation follows Armington (1969) such that there is imperfect substitution between foreign and domestic goods in each region. The most problematic region is Latin America and the Caribbean, where the model does not capture Brazil's dramatic output growth

subsidized through rural credit and price support mechanisms. However, market reforms introduced during the early 1990s reduced trade barriers in commodity markets (Chaddad and Jank 2006). There are also barriers to international trade in agricultural products, including poor quality domestic transport infrastructure, burdensome customs procedures, and poorly developed port facilities. These barriers to trade loom particularly large in Sub-Saharan Africa (Wilson et al. 2004) and have limited that region's engagement in the global trading system.

3 Validation at the Regional Level: The Case of Brazil

By focusing the validation exercise on a single region, it is possible to dig more deeply into region-specific data sources and policies. This section reports on the work of validating a region-specific model focused on gridded agriculture, land use, and the environment against region-level observations in Brazil (SIMPLE-G-Brazil, see Chap. 16 for additional details). After developing the initial version of SIMPLE-G-Brazil, we begin by hindcasting the model from its baseline (2017) to the year 2000 to compare the simulated crop output (converted to crop-equivalent and

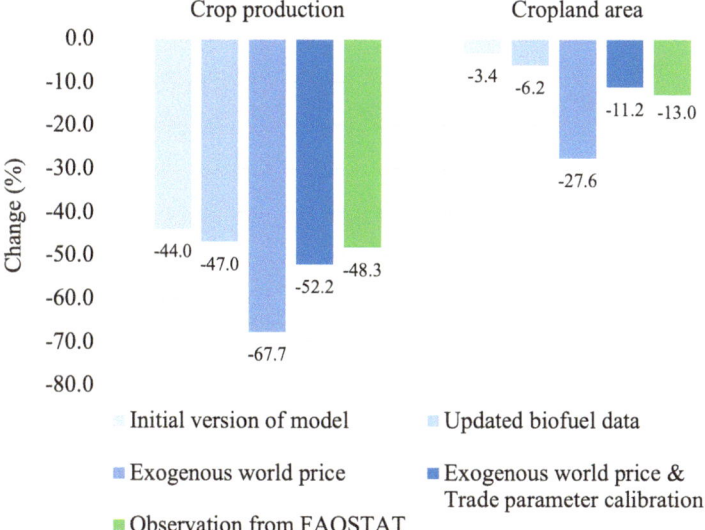

Fig. 9.3 Validation of crop output and cropland for Brazil
Model simulations (*blue*) are compared with observations reported by the United Nations Food and Agriculture Organization (*green*)

aggregated) and cropland area with observed data from FAOSTAT for Brazil. (Note that this validation exercise is backward-looking, in contrast to the forward-looking validation reported in Figs. 9.1 and 9.2.) Socioeconomic drivers used in the initial step of hindcasting include population and per capita GDP, calculated from population and GDP (in constant dollars) data from the World Bank (3-year averages are used to remove short-term volatility); total factor productivity (TFP), calculated from TFP growth rates (Fuglie 2022) for the crop, livestock (Ludena et al. 2007), and processed food (Griffith et al. 2004) sectors. The percentage change in demand for crops by the biofuel sector is calculated with the ratio of global biofuel demand in the transportation sector (representing biodiesel and ethanol from crops) between 2017 and 2000 from the World Energy Outlook 2018 (IEA 2018).

As is shown in the first bar of each grouping in Fig. 9.3, the initial version of the model underestimates the reduction in crop output (−44.0% in simulation versus −48.3% observed) and cropland area (−3.4% in simulation versus −13.0% observed) in Brazil. To identify potential issues that affect the performance of SIMPLE-G-Brazil, we apply a stepwise strategy to check dimensions of the model that may cause mismatches between simulations and observations.

According to the structure of SIMPLE-G (Fig. 4.1), demand for crop output can be divided into three categories: exogenous demand for biofuel production, endogenous demand for domestic consumption, and endogenous demand for foreign countries. First, we checked data sources on biofuel demand in the initial version and found that these data are based on the crop sales share for biofuel from GTAP-Bio database Version 6, with 2006 as the baseline. These data did not capture the

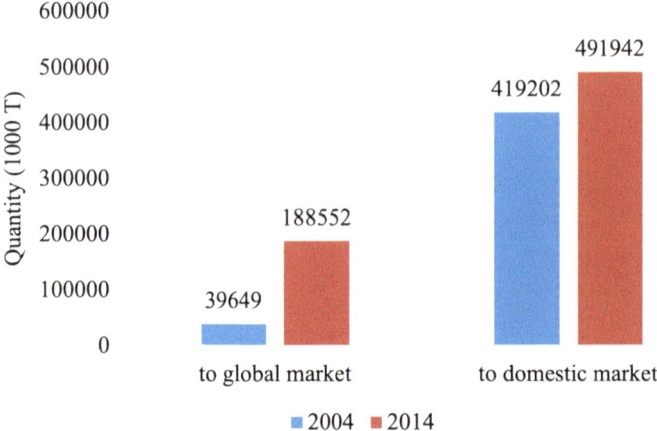

Fig. 9.4 Crop supply by market in Brazil, calculated using market share from GTAP database

rapid growth of biofuel production in Brazil during the 2000–2017 hindcasting period. To fix this problem, we update the biofuel demand data with the sales share from GTAP-Bio database Version 9 (with 2011 serving as the baseline). As a result, the validation outcomes (i.e., the "updated biofuel data" category in Fig. 9.3) show that the simulated change in crop output from 2017 to 2000 (−47.0%) becomes closer to the observed change (−48.3%), but the model still underestimates the (backward-looking) reduction of cropland area (−6.2% in simulation versus − 13.0% in observation). A smaller backward-looking reduction in area means that the model is not picking up all of the growth in cropland area from 2000 to 2017.

Second, we compared the simulated crop supply pattern (sales to domestic versus international markets) with the crop supply patterns calculated using the domestic and international market shares of crop supply from the GTAP version 10 database, with the baseline years of 2004 and 2014, respectively (Fig. 9.4).[1] We found that, compared with 2014, the GTAP database indicated that the crop supply in 2004 should have had a much greater reduction for the global market (−79%) than in the domestic market (−15%), indicating strong export growth over the 2004–2014 period. However, under the "updated biofuel data" category, the simulated reduction in crop supply (variable QSCROPr) was −60% for the global market and − 42% for the domestic market, which appears to greatly underestimate the role of export growth over this historical period. Given that the global crop price is simulated endogenously with socioeconomic drivers in this step, these findings indicate that there are some uncontrolled impacts on the global crop price, possibly from other regions, that cause mismatches in crop supply patterns.

---

[1] Although our simulation is between 2017 and 2000, GTAP version 10 database does not provide data on these two years, so we used the year closest to these two years to show the general patten of crop supply change.

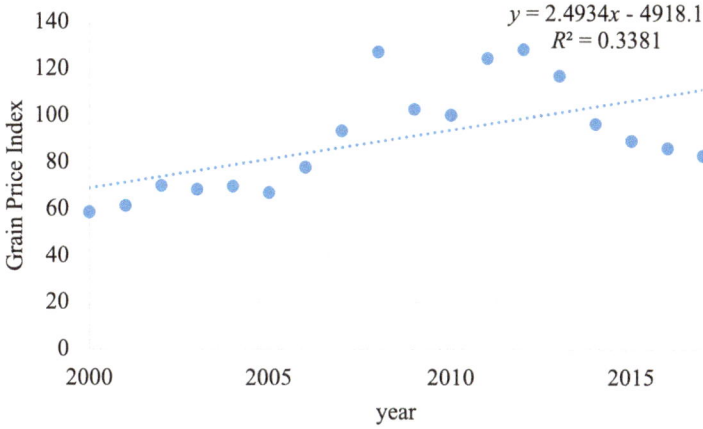

Fig. 9.5 Grain price index and its linear regression model

To control for potential bias originating from our modeling of the global crop market, we added another socioeconomic driver to the hindcast—the observed global crop price change over this historical period—to isolate Brazil from uncontrolled impacts emanating from non-Brazil regions via the global crop market. This allows us to focus more sharply on the response of crop production and land use within Brazil. The shock to the change in global crop prices was calculated with the annual indices for grains based on World Bank commodity price data (Fig. 9.5). We used the grain price index from 2000–2017 to fit a linear model between the grain price index and time and use the fitted value from the model between 2017 and 2000 to calculate the percentage shock for global grain price in order to smooth short-term variation in the data.

The simulation results with both socioeconomic drivers and additional global price shocks are reported under the "exogenous global price" category in Fig. 9.3. Although this simulation overestimates the reduction in both cropland and crop output, the simulated change in global crop supply is close to observations. The results suggest a 91% reduction of supply to the global market and a 59% decline in supply to the domestic market, which more closely mimics the observed crop supply pattern reported in the GTAP database. This result indicates that the global crop price is a partial source of the mismatch and that Brazil is overresponsive to global crop price changes in the current model.

To address the problem of overresponsiveness to global price changes, we further calibrated two key parameters in the Armington trade structure: the elasticity of substitution between crops demanded from domestic and international markets and the elasticity of transformation between crops supplied to domestic and international markets. The calibration was conducted by gradually changing these two parameters by the same amount and finding the parameter value (0.7) that produces the simulated results (both cropland and crop output) that most closely match the observations as reported in FAOSTAT. After calibration, the results are shown in

Fig. 9.3 as the group "exogenous global price & trade parameter calibration"; these results are not only more accurate than the initial validation results, but also exhibit more similar changes in the crop supply pattern (-73% to global market and -44% to domestic market) with the data from the GTAP database. This exercise highlights several useful strategies in model validation. It is important to break down the problem based on the theoretical foundation of this model, which can lead to fruitful scrutiny of the data and parameters used in the model. Proceeding in a stepwise fashion is important, as shown in Fig. 9.3.

4 Validation of Gridded Model for the United States

As above, we use backcasting as the main validation strategy for SIMPLE-G-US. This approach also provides insights into the strengths and weaknesses of a model in a specific application. In the case of the gridded model, the main challenge is obtaining accurate geospatial data as a reference for comparison and accurate geospatial data for drivers of the model. Here, we ask whether the model can replicate a movement to a new state of the agricultural economic systems from the period 2001–2002 to 2016–2017, an analysis that requires controlling many social, political, and economic variables.

4.1 Precision Assessment, 2001–2002 to 2016–2017

Here, we study the performance of the SIMPLE-G-US model in predicting cropland area and its changes in the continental United States over the long run using independently collected reference data. The validation results are reported first for the entire continental United States followed by tables for the USDA Farm Resource Regions. Given data availability, we draw on cropland reference data for 2001 and 2016 from the USGS National Land Cover Database (NLCD) and economic drivers for 2002 and 2017 from the USDA Census of Agriculture. Figure 9.6 illustrates the validation steps of SIMPLE-G-US.

Reference for Comparison Advances in satellite imagery over the past 2 decades have allowed researchers to compile new datasets with improved detail and geospatial accuracy. We draw on the National Land Cover Database (NLCD) for 2001–2016, a valuable resource for US land cover, as the reference dataset for this validation exercise. The NLCD is a collection of datasets created by the United States Geological Survey (USGS) that provides information on US land cover at a resolution of 30 meter. The maps are created by classifying USGS/NASA Landsat satellite imagery into different land cover categories. Here, we take cropland area from class #82 "Cultivated Crops," which includes annual crops and perennial woody crops. We aggregate these data to the SIMPLE-G model spatial resolution (5 arcmin) and then to the US county level, the level at which key USDA census data

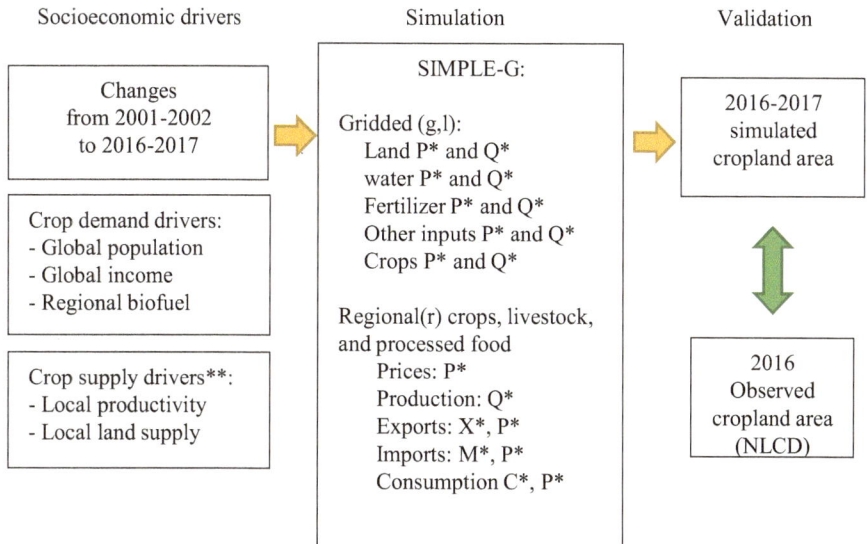

Fig. 9.6 Steps in the validation of the SIMPLE-G-US-Allcrops model, 2001–2002 to 2016–2017 When deployed in SIMPLE-G-US, the socioeconomic drivers of cropland change (*lefthand column*) result in gridded and regional outcomes (*center column*). For model validation, we compare simulated cropland area in 2016–2017 to observed area in 2016 (*righthand column*)

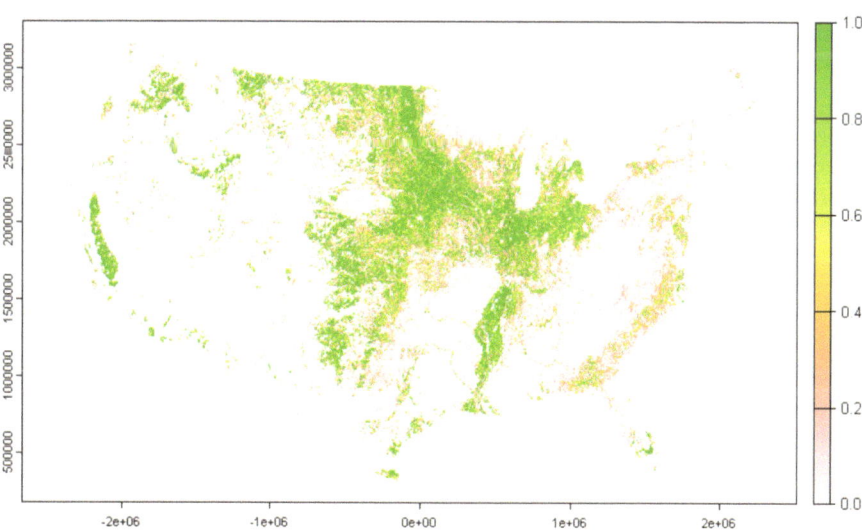

Fig. 9.7 Share of cropland in total grid cell area calculated from the National Land Cover Database, 2016

are reported. Fig. 9.7 illustrates the cropland share in total grid cell area calculated based on NLCD for the year 2016.

Global and Local Drivers of Land-Use Change from 2001–2002 to 2016–2017 - Historical validation is complicated when multiple biophysical and socioeconomic drivers interact at multiple scales. Hydroclimatic, agricultural, and economic systems are interconnected, and it would be impossible to factor in all the drivers of change in these systems. However, we find that including the most significant drivers in a well-structured model can give us a good representation of reality. The first step is the identification of these critical drivers. During the study period, four major changes are observed. First, population and income increased. We track these changes at the regional level. These increases lead to more demand for food and more demand for agricultural inputs, along with demand for land and water usage. Second, the United States mandated a sharp increase in ethanol production over this period, causing an increase in the production of corn. Third, enrollment in Conservation Reserve Programs declined significantly, such that more land became available for reversion to cropland. In addition, there has been a change in the rate of fallow/idle cropland, with significant implications for cropland supply curves. Finally, growth in agricultural productivity continued to be a dominant factor contributing to an increase in production in the United States (Clancy et al. 2016; Wang et al. 2020; Fuglie et al. 2022). We consider these drivers in the historical validation exercise. The drivers of the model are obtained from the USDA Census of Agriculture and FAO annual statistics, described below. The supply-side economic drivers are changes in county-level Hicks-neutral agricultural productivity for irrigated and rainfed crop production and county-level changes in the economic supply of cropland. The demand-side economic drivers are county-level changes in crop demand for biofuels and changes in global population and income.

Demand Drivers: Regional Population and Income The population and per capita income in 2001–2002 and 2016–2017 are obtained from the World Bank for each country and then aggregated to 16 SIMPLE regions (see Table 9.1). In this period, global population increased by 19.8% while per capita income increased by around 57.7%.

Demand Drivers: Crop Demand for Biofuels The biofuel boom has had important implications for corn-producing regions. However, the current SIMPLE-G model includes an aggregated measure of all crops produced. To accurately model the biofuel boom, we incorporate the increase in crop demand for biofuel for corn-producing subregions according to their share of corn in total crop production. For each county, the increase in demand is calculated using the following equation:

$$b_j = \theta_{corn,j} b_r, \tag{9.1}$$

where b_j is the percentage change in the demand for all crops produced in county j, $\theta_{corn,j}$ is the share of corn in county j's crop output, and b_r is the regional percentage

Table 9.1 Global drivers of land-use change, 2002–2017

	Population			Income		
	2002	2017	Growth (%)	2002	2017	Growth (%)
Eastern Europe	284,796	282,095	−0.9	4,678	7,842	66.0
North Africa	148,197	191,427	29.2	3,103	3,647	51.8
Sub Saharan Africa	648,302	985,597	52.0	929	1,378	125.5
South America	179,870	213,451	18.7	6,379	8,942	66.3
Australia, New Zealand	23,365	29,287	25.3	42,100	51,418	53.1
Europe	472,201	498,457	5.6	31,031	36,235	23.3
South Asia	1,419,749	1,755,775	23.7	832	1,778	164.2
Central America	354,091	420,080	18.6	6,958	8,400	43.2
South Africa	53,958	66,098	22.5	4,462	5,593	53.5
Southeast Asia	542,834	652,564	20.2	2,042	3,689	117.2
Canada	31,178	36,732	17.8	39,117	44,109	32.8
United States	287,279	325,085	13.2	49,184	58,330	34.2
China	1,336,765	1,452,625	8.7	2,600	8,901	272.0
Middle East	243,072	329,399	35.5	7,390	10,566	93.7
Japan, Korea	199,232	204,029	2.4	24,384	29,772	25.0
Central Asia	66,504	92,197	38.6	933	1,988	195.4
World	6,293,395	7,536,915	19.8	7,899	10,405	57.7

change in corn demand due to biofuel demand observed in the 2002–2017 period reported by the USDA (Chen et al. 2011).

Supply Drivers: Economic Supply of Cropland The cropland supply curve may shift with changes in the regulatory environment (e.g., conservation policy) or the relative return to alternative land use (e.g., pastureland rents decline). Here, the supply shifters (called "slack variables" in our model) are calculated based on changes in the average share of harvested cropland to total cropland (i.e., harvest rate) for each location:

$$H_g = \Psi_g L_g$$
$$h_g = \psi_g + l_g, \tag{9.2}$$

where H is the harvested area, Ψ is the harvest rate, and L is the cropland area defined for each grid cell g. Percentage changes are shown with lowercase letters. These data are taken from the USDA Census of Agriculture at the county level. Total cropland area includes harvested cropland, pastured cropland, idle cropland, and fallowed land.

Supply Drivers: Hicks-Neutral Agricultural Productivity TFP growth has been the major driver of US crop production in recent decades (Wang et al. 2020). Here, the productivity variables are represented by capital letter A (lowercase letter a for the percentage change). Hicks-neutral agricultural productivity is an economic term reflecting an increase in crop production using the same levels of land and nonland inputs. If we show the production volume by Q and the composite inputs by X, the percentage change in TFP can be shown as follows:

$$a = q - x. \tag{9.3}$$

Assuming no change in the production technology (input mixes), TFP growth can be approximated by changes in average yields. We estimate the changes in average yields using the following relationship:

$$\begin{aligned} V &= L.Y.P \\ v &= l + y + p \\ y &= v - l - p, \end{aligned} \tag{9.4}$$

where V is the value of crop sales; L is the cropland harvested area; Y is the average corn-equivalent yield; and P is the average crop price. For the validation exercise, P is calculated based on the average crop price index from the FAO for Farm Resource Regions. We obtain V and L from the USDA Census of Agriculture for 2002 and 2017 at the county level. Then, y is calculated for all counties using Eq. 9.4. Uniform values for productivity growth are assigned to all grid cells within a county.

Supply Drivers: Economic Supply of Water and Crop Water Demand Crop production is sensitive to water availability. Thus, changes in water conditions can affect the extent of cropland. In SIMPLE-G, the irrigation water supply curve may shift with changes in weather conditions (e.g., lower surface water availability) and changes in irrigation costs and expenditures (e.g., groundwater pumping costs). Ideally, we would calculate the shift in surface water supply for each grid cell. While the USGS reports the water conditions for each 5-year interval, it is not possible to decompose the demand drivers and supply drivers of changes in water withdrawal from those reports. As no other observational dataset is available, data collection can rely on outputs of a hydrologic model (e.g., the Water Balance Model). An additional complication is that measuring changes in groundwater supply requires converting changes in the groundwater table to equivalent changes in the supply shifter at each 5 arcmin grid cell in SIMPLE-G. However, as running a hydrology model requires multidisciplinary collaborations, we validate the model while excluding the supply shifters related to water. This may lead to less accurate cropland projections in highly irrigated areas where significant changes in water conditions have been observed (e.g., 2012 and 2015 drought and heat stress).

4.2   Agreement Between Observation (Satellite) and Simulation (SIMPLE-G)

At the national level, SIMPLE-G projects 128.68 million ha of cropland in the continental United States for 2016–2017. According to the NLCD, this number was actually 129.30 million ha for 2016. But how is this area distributed? For that, we focus on observations and simulation outcomes at the administrative units given by counties and states within the continental United States as well as the USDA Farm Resource Regions.

Agreement at the County Level Figure 9.8 illustrates projected cropland area (SIMPLE-G) and reference cropland area (NLCD) for 2016–2017 by county. Overall, the projected area is close to the reference data ($R = 0.98$).

Agreement at the State Level Figure 9.9 illustrates the cropland area projected by SIMPLE-G versus NLCD cropland area for major farming states around 2016–2017. Overall, SIMPLE-G can simulate total cropland area with only minor differences. However, in the case of Montana the difference is noticeable. Further investigation is required to explore the possible causes of this discrepancy.

Agreement at Farm Resource Regions Figure 9.10 illustrates the cropland area projected by SIMPLE-G versus NLCD cropland area for USDA Farm Resource Regions around 2016–2017. Visual inspection indicates that SIMPLE-G accurately simulates the total cropland area.

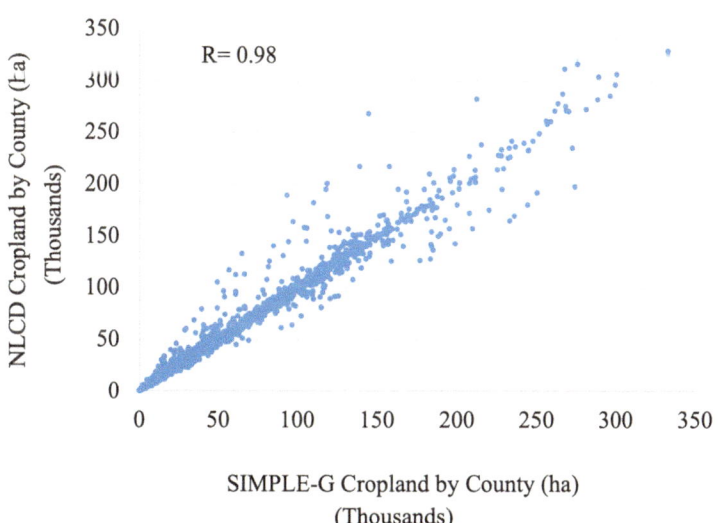

Fig. 9.8 Observed and simulated cropland area in the United States in 2016–2017 based external drivers, 2001–2002 to 2016–2017

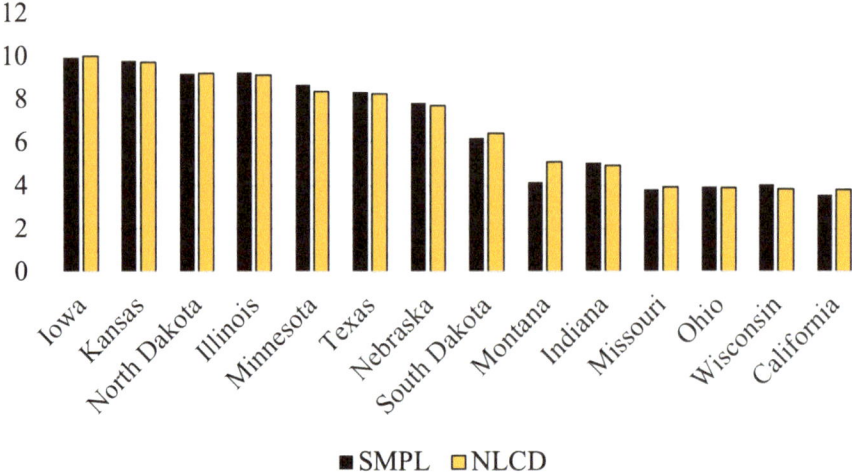

Fig. 9.9 Observed and simulated cropland area by major farming states, 2016–2017

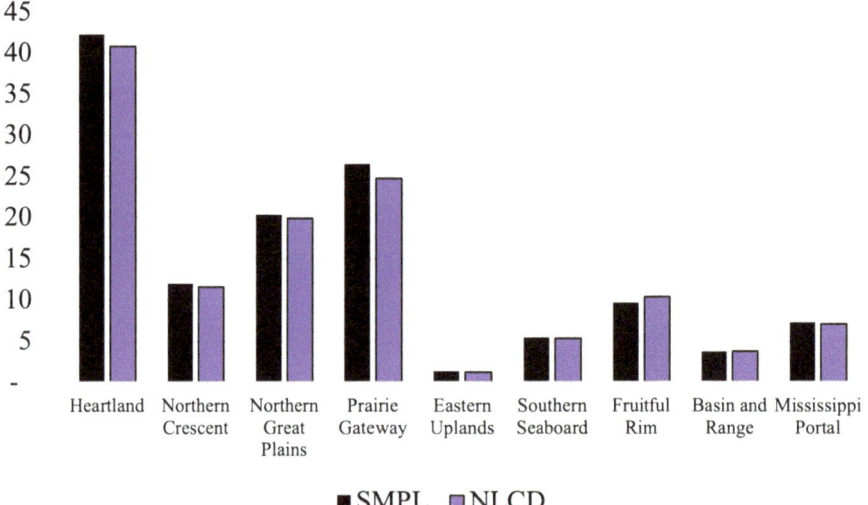

Fig. 9.10 Observed and simulated cropland area by USDA Farm Resource Regions in 2016–2017

To evaluate the significance of different drivers of cropland changes, Fig. 9.11 illustrates the role of supply (i.e., productivity and land supply) and demand drivers (i.e., global population, global income, local biofuels) by USDA Farm Resource Region. The results show that the supply and demand drivers are working in opposite directions in the Fruitful Rim and Basin and Range regions. This means that the final net change is sensitive to the magnitude of these drivers.

Kolmogorov–Smirnov Test The Kolmogorov–Smirnov test is a nonparametric test used to compare two continuous distributions. It is calculated as the maximum

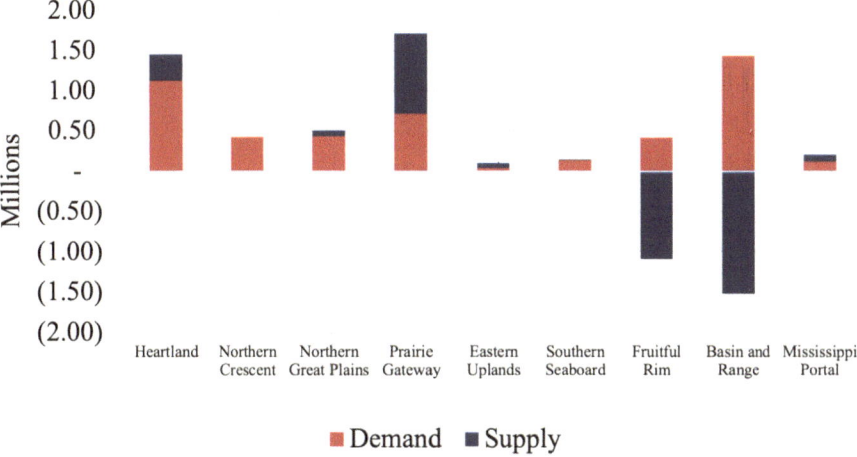

Fig. 9.11 Supply and demand drivers of change in cropland area by USDA Farm Resource Regions, 2016–2017

absolute difference between the cumulative distribution function of the reference distribution and the cumulative distribution function of a sample, estimated parameters, or simulated output:

$$D = \max_{x} |F_{SMPL}(x) - F_{NLCD}(x)|. \tag{9.5}$$

Figure 9.12 illustrates the cumulative distributions of cropland changes in the NLCD (reference) and SIMPLE-G. Overall, the distributions of changes are in agreement.

4.3 Discussion: Causes of Disagreement

Because the underlying human decisions that drive local changes in land use are rarely deterministic, these results can never be expected to be perfect when compared to empirical data. Additionally, many land-use changes are not the result of one simple process but rather a combination of biophysical and socioeconomic drivers that are mutually influential. Here, we review some of the most important confounding factors.

Government Policies Capturing changes in agricultural-related policies in the model can potentially improve the agreement between SIMPLE-G projections and NLCD data. Overall, government policies can introduce a significant amount of disagreement into coupled economic and environmental models, making it difficult to predict future outcomes. Local and national policies can change farmers'

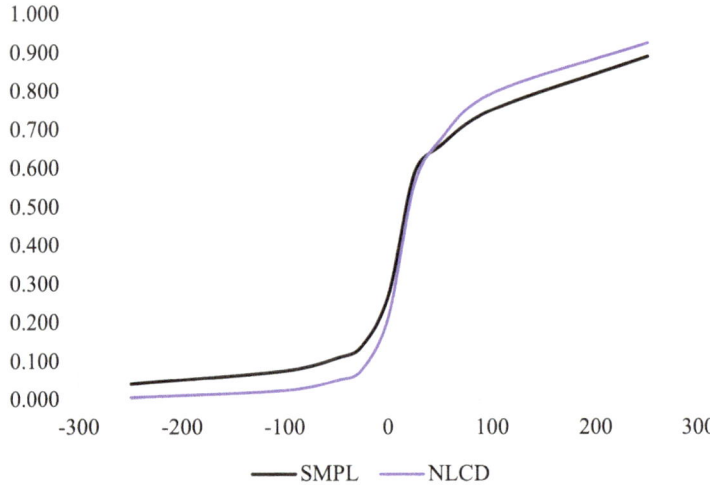

Fig. 9.12 Cumulative distribution of change in cropland (ha) by grid cell, 2001–2002 to 2016–2017

incentives and their decisions on the extent and intensity of farming. For example, policies that promote wetlands and improve water quality may increase the opportunity costs of cropland use and may increase the application of fertilizer on cropland, which will lead to changes in production patterns and application rates. Capturing these changes makes it difficult to accurately predict local conditions. This is important as some of these policies may have a significant local impact on economic and environmental outcomes and thus are necessary to be included in validation. In addition, the decision environment or policies can change over time. Farmers account for uncertainty about the future direction of policies in their land and water use decisions. However, government policies can be complex and difficult to implement. For example, many of the Conservation Reserve Programs (CRP) have multiple components and complicated details that may require different approaches for each location and program. Overall, acreage enrolled in CRPs declined from around 33 million acres in 2002 to around 24 million acres in 2017. Another example of policies with implications for land use is the Sustainable Groundwater Management Act, which requires local agencies to "form groundwater sustainability agencies (GSAs) for the high and medium priority basins." Prior expectations about such a policy might have implications for farmers' decisions about irrigation even before the act is passed.

Local Water Conditions We suspect that a major cause of disagreement between SIMPLE-G projections and NLCD data is related to changes in weather and water conditions between 2002 and 2017. One major determinant of the extent of irrigated cropland is hydroclimatic conditions. Higher temperatures increase irrigation requirements, and lower precipitation may reduce the availability of water resources. While SIMPLE-G does not include a hydrological model, it does contain an economic model for long-run economic demand and supply for water. Any changes to hydroclimatic conditions should be translated to economic demand and supply before entering the model as input. Obtaining gridded information on changes in water availability and water requirements is computationally expensive but can be done in the future with improved availability of these data.

Unobserved Local, Regional, and Global Drivers Many other factors related to agriculture can be sources of disagreement between SIMPLE-G and NLCD. These include, for example, the US–China trade war, which had significant implications for farmers in 2017 (Marchant and Wang 2018) and can be a negative demand driver for soy-producing grid cells. Additionally, the increase in energy prices observed in 2007–2008 led to increases in fertilizer prices and reductions in demand (Beckman and Riche 2015). This increase was a supply-side driver that affected production expenses. The grid cells with high-cost shares of commercial fertilizer will be more affected than other grid cells. Finally, the increasing trend in nonfarm investment in farmland has caused increases in land values and cropland rental rates (Burns et al. 2018). Overall, it is important to note that numerous external factors like trade wars, energy prices, and farmland investment create discrepancies between SIMPLE-G and NLCD's agricultural predictions. While it is impossible to consider all the changes, users need to consider the most important drivers.

Remote Sensing Accuracy While the overall accuracy of the NLCD 2019 data product is relatively high (90.3% ± 0.7%), the user accuracy and producer accuracy for forest loss and grass gain were > 70% and generally <50% for all other change themes (Wickham et al. 2021, 2023a, b), which means that we may be comparing model results to inaccurate observations. Therefore, caution should be exercised when directly comparing our model results to these observations. While this highlights the challenges of comparing model outputs to imperfect observations, it also underlines the importance of ongoing efforts to improve land cover change data collection and analysis methods.

5 Summary

Understanding the connections between food, trade, agriculture, land use, and water resources requires reliable quantitative insights. Gridded quantitative economic models can provide better insights into these relationships. However, validating these models is challenging because of the complex interactions, unprecedented changes, and uncertainties in human decisions. This chapter is about validating such gridded economic models by comparing their predictions to real-world data. It employs SIMPLE-G, a new economic model that can analyze land-use change, water management, and agricultural production. The goal is to demonstrate the potential of gridded economic models for understanding food systems, land use, and water management in sustainable development frameworks.

In this chapter, we draw on methods from different disciplines to validate the SIMPLE-G model. We first perform benchmark replication or calibration by comparing the results of the model to a set of observed data to ensure that the model can replicate a base reference condition. We then use backcasting to employ the model to make predictions about observed past events and compare them to actual observed outcomes from independent data sources not used in the calibration process. This helps ensure that the structural processes that can explain changes in the system are modeled correctly. We also call to use sensitivity analysis to vary the model's assumptions, parameters, and drivers to investigate how the predictions change (Haqiqi et al. 2023). This helps identify the assumptions and parameters of the model that are most important in determining outcomes and ensures the robustness of the model findings. Finally, we use uncertainty quantification and characterization to uncover sources of uncertainty in the results and reveal their implications for the major conclusions of each study. This approach to model validation can be applied to other geospatial economic models and can help uncover sources of uncertainty in the results and reveal their implications for the major conclusions of each study.

Acknowledgment and Competing Interests The authors acknowledge support from the U.-S. Department of Energy, Office of Science, Biological and Environmental Research Program, Earth and Environmental Systems Modeling, MultiSector Dynamics under Cooperative Agreement DE-SC0022141; the National Science Foundation award #2118329: "NSF Institute for Geospatial Understanding through an Integrative Discovery Environment (I-GUIDE)," the United States Department of Agriculture AFRI grant #2019-67023-29679, "Economic Foundations of Long Run Agricultural Sustainability," and the National Science Foundation INFEWS award #1855937, "Identifying Sustainability Solutions Through Global-Local-Global Analysis of a Coupled Water-Agriculture-Bioenergy System."

The findings and conclusions presented in this chapter are those of the authors and should not be construed to represent any official determination or policy of the US Department of Agriculture (USDA), the US government, the Department of Energy (DOE), or the National Science Foundation (NSF). Furthermore, we declare that there is no conflict of interest related to this work.

References

Armington, Paul S. 1969. A theory of demand for products distinguished by place of production. *Staff Papers* 16: 159–178. https://doi.org/10.2307/3866403.

Beckman, Jayson, and Stephanie Riche. 2015. Changes to the natural gas, corn, and fertilizer price relationships from the biofuels era. *Journal of Agricultural and Applied Economics* 47: 494–509. https://doi.org/10.1017/aae.2015.22.

Biondi, Daniela, Gabriele Freni, Vito Iacobellis, Giuseppe Mascaro, and Alberto Montanari. 2012. Validation of hydrological models: Conceptual basis, methodological approaches and a proposal for a code of practice. *Physics and Chemistry of the Earth, Parts A/B/C* 42–44: 70–76. https://doi.org/10.1016/j.pce.2011.07.037.

Burns, Christopher, Nigel Key, Sarah Tulman, Allison Borchers, and Jeremy Weber. 2018. *Farmland values, land ownership, and returns to farmland, 2000–2016*. Washington, DC: Economic Research Report 245. US Department of Agriculture, Economic Research Service. https://doi.org/10.22004/ag.econ.276249.

Chaddad, Fabio R., and Marcos S. Jank. 2006. The evolution of agricultural policies and agribusiness development in Brazil. *Choices* 21: 85–90. https://doi.org/10.22004/ag.econ.94415.

Chen, Xiaoguang, Haixiao Huang, Madhu Khanna, and Hayri Önal. 2011. Meeting the mandate for biofuels: Implications for land use, food, and fuel prices. In *The intended and unintended effects of US agricultural and biotechnology policies*, ed. Joshua S. Graff Zivin and Jeffrey M. Perloff, 223–267. Chicago: University of Chicago Press.

Clancy, Matthew, Keith Fuglie, and Paul Heisey. 2016. US agricultural R&D in an era of falling public funding. *Amber Waves*. https://doi.org/10.22004/ag.econ.249840.

Fuglie, Keith. 2022. International agricultural productivity. In *Economic research service, USDA*. Washington, DC: US Department of Agriculture, Economic Research Service. https://www.ers. usda.gov/data-products/international-agricultural-productivity/. Accessed 1 Oct 2023.

Fuglie, Keith, Srabashi Ray, Uris Lantz C. Baldos, and Thomas W. Hertel. 2022. The R&D cost of climate mitigation in agriculture. *Applied Economic Perspectives and Policy* 44: 1955–1974. https://doi.org/10.1002/aepp.13245.

Griffith, Rachel, Stephen Redding, and John Van Reenen. 2004. Mapping the two faces of R&D: Productivity growth in a panel of OECD industries. *The Review of Economics and Statistics* 86: 883–895. https://doi.org/10.1162/0034653043125194.

Haqiqi, I., Bowling, L., Jame, S., Baldos, U., Liu, J. and Hertel, T., 2023. Global drivers of local water stresses and global responses to local water policies in the United States. *Environmental Research Letters, 18*(6): 065007. https://doi.org/10.1088/1748-9326/acd269

Hansen, Lars Peter, and James J. Heckman. 1996. The Empirical foundations of calibration. *Journal of Economic Perspectives* 10: 87–104. https://doi.org/10.1257/jep.10.1.87.

Hertel, Thomas W., Navin Ramankutty, and Uris Lantz C. Baldos. 2014. Global market integration increases likelihood that a future African Green Revolution could increase crop land use and CO2 emissions. *Proceedings of the National Academy of Sciences* 111: 13799–13804. https:// doi.org/10.1073/pnas.1403543111.

Hertel, Thomas W., Lantz C. Uris, and Baldos. 2016. Attaining food and environmental security in an era of globalization. *Global Environmental Change* 41: 195–205. https://doi.org/10.1016/j. gloenvcha.2016.10.006.

IEA. 2018. *World Energy Outlook 2018*. Paris: IEA.

Kersebaum, K.C., K.J. Boote, J.S. Jorgenson, C. Nendel, M. Bindi, C. Frühauf, T. Gaiser, et al. 2015. Analysis and classification of data sets for calibration and validation of agro-ecosystem models. *Environmental Modelling & Software* 72: 402–417. https://doi.org/10.1016/j.envsoft. 2015.05.009.

Ludena, Carlos E., Thomas W. Hertel, Paul V. Preckel, Kenneth Foster, and Alejandro Nin. 2007. Productivity growth and convergence in crop, ruminant, and nonruminant production: Measurement and forecasts. *Agricultural Economics* 37: 1–17. https://doi.org/10.1111/j.1574-0862. 2007.00218.x.

Marchant, Mary A., and H. Holly Wang. 2018. Theme overview: US–China trade dispute and potential impacts on agriculture. *Choices* 33: 1–3. https://doi.org/10.22004/ag.econ.273328.

Ngo, The An, and Linda See. 2012. Calibration and validation of agent-based models of land cover change. In *Agent-based models of geographical systems*, ed. Alison J. Heppenstall, Andrew T. Crooks, Linda M. See, and Michael Batty, 181–197. Dordrecht: Springer Netherlands. https://doi.org/10.1007/978-90-481-8927-4_10.

Oreskes, N. 1998. Evaluation (not validation) of quantitative models. *Environmental Health Perspectives* 106: 1453–1460. https://doi.org/10.1289/ehp.98106s61453.

Razavi, Saman, and Hoshin V. Gupta. 2015. What do we mean by sensitivity analysis? The need for comprehensive characterization of "global" sensitivity in Earth and environmental systems models. *Water Resources Research* 51: 3070–3092. https://doi.org/10.1002/2014WR016527.

Razavi, Saman, Anthony Jakeman, Andrea Saltelli, Clémentine Prieur, Bertrand Iooss, Emanuele Borgonovo, Elmar Plischke, et al. 2021. The future of sensitivity analysis: An essential discipline for systems modeling and policy support. *Environmental Modelling & Software* 137: 104954. https://doi.org/10.1016/j.envsoft.2020.104954.

Rykiel, Edward J. 1996. Testing ecological models: The meaning of validation. *Ecological Modelling* 90: 229–244. https://doi.org/10.1016/0304-3800(95)00152-2.

van Vliet, Jasper, Arnold K. Bregt, Daniel G. Brown, Hedwig van Delden, Scott Heckbert, and Peter H. Verburg. 2016. A review of current calibration and validation practices in land-change modeling. *Environmental Modelling & Software* 82: 174–182. https://doi.org/10.1016/j.envsoft.2016.04.017.

Wang, Sun Ling, Roberto Mosheim, Richard Nehring, and Eric Njuki. 2020. *Productivity is the major driver of US farm sector's economic growth.* Amber Waves. https://doi.org/10.22004/ag.econ.303975.

Wickham, James, Stephen V. Stehman, Daniel G. Sorenson, Leila Gass, and Jon A. Dewitz. 2021. Thematic accuracy assessment of the NLCD 2016 land cover for the conterminous United States. *Remote Sensing of Environment* 257: 112357. https://doi.org/10.1016/j.rse.2021.112357.

Wickham, James, K. Anne Neale, Maliha Nash Riitters, Jon Dewitz, Suming Jin, Megan van Fossen, and D. Rosenbaum. 2023a. Where forest may not return in the western United States. *Ecological Indicators* 146: 109756. https://doi.org/10.1016/j.ecolind.2022.109756.

Wickham, James, Stephen V. Stehman, Daniel G. Sorenson, Leila Gass, and Jon A. Dewitz. 2023b. Thematic accuracy assessment of the NLCD 2019 land cover for the conterminous United States. *GIScience & Remote Sensing* 60: 2181143. https://doi.org/10.1080/15481603.2023.2181143.

Wilson, John S., Catherine L. Mann, and Tsunehiro Otsuki. 2004. *Assessing the potential benefit of trade facilitation: A global perspective.* Washington, DC: World Bank.

Part IV
Applications

Chapter 10
The R&D Cost of Climate Mitigation in Agriculture

Keith Fuglie, Srabashi Ray, Uris Lantz C. Baldos, and Thomas W. Hertel

1 Introduction

According to the International Panel on Climate Change (IPCC 2014), agriculture is responsible for about one-quarter of total global emissions of greenhouse gases (GHGs). According to FAOSTAT (FAO 2020), agriculture accounted for 7.21 Gt CO2e of emissions at the farmgate (i.e., from ongoing production and energy use) and another 3.50 Gt CO2e in net emissions from land-use changes in 2019.[1] Supply-side approaches for reducing agricultural emissions include investing in emissions-saving technological change and protecting carbon-rich natural lands from conversion. While technological and productivity improvements can reduce emissions intensity (GHG per unit of output) and save land overall, it could also lead to increased land conversion in some areas through competitiveness effects (Villoria 2019). If these local areas held large carbon sinks, then net emissions from land-use conversion could remain large even if the amount of agricultural land held globally

---

This chapter is a slightly revised version of a paper originally published as Fuglie, Keith, Srabashi Ray, Uris Lantz C Baldos, and Thomas W. Hertel. 2022. The R&D cost of climate mitigation in agriculture. *Applied Economic Perspectives and Policy* 44: 1955–1974. https://doi.org/10.1002/aepp.13245.

Data availability statement: The files needed to replicate this application are available at https://gtap.agecon.purdue.edu/simple-g/

---

[1] 1 Gt CO2e = 109 or one billion metric tons of CO2 equivalents (CO2e).

K. Fuglie (✉)
Economic Research Service, United States Department of Agriculture, Washington, DC, USA
e-mail: keith.fuglie@usda.gov

S. Ray · U. L. C. Baldos · T. W. Hertel
Center for Global Trade Analysis, Department of Agricultural Economics, Purdue University, West Lafayette, IN, USA

I. Haqiqi, T. W. Hertel (eds.), *SIMPLE-G*,
https://doi.org/10.1007/978-3-031-68054-0_10

remained unchanged or declined. Environmental policies can target and protect carbon-rich areas from conversion. However, the drawbacks of these policies include cost (assuming that landowners are compensated for forgone income from other land uses), lack of permanence, higher food prices, and possibly worsening global food insecurity (Baquedano et al. 2022).

This study uses a global economic model of the agri-food system (the Simplified International Model of agricultural Prices, Land use and the Environment, or SIMPLE) to compare outcomes from productivity policies in the form of higher agricultural research and development (R&D) spending and environmental policies that restrict agricultural land supply in the carbon-rich land most at risk to land-use change. Lobell et al. (2013) use SIMPLE model simulations to conclude that an additional US\$225 billion in agricultural R&D might save 15 Gt CO2e from avoided land-use change by 2050. However, they used a simplified framework to link R&D to productivity growth and did not include the effects of agricultural productivity growth on farmgate emissions. Our model linking R&D spending to productivity growth also takes into account time lags between R&D spending and the adoption of new technology by farmers, variation in the marginal response of productivity to R&D spending in different regions of the world, and possibilities for international technology spillovers, as informed by empirical studies (Fuglie 2018). This framework has been used to explore the R&D costs of adapting agriculture to climate change (Baldos et al. 2020) and potential gains from liberalizing trade in agricultural commodities and technology to improve global food security (Hertel et al. 2020).

For our policy simulations, the outcomes of interest are changes in global agricultural GHG emissions, land use, agricultural production, food prices, and the prevalence of undernutrition relative to policy cost. Simulations of the world agricultural economy over 2017–2050 are run for a set of policy scenarios involving R&D spending and land-use restrictions that compensate landowners for forgone income from avoided land-use change. We consider scenarios that target actions in less developed countries as well as some that include developed countries. In particular, we examine how more stringent environmental restrictions on agriculture envisioned by the European Union (EU) Green Deal could affect global outcomes and how unintended outcomes might be offset through accelerated R&D spending by EU countries.[2] We find that more R&D spending to accelerate productivity growth reduces GHG emissions from land-use change less effectively than targeted environmental policies. However, accelerated productivity growth also reduces the cost of environmental policies (by effectively making land less scarce and therefore compensation cheaper). Moreover, higher levels of productivity permanently lower farmgate GHG emission intensity and, by lowering global food prices, generally improve global nutrition and food security.

---

[2]The European Green Deal lays out specific targets for fertilizer and pesticide reductions in EU agriculture by 2030. Beckman et al. (2020) examine the productivity and market impacts of these input reductions but do not quantify how much R&D spending might be needed to offset anticipated productivity losses.

2   Agricultural Production, Productivity, and GHG Emissions Since 1990

Before presenting our model formulation and results, we first describe global and regional patterns of agricultural GHG emissions and output growth over the past three decades. Two types of GHG emissions are associated with agriculture: (1) GHGs released when forests are cut down, peatlands drained, or grasslands plowed to make way for more agricultural land and (2) direct "farmgate" emissions from existing agricultural production (e.g., methane from livestock and rice paddies, nitrous oxide from fertilizer applications, carbon dioxide from decomposition or burning of crop residues, and use of fossil fuels to power machinery and to light, heat, and cool farm buildings).

Figure 10.1 shows agricultural GHG emissions and outputs in 1990 and 2019 for major world regions. Sources of GHG emissions include land-use change and crop-related and animal-related production (Fig. 10.1a and c). Agricultural output is composed of crops, animal products, and products from aquaculture (Fig. 10.1b and d). Between 1990 and 2019, agricultural emissions intensity (i.e., kilograms of emissions per dollar of output) fell in all regions, but the total amount of emissions increased in Sub-Saharan Africa (SSA), South Asia, Northeast Asia, and Central and West Asia and North Africa (CWANA), while decreasing in Latin America (LAC), Southeast Asia, Europe, and other developed countries (other DC). Globally, emissions per unit of output (at constant prices) fell by nearly half between 1990 and 2019, as total emissions remained about constant while world agricultural output increased by 92%. This decline in global agricultural emission intensity was due to slowing rates of deforestation, a slowing rate of growth in emission-intensive factor inputs, and possibly to general improvements in agricultural total factor productivity (TFP). In this chapter, agricultural TFP is defined as the ratio of aggregate agricultural output to aggregate factor inputs (i.e., land, labor, capital, and intermediate inputs) as measured by the US Department of Agriculture Economic Research Service (USDA-ERS 2021).

The level and intensity of emissions from agriculture vary widely across the global landscape. GHG emissions per unit of output are exceptionally high in regions where forestland is being converted to agriculture (Sub-Saharan Africa, Latin America, and Southeast Asia). In Latin America, agricultural GHG emissions declined by nearly 20% between 1990 and 2019 (due to a decline in the rate of land-use change), even as output more than doubled. In Sub-Saharan Africa, both output and emissions increased, driven by an increase in land-use change. Remarkably, Northeast Asia was able to nearly triple agricultural output while increasing emissions by only 9%. The dramatic fall in emissions intensity of agricultural output in Northeast Asia was accompanied by rapid growth in TFP and a relatively small share of emission-intensive inputs—such as ruminant livestock—in production.

In developed country (DC) regions (i.e., Europe and other developed countries, including North America, Oceania, Japan, and South Korea), ruminant livestock account for the bulk of GHG emissions from agriculture. In Europe, there has been

Agricultural GHG Emissions

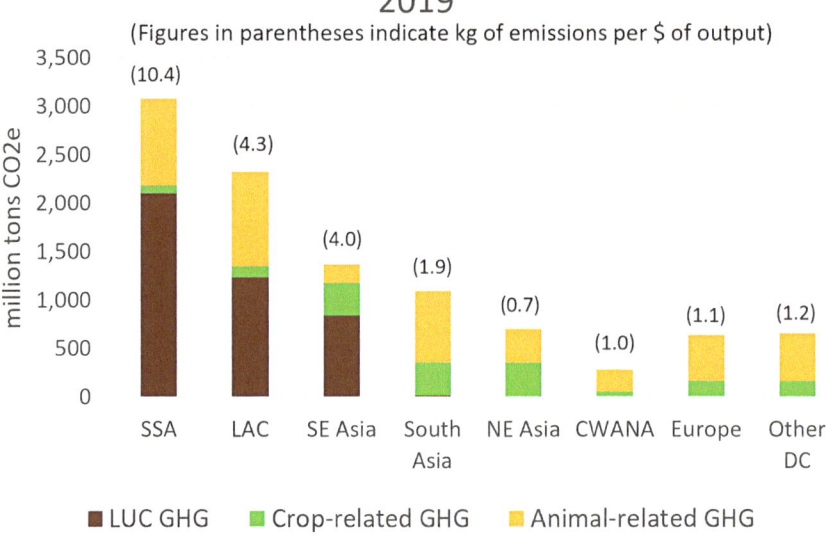

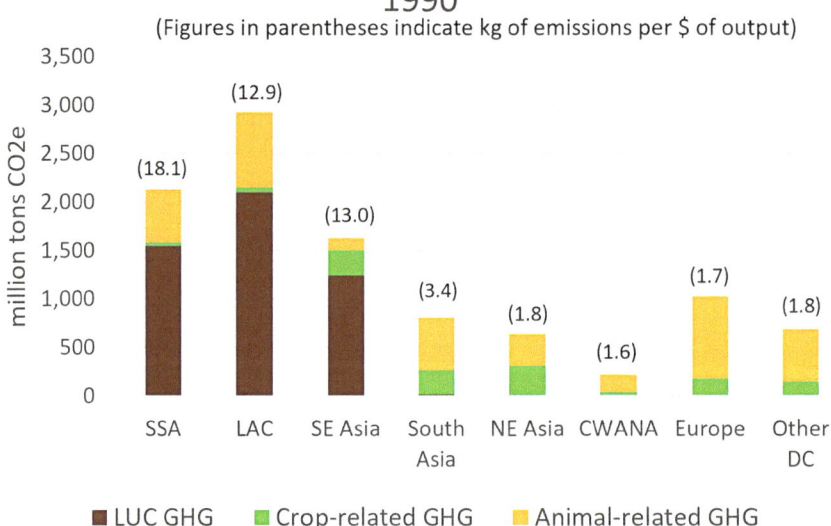

Fig. 10.1 Agricultural greenhouse gas emissions and output in world regions, 1990 and 2019 SSA, Sub-Saharan Africa; LAC, Latin America and Caribbean; CWANA, Central, West Asia, and North Africa; NE Asia, China, Mongolia, and North Korea; Europe includes all of Europe, the Russian Federation, and Kazakhstan; Other DC, North America, Oceania, Japan, and South Korea; and LUC GHG, GHG emissions from land-use change. Crop-related GHG emissions include emissions from rice cultivation, drained organic soils, fertilizer use, crop residue decomposition and the burning of crop residues; animal-related GHG emissions include emissions from enteric fermentation in ruminant livestock, manure, and savanna fires

Agricultural Output

2019

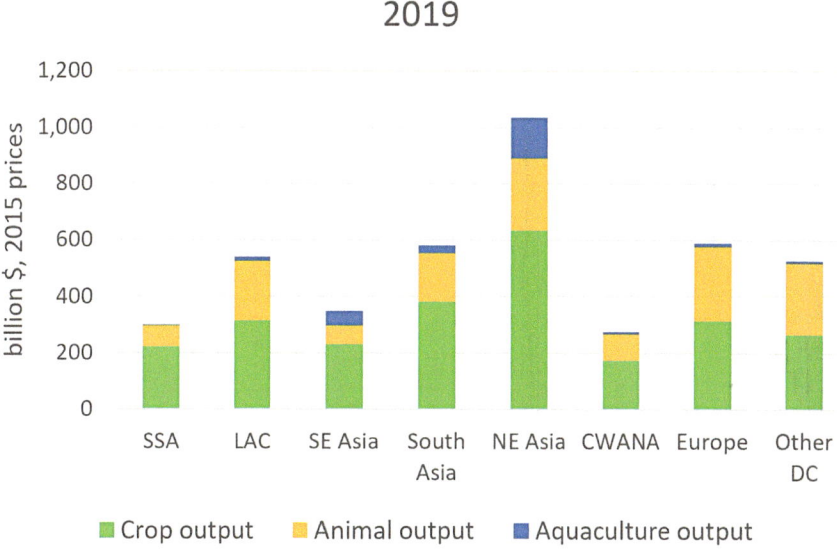

1990

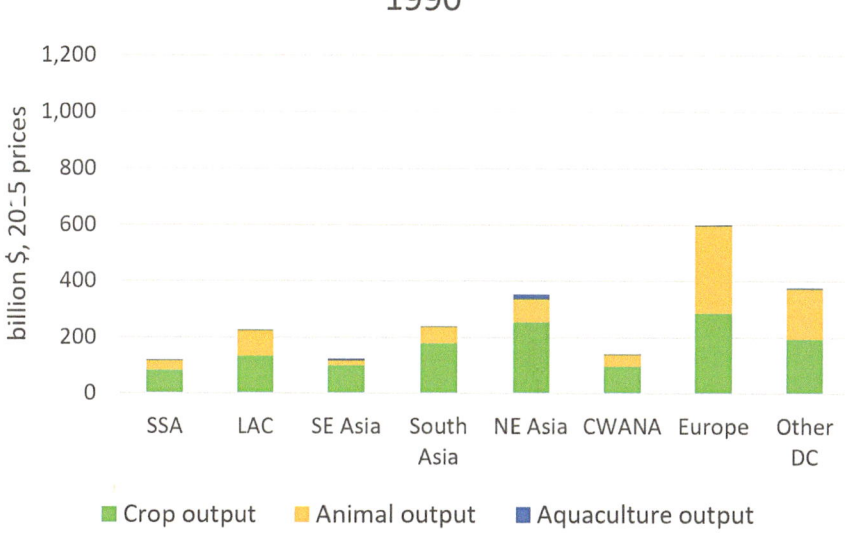

Fig. 10.1 (continued)

Table 10.1 Population and income growth assumptions (2017–2050)

Region	Population growth (% per year)	Income growth (% per year)
Eastern Europe	−0.16	2.59
North Africa	0.77	3.34
Sub-Saharan Africa	1.83	4.00
South America	0.59	2.47
Brazil	0.36	2.11
Australia + New Zealand	1.07	1.20
European Union	0.19	1.40
South Asia	0.82	4.01
Central America + Caribbean	0.57	2.32
South Africa	0.52	2.64
Southeast Asia	0.52	3.71
Canada	0.80	1.09
United States	0.62	1.07
China	−0.25	3.79
Middle East	1.13	2.05
Japan + Korea	−0.37	1.56
Central Asia	0.40	4.23

Growth is shown as compounded annual growth rates
Source: Middle-of-the-road shared socioeconomic pathway (SSP 2) projections, SSP database v2.0 (Riahi et al. 2017)

very little change in agricultural output over the past three decades, but productivity improvement has led to fewer inputs in the sector and less GHG emissions from agriculture. In other DC regions, total emissions (and total factor inputs) have remained roughly constant, while productivity improvement has expanded output. Thus, in both Europe and other developed countries, increases in TFP have led to reductions in GHG emissions intensity from agriculture.

The trends depicted in Fig. 10.1 strongly suggest that agricultural TFP growth has reduced agricultural GHG emission intensity. In existing agricultural areas, TFP growth reduces emissions intensity because fewer factor inputs are required to produce a given level of output. The relationship between agricultural TFP growth and land-use change is more nuanced. At the global level, TFP growth in agriculture almost certainly curtails land expansion into agriculture, but at the local level, it could increase incentives to expand agricultural land if local TFP growth improves trade competitiveness. This land expansion may arise in countries that achieve high TFP growth (relative to the global average) and are integrated into global markets so that their production costs fall relative to global commodity prices (Villoria 2019). But agricultural TFP is hardly the sole determinant of deforestation. National environmental policies also have a strong influence on land use, and recent policy changes in countries like Brazil and Indonesia may account for a significant part of those countries' reductions in GHG emissions from land-use change over the last three decades. Finally, consumer demand for agricultural products exerts an influence on emissions. The growing demand for meat and milk as populations and per capita incomes rise may shift the agricultural product mix toward more emissions-intensive ruminant livestock production.

3 Version of SIMPLE Used in This Application

To simulate how R&D and environmental policies could influence future GHG emissions from agriculture, we employ a nongridded version of the SIMPLE model (recall Part III).[3] The version of SIMPLE used here has a base year of 2017. Population and income growth to 2050 are assumed to follow the middle-of-the-road shared socioeconomic pathway (SSP 2) assumptions developed for integrated assessments of climate change (Riahi et al. 2017). Table 10.1 provides details on the population and income growth assumptions by global region used for this analysis. Based on estimated behavioral parameters, consumers in wealthy regions are assumed to be less responsive to price and income changes than those residing in low-income regions. Aggregated food commodities in SIMPLE include crops, livestock products, and processed foods, and consumption patterns evolve to reflect observed shifts in dietary preferences—moving away from crops toward livestock and processed foods as incomes rise. Population-wide caloric distributions (characterized by a lognormal distribution) in each region are calibrated to data from the Food and Agriculture Organization of the United Nations (FAO 2020) and shift over time based on prices and per capita incomes. The estimates of caloric consumption levels for different segments of a population permit us to estimate the proportion of a region's population that remains undernourished (i.e., minimum daily caloric intake of 1700–2000 kcal/day/person, depending on the age and gender structure of the population).

In this study, we pay attention to the effects of climate mitigation policies on both food and environmental security. By tracking impacts on food prices, we are able to estimate changes in the adequacy of dietary energy consumption in a given region (Baldos and Hertel 2014). Per capita food consumption is converted to caloric consumption equivalents. By applying minimum daily dietary energy requirements, it is possible to calculate hunger in terms of the average shortfall in caloric consumption among the undernourished population, given food prices and per capita income (FAO 2020). Lower food prices (resulting from greater R&D investment in response to climate change) reduce this shortfall and provide a food security benefit.

Environmental benefits are measured as avoided cropland expansion and carbon stock losses associated with the conversion of natural lands into cropland and

---

[3] Alternatively, a general equilibrium model could be used, which would give more explicit attention to how agricultural growth affects the rest of the economy. See Sands and Suttles (2022) for a recent example that explores questions similar to our own. A partial equilibrium approach is generally more tractable, however, and in this application gives similar results.

reduced emissions intensity from agricultural production. We rely on the global carbon stocks calculated by West et al. (2010) to quantify GHG emissions when cropland expands into natural lands. West et al. (2010) estimate these carbon stocks using spatially explicit datasets on potential vegetative and soil carbon. These are considered as one-time carbon emissions associated with bringing that land into crop production. The data on GHG emissions from agricultural production is based on the GTAP v.10 standard database (Aguiar et al. 2019), which reports CO_2 emissions from fossil fuel combustion and non-CO_2 GHG emissions, which rely on FAO (2020) for methane and nitrous oxide agricultural emissions (Chepeliev 2020).

4 Experimental Design

We turn now to the experimental design of this application of the SIMPLE model. This entails the construction of alternative future scenarios for R&D investments, the ensuing productivity growth, and policies aimed at achieving sustainability.

4.1 Constructing R&D Scenarios for Projections of Future Productivity Growth

R&D policy is modeled as the amount of government spending on agricultural R&D. This generates new technologies that, once adopted by farmers, raise agricultural TFP (Wang et al. 2022). As informed by past empirical studies (Fuglie 2018), the efficiency with which R&D generates TFP growth varies by region. In addition, R&D in developed countries may generate international technology spillovers— mostly to other developed countries but also to less developed countries—although this has historically been rather limited due to ecological and institutional constraints. The model also incorporates the effects of private sector R&D on TFP growth, although the growth rate in private R&D spending is held constant in the policy simulations.

Spending on agricultural R&D is assumed to affect agricultural TFP. We incorporate a lag to account for technology development and diffusion, with effects persisting for several years before eventually depreciating through technological obsolescence (Alene 2010; Alston et al. 2011; Andersen and Song 2013; Jin and Huffman 2016). We use historical R&D spending and capital stock estimates from Fuglie (2018), who assembles data on public and private agricultural R&D expenditures starting from the 1960s. We then project these spending pathways forward to 2050 under various R&D policy scenarios. The historical patterns of agricultural R&D spending show a sustained shift in the global total from developed to less developed countries and from the public to the private sector (Pardey et al. 2016). Since 2000, public agricultural R&D spending in high-income countries as a whole has been essentially flat in real terms (Heisey and Fuglie 2018), while it has continued to grow in less developed countries (LDC).

In this framework, growth in a region's agricultural TFP comes from technologies developed from four independent R&D sources (designated by the subscript s in Eqs. 10.1–10.3): public agricultural R&D by countries within a region, private R&D, R&D by the CGIAR (Consultative Group on International Agricultural Research), and technology spill-ins from public R&D done in other regions. Equations 10.1–10.3 summarize the key linkages under this framework:

$$\text{RDSTOCK}_{s,t} = \sum_{i=0}^{L} \beta_i (\text{RDEXPEND}_{s,t-i}) \tag{10.1}$$

$$\beta_i = \frac{(i+1)^{\delta/(1-\delta)} \lambda^i}{\sum_{i=0}^{L} (i+1)^{\delta/(1-\delta)} \lambda^i}, \text{ where } \sum_{i=0}^{L} \beta_i = 1 \tag{10.2}$$

$$\%\Delta\text{TFP}_t = \sum_{s=1}^{S} \alpha_s (\%\Delta\text{RDSTOCK}_{s,t}) \tag{10.3}$$

Starting with Eq. 10.1, the stock of R&D from source s at time t ($\text{RDSTOCK}_{s,t}$) is built up from the past L years of annual R&D expenditures ($\text{RDEXPEND}_{s,t}$), where β_i is the R&D lag weight at period i and total lag length L is the number of years in which R&D contributes to productivity until the R&D capital stock has fully depreciated. Equation 10.2 specifies a gamma distribution for the structure of the R&D lag weights (Alston et al. 2010). According to this distribution, R&D spending at time t initially contributes little to new R&D capital stock or TFP growth, but its effect builds over time as technology arising from that research is developed and disseminated to farmers. Eventually, the effects peak when technology is fully disseminated and then diminish due to technology obsolescence. We utilize separate R&D lag distributions for DC and LDC regions. For public R&D in developed countries, we impose an R&D lag structure of 50 years. The peak impacts of R&D spending on knowledge stocks (and productivity) occur after 26 years (δ, $\lambda = (0.90, 0.70)$ in Eq. 10.2). For R&D by less developed countries, the private sector, and the CGIAR, a total lag length of 35 years is imposed, with peak effects occurring at year 10 (δ, $\lambda = (0.80, 0.75)$ in Eq. 10.2). The longer lag structure for public R&D in developed countries reflects a greater focus on discovery R&D to push the science and technology frontier. The shorter lag length for less developed countries and private R&D reflects a greater emphasis on adaptive R&D, borrowing from global knowledge capital to close existing yield gaps.

Once the R&D capital stock is constructed from historical and projected future R&D spending by each R&D source, we link these stocks to future growth in agricultural TFP growth via elasticities (α_s) that describe the percentage rise in TFP given a 1% increase in R&D capital stock from source s (Eq. 10.3). The estimates of α_s are based on Fuglie's (2018) review of more than 40 studies that empirically estimated α_s from historical R&D spending and agricultural TFP growth

Table 10.2 Elasticities of agricultural research and development (R&D) capital stock, by region and technology provider (Fuglie 2018)

Region	Technology provider				
	Total—R&D from all sources	National public R&D	Private R&D	CGIAR R&D	Spill-ins from DC regions
Developed countries (DC)	0.67	0.27	0.20		0.21
Transition economies	0.07	0.07			
Less developed countries					
Latin America	0.77	0.23	0.13	0.05	0.36
East and South Asia	0.30	0.21	0.01	0.08	
West Asia & North Africa	0.19	0.15		0.04	
Sub-Saharan Africa	0.17	0.13		0.04	

Elasticities give the percentage change in agricultural total factor productivity (TFP) resulting from a 1% change in R&D capital stock

in various countries and regions, mostly using data since 1980 (see Table 10.2). The estimates vary by R&D source and by region, and the values of these elasticities are generally lower for LDC regions than for DC regions. Public R&D in DC regions is also more likely to generate international technology spillovers, but mainly to other developed countries.

4.2 Environment Policy

For this analysis, environmental policies are defined as land policies that restrict land areas rich in carbon stocks from agricultural use. The cost of this environmental policy is the amount of compensation paid to landowners to replace forgone income from agricultural uses of the land. We call this a land set-aside payment; it is equivalent to an annual rental payment for agricultural land. Environmental policy is modeled as a backward shift in the land supply function in regions where there is considerable potential for agriculture to expand into carbon-rich natural areas (i.e., forests and grasslands). These areas are identified from global grid cells of areas (1) in the top 80th percentile of carbon intensity (West et al. 2010) and (2) in which less than 70% of the area is currently in cropland. This scheme assures that land restriction policies are imposed in areas with high carbon sinks near agricultural frontiers. The policy removes a specified share of land from agricultural production in these areas.

4.3 Policy Scenarios

Table 10.3 describes a set of R&D and environmental policy scenarios for reducing GHG emissions from agriculture that form the basis for the SIMPLE model simulations of the global agricultural economy from 2017 to 2050. Scenario S1 is the business-as-usual (BAU) case and assumes that public and private spending on agricultural R&D will continue at the same (real) annual growth rate over 2017–2050 as it did in 2000–2017. For DC regions as a whole, the growth rate in public agricultural R&D was virtually zero (Heisey and Fuglie 2018), while the annual growth in public R&D spending in less developed countries averaged 4.6%, ranging from 2.1% in Sub-Saharan Africa to 9.1% in China (Fuglie 2018). The BAU case assumes that these rates will continue into the future except for in China, which is scaled down to 3% annual growth in R&D spending, since it is unlikely to continue at such a high rate indefinitely. In the BAU case, R&D spending in less developed countries would more than double over the simulation period and would increase as a percentage of gross agricultural output from 1.0% to 1.5%. R&D spending by the private sector is assumed to grow at 4% per year, and CGIAR spending is assumed to grow by 5.8% per year (reflecting their respective 2000–2017 average growth rates).

Scenario S2:ENV adds an environmental policy that places limits on agricultural land use in frontier areas where forests, peatlands, and grasslands provide carbon-rich sinks. In this scenario, we target a 5% reduction in cropland in areas with high potential to serve as carbon sinks and provide compensation to landowners for forgone income from agricultural uses of this land. Compensation costs are estimated from regional average agricultural land rental rates derived from SIMPLE. Most of the set-aside area lies on the agriculture–forest fringe in less developed countries but also includes some areas in developed countries, such as the southeastern United States (Fig. 10.2).

Under scenarios S3, S4, and S5, growth in public agricultural R&D spending is accelerated beyond that in the BAU case for the 2017–2040 period. Due to lags between R&D spending and TFP growth, higher spending during 2017–2040 will begin to nudge TFP growth rates upward in the late 2020s, with effects persisting for several decades; however, R&D spending beyond 2040 will not have much effect on TFP growth before midcentury, so only R&D costs between 2017 and 2040 are considered. Under Scenario S3:LDC, agricultural R&D spending is assumed to grow uniformly in all LDC regions by 6.3% per year but is held at BAU levels (zero growth) in DC regions. At this aggressive rate of growth, R&D spending in less developed countries would grow from US$28.0 billion per year in 2020 to US$126.1 billion per year by 2040 (or to 3.2% of the projected value of agricultural output, compared to 1.5% of the output in the BAU case). Scenario S4:ENV-LDC combines this R&D growth path with the environmental policy under S2:ENV, which protects carbon-rich land from conversion to cropland. Finally, under scenario S5:GLOBAL, R&D spending is accelerated in both LDC and DC regions but for the same global total as S3 and S4. For S5:GLOBAL, we assume that R&D spending in LDC regions

Table 10.3 Research and development (R&D) and environmental policy scenarios for the simulations

Policy scenarios		Total global agricultural R&D spending in 2040 (billion PPP$)	Public agricultural R&D spending in 2040[a]		Total global agricultural R&D spending, 2017–2040 (billion PPP$)	Total global land set-aside payments, 2017–2050 (billion PPP$)	Total global policy cost, 2017–2050 (billion PPP $)
			Less developed countries (LDC) (billion PPP$)	Developed countries (DC) (billion PPP $)			
Global policy scenarios S1 to S5							
S1:BAU	Business-as-usual (BAU) case: continue current spending growth in global agricultural R&D	131.7	59.1	19.6	2296	0	2296
S2:ENV	Enact environmental policies that protect carbon-rich land from conversion with BAU R&D growth	131.7	59.1	19.6	2296	1113	3409
S3:LDC	Accelerate public agricultural R&D spending in LDC	199.5	126.5	19.6	2907	0	2907
S4:ENV-LDC	Combines S2 & S3: Accelerate R&D spending in LDC and protect carbon-rich land from conversion	199.5	126.5	19.6	2907	1041	3948
S5: GLOBAL	Accelerate public agricultural R&D spending in both LDC to DC (for same global total as S2 and S3)	192.6	99.9	40.2	2907	0	2907
Regional policy scenario S6 EU "Green Deal"							
S6A:EU-ENV	S5 plus EU imposes reduction of nonland agricultural inputs by 20%	192.6	99.9	40.2	2907		
S6B:EU-RD	S6A plus additional increase in EU R&D spending	203.1	100	50.5	2999		

Dollar amounts are measured in purchasing power parity (PPP)

[a] R&D policy is set by assuming annual growth rates in spending over 2017–2040 and projecting total factor productivity (TFP) growth to 2050. Due to the R&D lag structure, R&D spending after 2040 has minimal effect on TFP growth before 2050 so only R&D costs over 2017–2040 are considered. Total global R&D spending includes private and CGIAR R&D

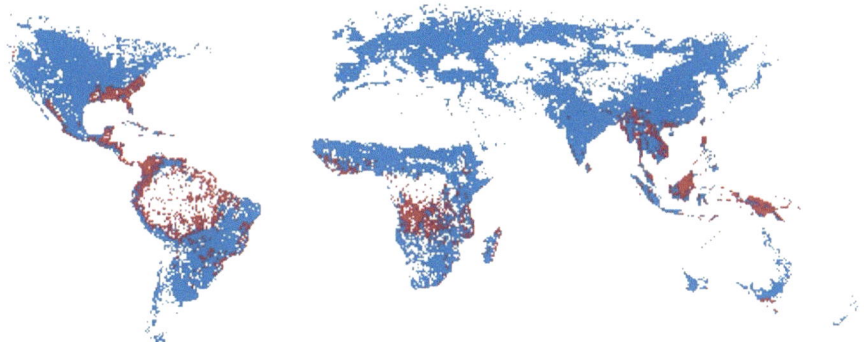

Fig. 10.2 Areas targeted for cropland restrictions in the policy simulations (5% of global cropland area)
Red grid cells lie on or above the 80th percentile in carbon intensity and presently have 70% or less area in cropland

grows by 5.4% per year, while R&D spending in DC regions grows by 3.1% per year. Public R&D as a percentage of agricultural output would reach 3.6% in developed countries and 2.8% in less developed countries by 2040. This scenario exploits the greater efficiency of R&D systems in developed countries to push out the science and technology frontier and produce international R&D spillovers. In all the scenarios, private R&D grows at a constant rate of 4.1% per year (the BAU rate), while under scenarios S2 through S5, CGIAR spending is assumed to grow at the same rate as LDC public R&D.

Finally, scenario S6:EU considers policies that seek to reduce the environmental footprint of agriculture in a developed region. Modeled loosely on the EU Green Deal and its agricultural Farm to Fork Strategy, scenario S6A:EU-ENV deintensifies agricultural production in the European Union by imposing a 20% reduction in the use of nonland inputs.[4] Scenario S6B:EU-RD adds additional R&D spending to offset production losses from the input-use restriction. While we are not able to present the full cost of these policies (in Table 10.3 we show only R&D cost and not the environmental policy cost), scenario 6 is nonetheless useful for exploring the effects of regional policies on global environmental and food security outcomes.

---

[4]The EU Green Deal policy (European Commission 2020) lays out specific goals for agricultural input use reduction by 2030: reduce pesticide use by 50%, fertilizer use by 20%, sales of antimicrobials in farmed animals and aquaculture by 50%, and place 25% of total farmland under organic farming. Given that nonland inputs account for about 90% of total agricultural costs and material inputs about half of nonland costs (Ball et al. 2010; Fuglie 2015), it is reasonable to assume that these restrictions might mean a 20% overall reduction in nonland inputs. The EU Green Deal policy also commits €10 billion to research and innovation related to food, bioeconomy, natural resources, agriculture, fisheries, aquaculture, and environment. However, our policy scenarios are not directly comparable to the EU Green Deal since our scenarios envision agriculture supplying global demand in 2050 rather than 2030. Given R&D lags, our scenarios also give a more reasonable timeframe for modeling the effects of new R&D spending on productivity growth.

5   Results

Table 10.4 summarizes the implications of the policy scenarios for global agricultural production and prices, GHG emissions from agriculture and land-use change, and the number of undernourished people. Under scenario S1:BAU, the projected growth in agricultural TFP over 2017–2050 is 42.3% (for an average annual rate of 1.07%), and global agricultural output rises by 51.6%. Our finding that output will grow more than TFP indicates that the use of agricultural (land and nonland) inputs also increased. The increase in TFP causes average global crop prices to fall by an average of 21.6% (but regional price changes vary due to market segmentation, as seen in Fig. 10.3). Global cropland expands by 6%, or by about 95 million hectares. This land-use change causes a release of 31.8 Gt CO_2e of terrestrial GHGs into the atmosphere. At the same time, farmgate GHG emissions rise to 8.3 Gt CO_2e per year, an increase of nearly 3.0 Gt CO_2e over 2017 farmgate emissions. Exogenous growth in per capita income and the fall in crop prices reduce the undernourished population by 146 million compared to 2017 levels.

Under Scenario S2:ENV, imposing an environmental policy that protects carbon-rich land from agricultural use reduces the increase in global cropland from 6% to just 1.4%. Net GHG emissions from land-use change are only 2.9 Gt CO_2e (compared to 31.8 Gt CO_2e under S1:BAU). The increase in global cropland masks a larger shift in cropland from areas with high carbon sink potential to other parts of the world. Even though there is no extra gain in productivity compared with S1:BAU, the projected farmgate GHG emissions in 2050 are cut from 8.3 to 7.4 Gt CO_2e. A major contributing factor to the reduction in farmgate emissions is the avoidance of crop production in former peatlands, which continue to emit GHG emissions from peat decomposition as they are farmed. Imposing land-use restrictions without adding productivity growth, however, exacerbates global food insecurity: There are four million more undernourished people compared with the S1:BAU scenario (i.e., the number of undernourished people falls by 142 million under S2:ENV compared to 146 million under S1:BAU).

Increasing R&D spending in less developed countries (scenario S3:LDC) above BAU levels accelerates global growth in agricultural TFP and output while further lowering crop prices. Global cropland expansion falls by nearly half (from 6.0% to 3.4%), but not by as much as the environmental policy under S2:ENV (1.6%). GHG emissions per unit of output are lower under S3:LDC than S2:ENV (see Fig. 10.4); however, total farmgate emissions are higher under S3:LDC than under S2:ENV because output is larger. Lower crop prices under S3:LDC have major benefits for global food security: The prevalence of undernutrition falls by 208 million people by 2050.

Scenario S4:ENV-LDC, which combines increased R&D spending with environmental policies that protect carbon-rich land, has the largest impact among these scenarios at curtailing GHG emissions from agriculture while preserving gains in global food security. In fact, global cropland declines by 1.2% under this scenario. This represents a reduction of 18 million hectares of cropland from 2017 levels and

Table 10.4 Effects of research and development (R&D) and environmental policy scenarios on global agricultural greenhouse gas (GHG) emissions, prevalence of malnutrition, food prices, cropland, and agricultural production

Policy scenario	World agricultural TFP in 2050 (Index, 2017 = 100)	World agricultural output in 2050 (Index, 2017 = 100)	Average annual GHG emissions from land-use change, 2017–2050 million tons CO2e/year	Average annual farmgate GHG emissions million tons CO2e/year	Total GHG emissions from land-use change, 2017–2050 million tons CO2e	Total change in agricultural GHG emissions, 2017–2050 million tons CO2e	Difference between 2050 and 2017 global levels Farmgate GHG emissions million tons CO2e/year	Under-nutrition million persons	Crop price (%)	Crop land (%)
S1:BAU	142.3	151.6	963	8258	31,764	272,510	2985	−146	−21.6	6.0
S2:ENV	142.3	151.4	86	7357	2851	242,791	2060	−142	−20.8	1.4
S3:LDC	156.8	157.7	649	7719	21,403	254,733	2221	−208	−32.2	3.4
S4:ENV-LDC	156.8	157.4	−205	6842	−6756	225,787	1320	−205	−31.4	−1.2
S5: GLOBAL	164.8	159.5	463	7446	15,264	245,703	1860	−212	−42.1	1.9
S6A:EU-ENV	164.8	159.2	499	7479	16,462	246,798	1890	−210	−41.1	2.2
S6:EU-RD	171.1	161.4	337	7251	11,111	239,280	1597	−223	−46.7	0.9

GHG emissions intensity in agriculture
(kg CO2 eq per USD Y2017)

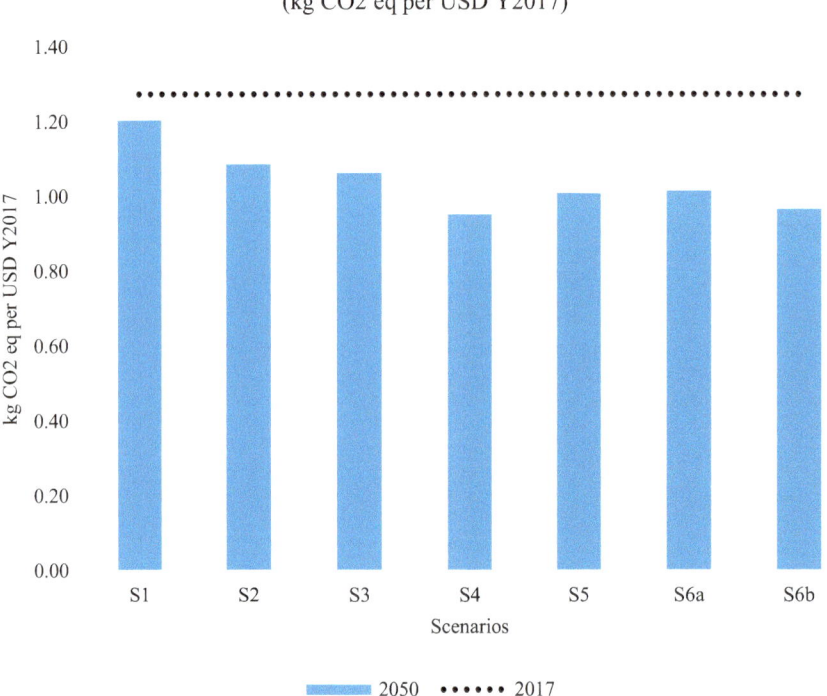

Fig. 10.3 Effect of research and development and land policy scenarios on farmgate greenhouse gas emissions intensity (kg CO2e per constant 2017 USD)
Results for S1:BAU, S2:ENV, S3:LDC, S4:ENV-LDC, and S5:GLOBAL are displayed left to right in the bar graphs

113 million hectares less cropland in 2050 than under S1:BAU projections. The reduction in cropland (and the accompanying expansion of forest area) sequesters 6.8 Gt CO2e. A further advantage of S4:ENV-LDC is that it lowers the cost of the environmental policy. Because faster productivity growth reduces commodity prices, which in turn reduces land rents, compensation to landowners for forgone earnings from agricultural land use also falls (from US$1113 billion to US$1041 billion over the 2017–2050 period, see Table 10.3). Under S4:ENV-LDC, farmgate emissions are also considerably lower compared with those under other scenarios due both to the rise in productivity of existing farmland and the prevention of agricultural expansion into environmentally sensitive peatlands, which continue to emit GHGs for decades following land-use change.

Scenario S5:GLOBAL illustrates the efficiency gains that could be achieved under an R&D strategy that includes additional investments in both DC and LDC regions. Using the same global R&D spending as S3:LDC, this scenario demonstrates that shifting some of this spending from less developed to developed

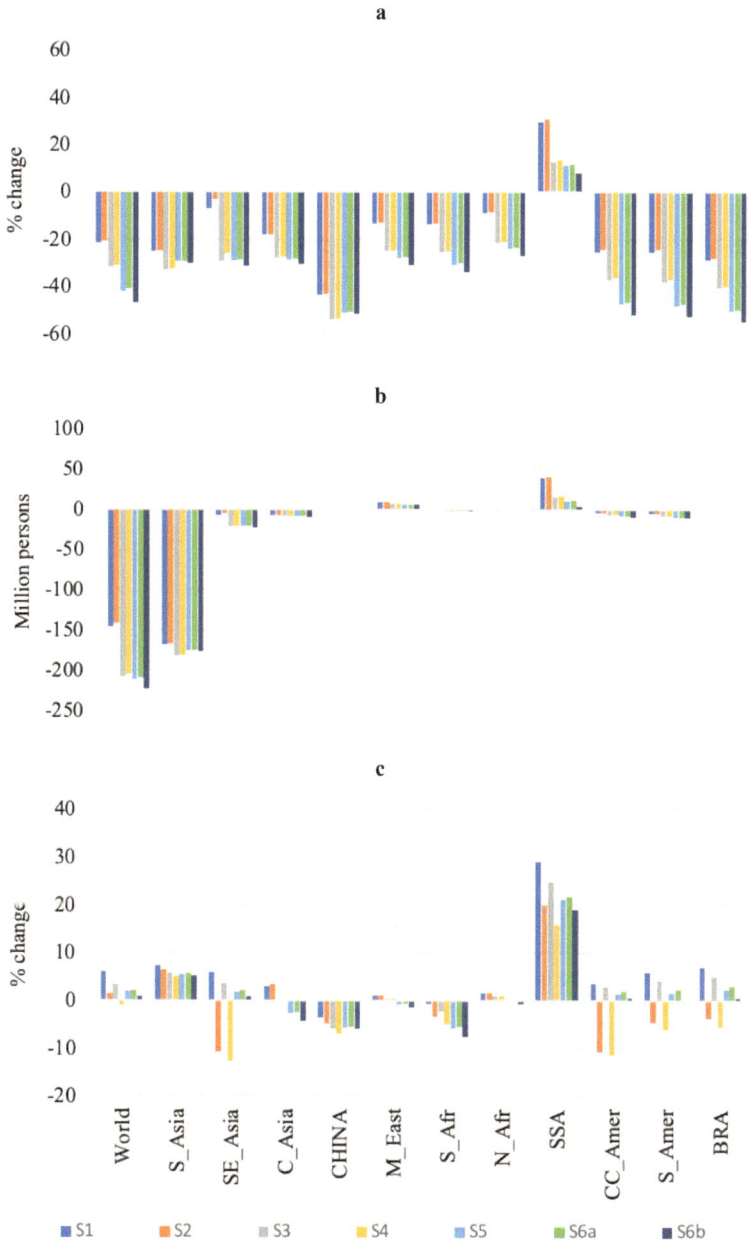

Fig. 10.4 Change in (**a**) crop prices, (**b**) undernutrition, and (**c**) cropland use as a result of global policy scenarios: differences between 2017 and 2050, by region

countries achieves more rapid gains in global agricultural TFP and output. This further reduces food prices, GHG emissions, and undernutrition. Even without the environmental policy, global cropland expands by only 1.9%. The higher return on R&D in DC regions stems from the assumption, based on historical evidence, that these R&D systems generate more technological breakthroughs with potential for international spillovers.

Figure 10.4 illustrates the regional implications of policy scenarios S1–S5 for crop prices, undernutrition, and cropland use. In all policy scenarios, crop prices and undernutrition are projected to decline between 2017 and 2050 except in Africa, especially sub-Saharan Africa (SSA). SSA is the region least integrated into global agricultural markets; as a result of slow agricultural productivity growth coupled with rapidly rising population, the model projects that food prices, cropland use, and the total number of undernourished people in this region will increase even as they fall in other parts of the world. In all regions, the policy scenarios that increase R&D spending (S3:LDC, S4:ENV-LDC, and S5:GLOBAL) have the largest impact on reducing crop prices and undernutrition, while the scenarios with environmental policies (S2:ENV and S4:ENV-LDC) have the largest impact on reducing cropland and therefore avoiding GHG emissions from land-use change. Not shown in Fig. 10.3 are impacts on developed countries. Under BAU productivity gains, cropland is projected to fall in Europe and rise in North America and Oceania. With accelerated R&D in developed countries, cropland is projected to decline in all DC regions except the United States, where cropland is projected to remain unchanged.

The estimates of policy costs and GHG emissions in Tables 10.3 and 10.4 can be used to approximate the marginal cost of GHG abatement. Table 10.5 shows the estimated total costs of R&D and environmental policies incurred over the 33 years from 2017 to 2050, in constant 2017 (undiscounted) dollars, and total agricultural GHG emissions over the same period. Marginal abatement costs are found by comparing the differences in emissions and policy costs with those under the BAU scenario (S1). The ratio of the increase in policy cost to the reduction in emissions then gives an approximate marginal abatement cost in USD per metric ton of CO_2e emissions averted. These estimates are approximate because they do not account for the time path of policy costs and environmental benefits but simply make a static comparison of total costs (in constant dollars) and emissions changes over 2017–2050. They also ignore post-2050 benefits from productivity gains due to R&D spending in 2017–2040. The results suggest abatement costs of US$19–US $22 per ton of CO_2e for policies S2:ENV, S3:LDC, and S4:ENV-LDC and a substantially lower abatement cost of US$14 per ton for S5:GLOBAL due to the efficiency gains from shifting some R&D spending from LDC to DC regions. By comparison, Lobell et al. (2013) estimate that using greater agricultural R&D spending to mitigate GHG emissions from land-use change would cost about US $15 per ton of CO_2e. Our results consider not only emissions saving from avoided land conversion but also changes in farmgate emissions. We also use a more refined model of R&D spending in which the lag between investments and productivity realizations is taken into account. Nonetheless, our results are quite similar to those of Lobell et al. (2013).

Table 10.5 Marginal costs of agricultural greenhouse gas (GHG) emissions abatement costs from the global policy scenarios

Policy scenario	R&D policy cost (2017–2040) billion PPP$	Environmental policy cost (2017–2050) billion PPP$	Total policy cost (2017–2050) billion PPP$	Cost change from BAU billion PPP$	Total agricultural GHG emissions (2017–2050) Gt CO2e	GHG emissions change from BAU (2017–2050) Gt CO2e	Approximate marginal cost of GHG abatement PPP$/ton CO2e
S1:BAU	2296	0	2296	0	304	0	–
S2:ENV	2296	1113	3409	1113	246	–59	18.97
S3:LDC	2907	0	2907	611	276	–28	21.71
S4:ENV-LDC	2907	1041	3948	1652	219	–85	19.38
S5: GLOBAL	2907	0	2907	611	261	–43	14.11

Table 10.6 Effects of regional European Union (EU) policy scenarios on agricultural productivity, cropland, and greenhouse gas (GHG) emissions

Policy scenario		Agricultural total factor productivity, 2050 (Index, 2017 = 100)		Change in cropland, 2017–2050 (%)		Agricultural GHG emissions, 2017–2050 (million tons CO2e)	
		EU	Global	EU	Global	EU	Global
S5: GLOBAL	Accelerate public agricultural R&D spending in both less developed and developed countries	164.6	164.6	(1.3)	1.9	15,544	245,703
S6A:EU-ENV	S5 plus EU imposes reduction of nonland agricultural inputs by 20%	164.6	164.8	(1.3)	2.2	15,232	246,798
S6B:EU-RD	S6A plus additional increase in EU R&D spending	195.5	171.1	(0.9)	`0.9	14,938	239,280

Finally, Table 10.6 reports the implications of policy scenarios that focus on developed countries. Here, we consider a policy imposed on EU countries that deintensifies the use of nonland inputs to reduce emissions and other environmental costs from agriculture. Using the productivity growth assumed under scenario S5: GLOBAL, scenario S6A:EU-ENV adds a mandate to reduce nonland inputs in EU agriculture by 20%. This reduces EU GHG emissions but also reduces output and raises global prices (see Table 10.4). More cropland enters agriculture in other parts of the world and total global GHG emissions increase. However, additional R&D spending by the European Union (scenario S6B:EU-RD) could offset these adverse global outcomes. This would raise EU agricultural production and keep more area in cropland in the European Union while reducing the growth in cropland area in the rest of the world. Higher R&D spending by the European Union would also raise agricultural productivity in other parts of the world due to international technology spillovers.

6 Discussion

This chapter has investigated the role of productivity investment policies in reducing GHG emissions from agriculture and compares these outcomes with those obtained through environmental policies that restrict agricultural land and input use. Using the SIMPLE partial equilibrium model of the global agri-food economy, we carried out a series of simulations of world agricultural supply and demand to 2050 that included various combinations of R&D spending and environmental restrictions on agriculture in LDC and DC regions. Our baseline, or business-as-usual (BAU), scenario assumes that R&D spending patterns continue on their present trajectory and that world population and income grow according to SSP 2 middle-of-the-road assumptions. Under the BAU scenario, global cropland increases by 6% (about 95 million hectares) from 2017 levels and agriculture releases a total of 272.5 Gt CO2e from land-use change and farmgate production. A simulation of an environmental policy that reduces agricultural cropland by 5% in carbon-rich areas reduces overall agricultural emissions by 29.7 Gt CO2e but causes food prices to rise and malnutrition to increase. It is also costly: We estimate that compensation to landowners for forgone income will total US$1.11 trillion over 2017–2050 (and continue indefinitely after that). A simulation with a productivity policy scenario that adds US$0.61 trillion to R&D spending in LDC regions over 2017–2050 results in lower food prices and reduces the number of undernourished people worldwide but is only about 60% as effective as the environmental policy in reducing emissions. A simulation that combines these approaches cuts global agricultural emissions by 46.7 Gt CO2e and preserves the reduction in undernutrition at a cost of US$1.65 trillion. Higher productivity results in lower land rents (because output prices are lower) which reduce the cost of the environmental policy by about US$100 billion. All three policy options imply roughly the same marginal abatement cost of US$19–US$22 per ton of CO2e.

A further scenario considers a variation in productivity policy that shifts part of the US$0.61 trillion increase in R&D spending from LDC to DC regions. Because R&D systems in DC regions are assumed, based on historical evidence, to be more efficient at generating international technology spillovers, this scenario achieves more rapid gains in global agricultural productivity and reduces total GHG emissions from agriculture by 27 Gt CO2e, for a unit abatement cost of US$14 per ton CO2e. This scenario also achieves a greater reduction in the size of the undernourished population. A final scenario imposes more stringent environmental restrictions on developed countries—in this case, it targets a 20% reduction in the use of nonland inputs in EU agriculture (roughly what the EU Green Deal envisions). While this reduces GHG emissions in the European Union, it slightly raises agricultural emissions worldwide because lower agricultural outputs in the European Union raise world prices and cause agriculture to expand in the rest of the world. However, an aggressive EU policy of R&D spending could offset those effects by further increasing agricultural productivity not only in the European Union but also in other countries through international technology spillovers.

It is important to note that none of the policy scenarios considered in this chapter resulted in an actual reduction in global GHG emissions from agriculture by 2050. Rather, they curbed the growth in emissions as agricultural output expanded to better nourish the growing world population. Another aspect of our modeling approach is that improvements in productivity are assumed to be factor neutral, meaning that they reduce the amount of all factor inputs (e.g., land, labor, capital) required to produce a unit of output in equal proportions. Because it unambiguously reduces unit costs of production, factor-neutral technical change is profitable for farmers and thus can be expected to face few hurdles to widespread adoption. A suggestion for future work is to consider factor-biased technical change, wherein technologies are designed to save the production factors most closely associated with GHG emissions. Factor-biased technologies that target emissions reductions may require fewer R&D expenditures than factor-neutral technologies for similar savings in farmgate emissions intensity. However, factor-biased technical change may increase the use of other inputs and may not be profitable for farmers to adopt without additional incentives.

Generally, it appears that productivity and environmental policies to curb GHG emissions from agriculture are likely to complement each other. As Stevenson et al. (2013) note, higher agricultural productivity by itself is likely to be too blunt an instrument for protecting environmentally sensitive land from conversion to cropland. However, environmental policies that protect such land from agricultural use may be difficult to monitor and enforce, lack permanence, and exacerbate hunger. Combining these policies may better achieve the dual societal goals of climate change mitigation and global food security.

Acknowledgment and Competing Interests This work was supported by the US Department of Agriculture, Economic Research Service. The findings and conclusions presented in this chapter are those of the authors and should not be construed to represent any official determination or policy of the US Department of Agriculture (USDA), or the National Science Foundation (NSF). Furthermore, we declare that there is no conflict of interest related to this work.

References

Aguiar, Angel, Maksym Chepeliev, Erwin L. Corong, Robert McDougall, and Dominique van der Mensbrugghe. 2019. The GTAP data base: Version 10. *Journal of Global Economic Analysis* 4: 1–27. https://doi.org/10.21642/JGEA.040101AF.

Alene, Arega D. 2010. Productivity growth and the effects of R&D in African agriculture. *Agricultural Economics* 41: 223–238. https://doi.org/10.1111/j.1574-0862.2010.00450.x.

Alston, Julian M., Matthew A. Andersen, Jennifer S. James, and Philip G. Pardey. 2010. *Persistence pays: U.S. Agricultural productivity growth and the benefits from public R&D spending.* New York: Springer. https://doi.org/10.1007/978-1-4419-0658-8.

———. 2011. The economic returns to U.S. public agricultural research. *American Journal of Agricultural Economics* 93: 1257–1277. https://doi.org/10.1093/ajae/aar044.

Andersen, Matthew A., and Wenxing Song. 2013. The economic impact of public agricultural research and development in the United States. *Agricultural Economics* 44: 287–295. https://doi.org/10.1111/agec.12011.

Baldos, Uris Lantz C., and Thomas W. Hertel. 2014. Global food security in 2050: The role of agricultural productivity and climate change. *Australian Journal of Agricultural and Resource Economics* 58: 554–570. https://doi.org/10.1111/1467-8489.12048.

Baldos, Uris Lantz C., Keith Fuglie, and Thomas W. Hertel. 2020. The research cost of adapting agriculture to climate change: A global analysis to 2050. *Agricultural Economics* 51: 207–220. https://doi.org/10.1111/agec.12550.

Ball, Veldon, Jean-Pierre Butault, Carlos San Mesonada, and Ricardo Mora. 2010. Productivity and international competitiveness of agriculture in the European Union and the United States. *Agricultural Economics* 41: 611–627. https://doi.org/10.1111/j.1574-0862.2010.00476.x.

Baquedano, Felix, Jeremy Jelliffe, Jayson Beckman, Maros Ivanic, Yacob Zereyesus, and Michael Johnson. 2022. Food security implications for low- and middle-income countries under agricultural input reduction: The case of the European Union's farm to fork and biodiversity strategies. *Applied Economic Perspectives and Policy* 44: 1942–1954. https://doi.org/10.1002/aepp.13236.

Beckman, Jayson, Maros Ivanic, Jeremy L. Jelliffe, Felix G. Baquedano, and Sara G. Scott. 2020. *Economic and food security impacts of agricultural input reduction under the European Union Green Deal's Farm to Fork and biodiversity strategies*, Economic Brief 30. US Department of Agriculture, Economic Research Service.

Chepeliev, Maksym. 2020. *Development of the non-CO2 GHG emissions database for the GTAP 10A data base (GTAP)*. Research Memorandum 32. Department of Agricultural Economics, Purdue University, West Lafayette: Global Trade Analysis Project (GTAP). https://doi.org/10.21642/GTAP.RM32.

European Commission. 2020. *From Farm to Fork: Our food, our health, our future*. Fact Sheet. Brussels: European Union. https://doi.org/10.2875/653604.

FAO. 2020. *FAOSTAT: Food and agriculture data*. Rome: Food and Agriculture Organization of the United Nations. http://faostat.fao.org/.

Fuglie, Keith. 2015. Accounting for growth in global agriculture. *Bio-Based and Applied Economics Journal* 4: 221–254. https://doi.org/10.22004/ag.econ.231887.

———. 2018. R&D capital, R&D spillovers, and productivity growth in world agriculture. *Applied Economic Perspectives and Policy* 40: 421–444. https://doi.org/10.1093/aepp/ppx045.

Heisey, Paul W., and Keith Fuglie. 2018. *Agricultural research investment and policy reform in high income countries*, Economic Research Report 249. Washington, DC: US Department of Agriculture, Economic Research Service.

Hertel, Thomas W., Uris Lantz C. Baldos, and Keith Fuglie. 2020. Trade in technology: A potential solution to the food security challenge of the 21st century. *European Economic Review* 127: 103479. https://doi.org/10.1016/j.euroecorev.2020.103479.

IPCC. 2014. *Climate change 2014 mitigation of climate change. Working Group III contribution to the fifth assessment report of the intergovernmental panel on climate change*. Cambridge, UK: Cambridge University Press.

Jin, Yu, and Wallace E. Huffman. 2016. Measuring public agricultural research and extension and estimating their impacts on agricultural productivity: New insights from U.S. evidence. *Agricultural Economics* 47: 15–31. https://doi.org/10.1111/agec.12206.

Lobell, David B., Uris Lantz C. Baldos, and Thomas W. Hertel. 2013. Climate adaptation as mitigation: The case of agricultural investments. *Environmental Research Letters* 8: 015012. https://doi.org/10.1088/1748-9326/8/1/015012.

Pardey, Philip G., Connie Chan-Kang, Steven P. Dehmer, and Jason M. Beddow. 2016. Agricultural R&D is on the move. *Nature* 537: 301. https://doi.org/10.1038/537301a.

Riahi, Keywan, Detlef P. Van Vuuren, Elmar Kriegler, Jae Edmonds, Brian C. O'Neill, Shinichiro Fujimori, Nico Bauer, et al. 2017. The shared socioeconomic pathways and their energy, land use, and greenhouse gas emissions implications: An overview. *Global Environmental Change* 42: 153–168. https://doi.org/10.1016/j.gloenvcha.2016.05.009.

Sands, Ronald D., and Shellye A. Suttles. 2022. World agricultural baseline scenarios through 2050. *Applied Economic Perspectives and Policy* 44: 2034–2048. https://doi.org/10.1002/aepp.13309.

Stevenson, James R., Nelson B. Villoria, Derek Byerlee, Timothy Kelly, and Mywish Maredia. 2013. Green revolution research saved an estimated 18 to 27 million hectares from being brought into agricultural production. *Proceedings of the National Academy of Sciences* 110: 363–368. https://doi.org/10.1073/pnas.1208065110.

USDA-ERS. 2021. *Agricultural productivity in the U.S.* Washington, DC: US Department of Agriculture, Economic Research Service. https://www.ers.usda.gov/data-products/agricultural-productivity-in-the-u-s/.

Villoria, Nelson B. 2019. Technology spillovers and land use change: Empirical evidence from global agriculture. *American Journal of Agricultural Economics* 101: 870–893. https://doi.org/10.1093/ajae/aay088.

Wang, Sun Ling, Nicholas Rada, Ryan Williams, and Doris Newton. 2022. Accounting for climatic effects in measuring U.S. field crop farm productivity. *Applied Economic Perspectives and Policy* 44: 1975–1994. https://doi.org/10.1002/aepp.13289.

West, Paul C., Holly K. Gibbs, Chad Monfreda, John Wagner, Carol C. Barford, Stephen R. Carpenter, and Jonathan A. Foley. 2010. Trading carbon for food: Global comparison of carbon stocks vs crop yields on agricultural land. *Proceedings of the National Academy of Sciences* 107: 19645–19648. https://doi.org/10.1073/pnas.1011078107.

Chapter 11
Gridded Implications of Total Factor Productivity Growth

Elizabeth A. Fraysse, Thomas W. Hertel, and Srabashi Ray

This chapter introduces novel techniques for undertaking gridded analyses within the Simplified International Model of agricultural Prices, Land use, and the Environment (SIMPLE)-G framework. It builds on the baseline experiment developed in Chap. 10, which projected the global economy from 2017 to 2050 using the nongridded SIMPLE model. In this chapter, we replace the single, national production function in SIMPLE with a gridded representation of US agriculture, following the methodology laid out in Chaps. 3 and 5 and using SIMPLE-G1, a version of SIMPLE-G that does not distinguish between rainfed and irrigated agriculture. It also follows Chap. 10 in having only two inputs: land (natural resources) and nonland (human and produced) inputs. We begin by exploring the impacts of a marginal perturbation in US total factor productivity (TFP) growth, demonstrating that the model results mirror the results obtained by using the theoretical equations from Chap. 3. We then run the same global baseline scenario as in Chap. 10 and show that the equivalent gridded results can be obtained from a "minimodel." which takes as its inputs the US crop price change and exogenous grid-level shocks (in this case, TFP). This minimodel is highly amenable to in-depth analysis, one grid cell at a time. The chapter concludes by demonstrating how to attribute local grid-cell outcomes to global change drivers—including income, population, and technology—in non-US regions.

Data availability statement: The files needed to replicate this application are available at https://gtap.agecon.purdue.edu/simple-g/.

E. A. Fraysse (✉) · T. W. Hertel · S. Ray
Center for Global Trade Analysis, Department of Agricultural Economics, Purdue University, West Lafayette, IN, USA
e-mail: efraysse@purdue.edu

1 Introduction

In Chap. 10, which utilized the (nongridded) SIMPLE, the authors analyzed the impacts of environmental policies at a regional level (i.e., groups of countries). This approach allowed them to evaluate how national and regional investments in agricultural research and development (R&D) interact with environmental policies to achieve the dual goals of restricting harmful greenhouse gas emissions while minimizing adverse impacts on food affordability. The regional-scale SIMPLE model is well positioned for this analysis as it allows for complex regional interactions via food markets and spillover effects from technological advances. However, using SIMPLE at the regional level lacks the nuances of geographic specificity in the policy impacts that are determined by local-level biophysical and socioeconomic factors. For example, the conservation policies implemented in the previous chapter aim to conserve terrestrial carbon and biodiversity, yet these features of the landscape vary widely within each SIMPLE region. This calls for analysis at a more granular scale. This chapter employs the simplest possible version of the SIMPLE-G model developed in Part III, called SIMPLE-G1. Furthermore, we introduce the SIMPLE-G "minimodel" approach to gridded analysis—previously discussed in Chap. 3—which will allow us to focus on specific grids of interest to understand the determinants of individual grid-cell-level outcomes. This level of economic analysis has been missing from the integrated assessment models that are currently used to evaluate regional and global-scale environmental and conservation policies.

2 SIMPLE-G Version Employed

To highlight the benefits of the minimodel approach, we build on Chap. 10 and replicate the economic growth and TFP shocks from that chapter, using the gridded SIMPLE-G1 model. As a prelude to the next few chapters, we focus on the United States as the gridded region of interest. Our goal is to demonstrate how we can use the minimodel approach to delve into grid-cell-level results, relating these outcomes back to key economic and biophysical parameters.

As with the SIMPLE model utilized in the previous chapter, SIMPLE-G1 is based on two aggregated inputs: natural resources, a category that includes land and water inputs, and human resources, which includes all manufactured inputs and human labor. This two-input approach allows us to leverage the foundational economic theory laid out in Chap. 3 to understand the model results. After running the TFP shocks in the SIMPLE-G1-US model, we explain the model results and how they relate to the foundational theory, data, and parameters in the SIMPLE-G1 model. This deeper understanding of the SIMPLE-G1-US model provides the reader with a strong basis for exploring the more complex versions of the SIMPLE-G model featured in subsequent applications.

3 Experimental Design

Many factors can induce shifts in TFP, but one of the biggest contributors to changes in TFP is an investment in agricultural R&D. As in Chap. 10, R&D-induced TFP growth is assumed to uniformly affect the productivity of all inputs in all grid cells and is represented in this model by the variable p_AOCROP, which is the percentage change in a productivity index. With an increase in TFP ($p\_AOCROPr > 0$), the amount of each input required in region r to produce a given amount of output decreases. Therefore, to produce the same amount of food, producers need fewer human and resource inputs. However, if the region where TFP improvement occurs becomes more competitive, input use might increase due to an overall expansion in output level. In Chap. 2, we referred to this as Jevons' paradox and discussed the conditions under which this phenomenon will arise.

In this chapter, we consider three experiments. The first, which we use for illustrative purposes, simulates a 1% improvement in TFP in US grid cells. This marginal perturbation of the model allows us to use the analytical equations from Box 3.1 in Chap. 3 to explore in more depth the model theory, data, and parameters that drive the observed changes in both resource and human input use. In the second experiment, we implement the full set of population, income, TFP, and biofuel shocks, as implemented in the baseline R&D scenario in Chap. 10. Finally, we show how the exact same gridded outcomes can be obtained in a two-step simulation. In this case, we shock only the TFP in the US grid cells. However, we accompany this TFP shock with a market price shock that carries all the necessary information about global demands and supplies. We show that the US market price is a "sufficient statistic" embodying the boundary conditions facing any given grid cell in the United States. In other words, it is a complete summary of market developments elsewhere in the global economy. Once producers in a grid cell know this price, they do not need any other information to make informed economic decisions about production practices on their farms. We leverage this knowledge to facilitate grid-by-grid analysis of the results.

4 Selected Grids

For the minimodel analysis, we selected 12 grids to represent the range of important SIMPLE-G1 parameters (i.e., the land supply elasticity, cost share of land, and land–nonland input substitution elasticity). Table 11.1 lists the values of these parameters, and Fig. 11.1 shows the grid cells' location in the United States. In some grid cells—such as grid I04106 in the state of Washington (WA)—resource supply is largely unresponsive to price (i.e., it is inelastic); in others—such as grid I06314 in the state of Montana (MT)—resource supply is considerably more responsive to price. In contrast, the supply elasticity of human inputs is elastic and assigned a uniform value, given the lack of grid-cell-specific estimates of this parameter. The value of

Table 11.1 SIMPLE parameters for 12 selected grid cells for in-depth analysis of model results

Grid ID	Land supply elasticity	Nonland supply elasticity	Land cost share	Substitution elasticity
WA: I04106	0.003	1.340	0.2906	1.00
NV: I04259	0.003	1.340	0.1179	1.00
ID: I06003	0.102	1.340	0.2424	0.86
MT: I06314	0.255	1.340	0.0971	0.20
OK: I24220	0.111	1.340	0.1300	1.00
TX (1): I27726	0.326	1.340	0.1243	1.00
MN: I33495	0.004	1.340	0.2623	0.22
TX (2): I36312	0.350	1.340	0.2475	1.00
WV: I51326	0.368	1.340	0.2542	0.18
IN: I56025	0.129	1.340	0.2796	0.22
AL: I58595	0.144	1.340	0.1090	0.18
PA: I68537	0.300	1.340	0.0956	0.20

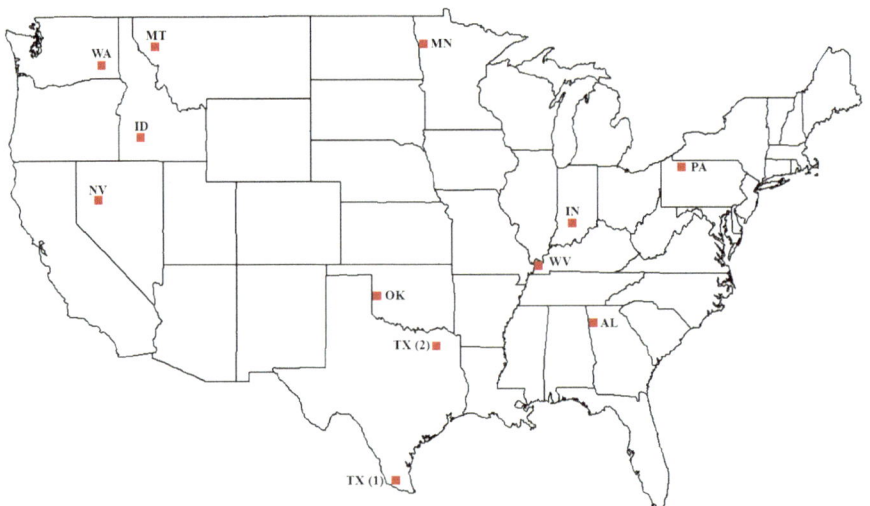

Fig. 11.1 Map of selected grid cells used in this chapter

the elasticity of substitution varies greatly across grid cells, implying that some grid cells have significant potential for boosting production at the intensive margin (i.e., substituting nonland for land inputs). For example, the elasticity of substitution is greater than 0.6 in Nevada (NV) and Idaho (ID) but much smaller in Indiana (IN) and Alabama (AL). A small value suggests that the application of more variable inputs (e.g., fertilizer) will not greatly increase yields. In Chap. 7, we discussed further details about how this parameter is calculated. Note also that the grid-cell-level share of land rents in total production costs is also quite variable, ranging between 10% and 30%. This will prove to be important in driving the results.

5 Results

For expository purposes, we begin with a simple 1% perturbation of TFP in the United States and examine how this plays out at the level of individual grid cells. This is a good opportunity to utilize the grid-cell-level theory developed in Chap. 3. We then analyze the impacts of the full baseline scenario from the preceding chapter. This approach allows us to explore the gridded impacts of this forward-looking baseline within the continental United States. Finally, we show how the minimodel can be used to replicate the results from the full model and explore them more deeply.

5.1 SIMPLE-G1 with a 1% TFP Shock in the United States

A 1% increase in TFP in the United States leads to a 1.3% increase in crop production with a 0.6% reduction in crop price. The improvement in TFP in the United States makes crops produced in the United States more competitive in the national and international markets and leads to an increase in overall production, which dominates the resource-conserving aspect of the TFP improvement. In contrast, the TFP boost in the United States is resource-conserving in the rest of the world (ROW): Production declines by 0.01–0.22% in the other 16 SIMPLE regions.

We present the results of the 1% US TFP shock for the 12 focus grids in the United States in Table 11.2. Columns 2–4 show the impacts of the 1% TFP shock on input use and crop production. Columns 5–7 show the same results calculated using the theoretical equations developed in Box 3.1. (We have omitted the calculations of input price changes, which can be explored using Eqs. 3.7 and 3.8.) By comparing these two sets of columns in Table 11.2, we show that the simulation model results can be replicated (for a marginal perturbation of the model) with our analytical model using the relevant parameter values. Note that, for larger (nonmarginal) changes, the analytical formulae may yield different results as these formulae are based on a local linearization of the underlying nonlinear model.

In the following, we reproduce the relevant equations from the foundational two-input analytical model presented in Box 3.1 to calculate the impacts of a 1% TFP change in the system. Recall that the change in TFP is captured with the a parameter, which refers to factor-neutral technical change in the production function. As the US crop price is also changing, we need to include the change in p as well. However, we omit the conservation policy shock ($\phi_R = 0$) that was the focus of discussion in Chap. 3.

Percentage change in input H use when $\phi_R = 0$:

Table 11.2 Simulated and calculated impacts of a 1% TFP shock to 12 selected grids in SIMPLE-G1

Grid ID	Simulated results for 1% AOCROP$_g$ shock			Calculated results for $a = 1$ using Eqs. 11.1, 11.2, and 11.5		
	Nonland input	Land input	Output	Nonland input (Eq. 11.1)	Land input (Eq. 11.2)	Output (Eq. 11.5)
(1)	(2)	(3)	(4)	(5)	(6)	(7)
WA: I04106	0.37	0.002	1.27	0.38	0.002	1.27
NV: I04259	0.45	0.002	1.4	0.45	0.002	1.40
ID: I06003	0.40	0.069	1.32	0.40	0.069	1.32
MT: I06314	0.31	0.004	1.29	0.42	0.002	1.41
OK: I24220	0.45	0.079	1.41	0.46	0.080	1.41
TX (1): I27726	0.47	0.203	1.44	0.48	0.205	1.44
MN: I33495	0.20	0.004	1.15	0.20	0.004	1.15
TX (2): I36312	0.44	0.199	1.38	0.44	0.200	1.38
WV: I51326	0.36	0.272	1.34	0.36	0.274	1.34
IN: I56025	0.26	0.113	1.22	0.27	0.114	1.22
AL: I58595	0.37	0.186	1.35	0.37	0.188	1.35
PA: I68537	0.43	0.298	1.42	0.44	0.301	1.42

$$q_H = v_R \frac{(p^* + a)}{\theta_R} \Gamma_H + \sigma \frac{(p^* + a)}{\theta_R} \Gamma_H; \tag{11.1}$$

Percentage change in input R use when $\phi_R = 0$:

$$q_R = (p^* + a) \frac{v_R}{\theta_R} \Gamma_R; \tag{11.2}$$

Percentage change in price of input H when $\phi_R = 0$:

$$p_H = \frac{p^* + a}{\theta_H} (1 - \Gamma_R); \tag{11.3}$$

Percentage change in price of input R when $\phi_R = 0$:

$$p_R = \frac{p^* + a}{\theta_R} \Gamma_R; \tag{11.4}$$

Percentage change in production when $\phi_R = 0$:

Table 11.3 Decomposition of total change in crop production as a percentage change

Grid ID	Total change in crop production Result of Eq. 11.5.	Direct impact of TFP increase First element of Eq. 11.5	Extensive margin response to price change Second element of Eq. 11.5	Intensive margin response to price change Third element of Eq. 11.5
(1)	(2)	(3)	(4)	(5)
WA: I04106	1.27	1	0.002	0.266
NV: I04259	1.40	1	0.002	0.396
ID: I06003	1.32	1	0.069	0.249
MT: I06314	1.14	1	0.002	0.136
OK: I24220	1.41	1	0.080	0.328
TX (1): I27726	1.44	1	0.205	0.238
MN: I33495	1.15	1	0.004	0.147
TX (2): I36312	1.38	1	0.200	0.182
WV: I51326	1.34	1	0.274	0.064
IN: I56025	1.22	1	0.114	0.109
AL: I58595	1.35	1	0.188	0.165
PA: I68537	1.42	1	0.301	0.122

$$q = a + (p^* + a)\frac{v_R}{\theta_R}\Gamma_R + (p^* + a)\frac{\sigma\theta_H}{\theta_R}\Gamma_\sigma, \qquad (11.5)$$

where $\Gamma_H = \frac{\theta_R v_H}{denom}$, $\Gamma_\phi = \frac{\theta_R(v_H+\sigma)+\theta_H\sigma}{denom}$, $\Gamma_\sigma = \frac{\theta_R(v_H - v_R)}{denom}$, and denom $=$ $\sum_{j,k}\theta_j(v_k + \sigma), j, k = H, R$ for $v_R < v_H$.

It is clear from Eqs. 11.1, 11.2, 11.3, 11.4, and 11.5 and the results in Table 11.3 that SIMPLE-G results are firmly underpinned by economic theory. This is important, as it allows for a thorough explanation of the model results, and the results can be related back to key parameters. In addition, the theory allows us to decompose the simulation results to understand the biophysical and socioeconomic grid-cell characteristics that drive those results. For example, using Eq. 11.5, we can explore the elements that contribute to the total change in crop production due to a change in TFP. Table 11.3 decomposes the total change in crop production into the direct impact of a 1% increase in TFP (first element of Eq. 11.5) and the supply response to the ensuing price change, which comprises both an extensive margin (second element of Eq. 11.5) and an intensive margin (third element of Eq. 11.5) of the supply response. Column 4 of Table 11.4 shows that the extensive margin response

Table 11.4 Minimodel analysis of grid-cell impacts of the baseline scenario, incorporating population, income, biofuels, and technology growth worldwide, 2017–2050

Grid cells	Baseline	Minimodel		
		Total	Market price	Own-TFP
(1)	(2)	(3)	(4)	(5)
Panel A. Cropland change by grid cell (QLAND)				
WA: I04106	0.08	0.08	−0.08	0.17
NV: I04259	0.10	0.10	−0.10	0.20
ID: I06003	2.98	2.99	−2.91	5.90
MT: I06314	0.14	0.14	−0.13	0.27
OK: I24220	3.45	3.45	−3.36	6.81
TX (1): I27726	9.10	9.10	−8.92	18.02
MN: I33495	0.17	0.17	−0.16	0.33
TX (2): I36312	8.88	8.89	−8.71	17.59
WV: I51326	12.12	12.12	−11.87	23.99
IN: I56025	4.77	4.78	−4.64	9.41
AL: I58595	7.92	7.92	−7.70	15.62
PA: I68537	13.37	13.37	−13.11	26.48
Panel B. Crop output by grid cell (QCROP)				
WA: I04106	55.42	55.43	−14.24	69.66
NV: I04259	64.31	64.32	−21.94	86.26
ID: I06003	58.72	58.73	−17.07	75.80
MT: I06314	54.57	54.57	−13.15	67.72
OK: I24220	64.95	64.96	−22.50	87.46
TX (1): I27726	67.46	67.47	−24.69	92.16
MN: I33495	47.00	47.00	−6.87	53.87
TX (2): I36312	63.18	63.19	−20.95	84.14
WV: I51326	59.63	59.63	−17.78	77.41
IN: I56025	51.72	51.72	−10.93	62.65
AL: I58595	59.80	59.81	−17.80	77.61
PA: I68537	65.39	65.40	−22.77	88.17

These impacts are conveyed to grid cells via the US market price for crops combined with the local change in technology. This approach allows for accurate prediction of local impacts as derived from the full model

to the price change closely follows the magnitude of the price elasticity of resource supply reported in Table 11.1. In grids like that in Washington, where the supply elasticity is very small (0.003), the extensive margin response is also negligible (0.002). However, in West Virginia, the supply elasticity of resources is 0.37, leading to the largest extensive margin supply response in this table (0.27). Similarly, in grids where the substitution elasticity between the two inputs is large (e.g., 1), such as in Nevada, the intensive margin supply response is much larger. In Nevada, this parameter reaches its largest value in this table: 0.396.

5.2   SIMPLE-G1 with a Complete Baseline

In this section, we repeat the baseline experiment from Chap. 10, shocking not only the supply side (i.e., TFP in the United States and TFP in other world regions) but also the demand side, including population, income, and changes in demand for biofuels for 2017–2050. This experiment generates regional outcomes similar to those in Chap. 10, but it also allows us to explore the impacts of demand and supply side shocks in the ROW on individual grid cells across the United States. The United States is a major producer of food consumed across the world. This implies that both changes in the productivity of the agricultural sector and changes in food demand in the ROW have a substantial impact on the US agricultural sector. To better investigate this phenomenon, we replicate the set of shocks from Chap. 10 using the SIMPLE-G1 model while utilizing the subtotal feature of GEMPACK to further decompose the impacts of those shocks into ten groups of drivers:

1. Change in population in ROW.
2. Change in income in ROW.
3. Change in biofuel demand in ROW.
4. Change in crop TFP in ROW.
5. Change in TFP in the crop-using sectors (livestock and processed foods) in ROW.
6. Change in population in the United States.
7. Change in income in the United States.
8. Change in biofuel demand in the United States.
9. Change in crop TFP in the United States.
10. Change in TFP in the Crop-Using Sectors (Livestock and Processed Foods) in the United States

We further show that all the information about these developments across the global economy is conveyed to US producers locally through changes in crop prices. Therefore, changes in grid-cell-level production can be replicated by simply implementing a US crop price shock along with any local-level TFP changes affecting individual producers.

5.2.1   Impact of Baseline Scenario on US Crop Price

Under this baseline scenario for 2017–2050, growth in global supplies outpaces growth in global demands, and the crop price declines globally and in the United States where it falls by 14.90%. However, TFP grows—hence unit costs fall. The cost decrease is greater than the fall in prices, so there are benefits to be gained at the farm level. A large share of these gains is captured by the inelastically supplied natural resource input, the rents of which increase by 55.44%. However, returns to human and manufactured inputs (the more elastically supplied input) increase by a relatively smaller amount, 9.87%.

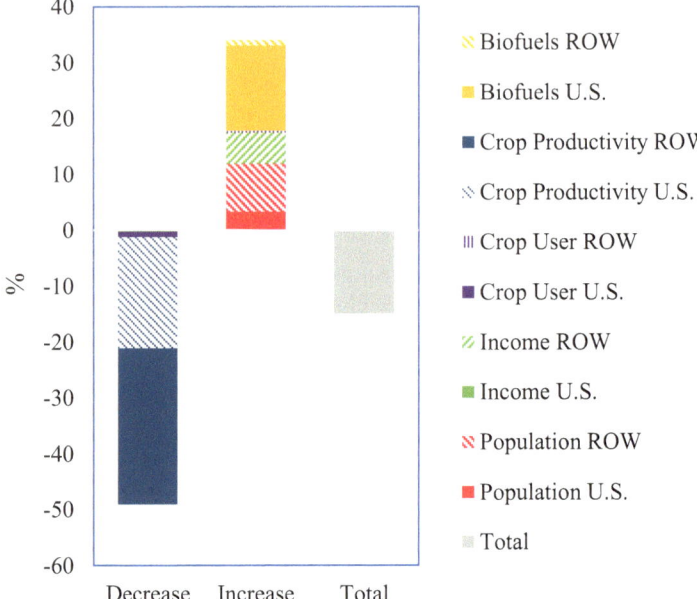

Fig. 11.2 Decomposition of changes in US crop price by ten sources of change, 2017–2050

Figure 11.2 utilizes the subtotal feature of **GEMPACK** to decompose the total change in US food prices into the ten sources of global change drivers identified above. The main factor that leads to the reduction in crop prices in the United States is the technological improvements in crop productivity due to agricultural R&D investments in the United States and the ROW. Improved productivity in crop-using sectors (i.e., livestock and processed foods) also generates a small amount of downward pressure on prices. Increases in global population and income lead to increased demand and therefore put upward pressure on crop prices. Demand for crops from the US biofuel sector is also a major contributor to upward pressures on US crop prices.

5.3 SIMPLE-G1-Mini

The SIMPLE-G1-Mini model is designed to replicate individual grid-level results produced by a SIMPLE-G1 model considering only the equilibrium price changes from SIMPLE-G1. This is possible because the price changes carry all the relevant market information determined by the global model. The SIMPLE-G1-Mini model can be run on a few selected grids (e.g., the 12 grids discussed above). These grid cells are typically selected due to their (perhaps surprising) outcomes as well as their potential relevance to local policymakers. Of course, any local policy or technology changes must also be communicated to the gridded model. Therefore, in the case of this SIMPLE-G1 experiment, each grid cell will be shocked by both the crop price (which is deemed exogenous to local producers) and the TFP improvement experienced by local producers.

Column 2 in Table 11.4 (baseline) reports the gridded impacts from the baseline scenario, run with the full set of shocks to population, income, biofuels, and technology around the world in a model with thousands of grid cells. Columns 3–5 report the results from the minimodel, which is run for individual grid cells by shocking the market price as well as grid-cell-level technology (own-TFP). Note that the total impact agrees with the baseline results to the first decimal point. We have successfully transferred global supply and demand information to the grid cells via the crop price signal. The impact of this change in market conditions is reported in column 4. If producers in the WA grid cell do not experience a TFP improvement over the 2017–2050 time horizon, they will reduce cropland (QLAND $= -0.08$) and output (QCROP $= -14.24$). However, this model assumes that producers experience the average rate of TFP growth in the entire US region. As a consequence, their costs fall, and they end up expanding cropland cover as well as output. The remaining entries in this table report the market price and own-TFP impacts on cropland and output for the other 11 grid cells illustrated in Fig. 11.1. Note that the responses vary widely due to the widely varying grid-cell-level parameter values reported in Table 11.1.

We see that the grid cell labeled TX (1) has the largest increase in crop production, 67.46%, driven by the relatively high supply elasticity of land (0.326) and substitution elasticity (1) as shown by the parameters in Table 11.1. Both of these factors ensure that the producers in the grid cell can respond to the TFP shock with more flexibility. On the contrary, the grid cell labeled MN has a combination of parameters (small land supply elasticity and small elasticity of substitution between human and natural resource inputs) that limits producers' ability to respond to the TFP shock; it therefore reports the smallest increase in crop production.

The grid cell labeled PA shows the largest increase in cropland area, the result of a large increase in the price of land. The magnitude of this price increase is inversely proportional to the cost share of land (recall Eq. 11.4). The reason is that as unit costs fall due to the TFP shock, the associated benefits that are not passed forward to consumers in lower prices are passed backward to input suppliers, with the relatively inelastic input (land, in this case) garnering more of the gains. As the land resources currently account for only a small share of total costs (less than 10% in the case of the PA grid cell), the resulting increase in percentage returns must be very large. This increase in land returns operates on the land supply elasticity, which is also relatively large (0.3, see Table 11.1), resulting in an increase of more than 13% in cropland area. In short, each grid cell has its own story, even though they all face the same changes in national price and TFP. Their outcomes are varied, reflecting the interplay of cost shares, factor supply elasticities, and input substitution possibilities as seen in Eqs. 11.1, 11.2, 11.3, 11.4, and 11.5.

5.4   A Global-Local-Global Analysis for Selected Grid Cells

With this minimodel in hand, we can also quantify the impact of diverse global change drivers on local outcomes in individual grid cells. Figure 11.3 illustrates this global-to-local linkage in the case of crop output growth in the WA grid cell. Table 11.4 reports the growth in output (Grid Total) in the WA grid cell as 55.4%. How much of this is due to economic developments in the non-US, ROW regions? By combining the results in Fig. 11.2 with those in Table 11.4, we can calculate a full attribution of local output changes to the underlying global drivers. Figure 11.3 reports these results. Crop productivity growth—both in the United States and in the ROW—depresses prices and discourages output in the WA grid cell. However, biofuel production, income, and population growth boost prices and hence output. The net impact of these market price effects on crop output is negative (Table 11.4). However, when combined with the own-TFP impacts, which lower producer costs, output in the WA grid cell is predicted to rise.

These findings are obtained by combining results from Fig. 11.2 with those from Table 11.4.

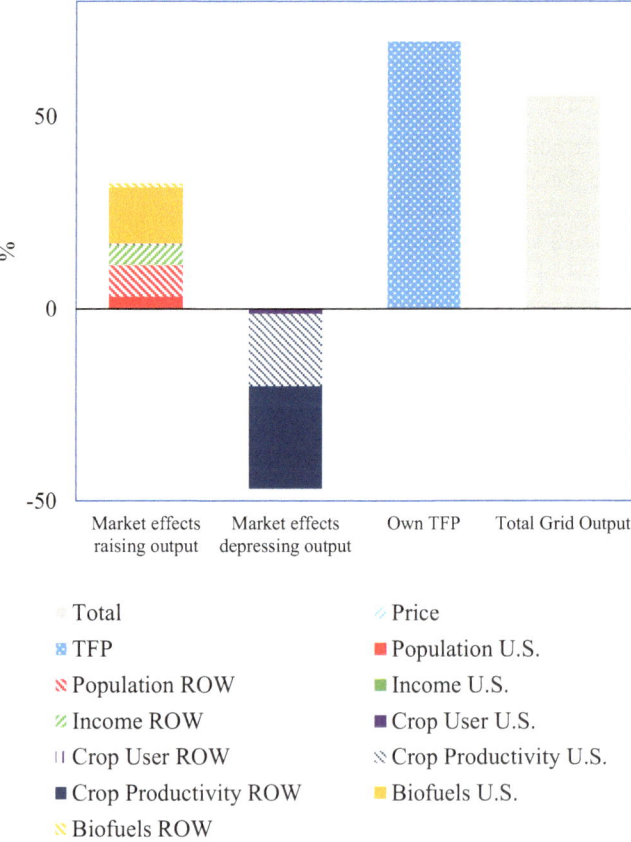

Fig. 11.3 Decomposing the global and local drivers of crop output in grid WA

6 Conclusions

This chapter is key for those who wish to evaluate the critical drivers and processes in sustainability analyses using SIMPLE-G. The SIMPLE-G1 and minimodel are useful tools when a deeper analysis is required to understand—and explain—spatially heterogeneous results. As has been shown here, to understand what is behind a given set of results, it is essential to refer back to the underlying parameters and explore the connection to regional and global markets.

The tools introduced in this chapter are also helpful for uncertainty analyses and robustness checks. It is worthwhile to conduct a systematic sensitivity analysis of model results with respect to the uncertainty in the underlying parameters and drivers. Building meaningful confidence intervals around the results is an important part of the scientific endeavor.

The beauty of these SIMPLE-G tools is that they allow researchers to improve transparency and traceability of global economic analysis of food security and environmental sustainability while also drawing out the implications of these global developments for local outcomes. Exploring the links among global drivers, local stresses and responses, and feedback at the global level is a central theme of this book.

Acknowledgments and Competing Interests The authors acknowledge support from the National Science Foundation, including INFEWS award #1855937, "Identifying Sustainability Solutions through Global-Local-Global Analysis of a Coupled Water-Agriculture-Bioenergy System" and AccelNet Award #NSF-OISE-2020635, GLASSNET, from the US Department of Agriculture NIFA grants #2019–67,023-29,679, "Economic Foundations of Long Run Agricultural Sustainability," and 2022–67,023-36,403, "Labor Markets and the Impacts of Environmental Stresses and Conservation Policies on US Agriculture," as well as USDA Hatch Project 1,003,642.

The findings and conclusions presented in this chapter are those of the authors and should not be construed to represent any official determination or policy of the US Department of Agriculture (USDA) or the National Science Foundation (NSF). Furthermore, we declare that there is no conflict of interest related to this work.

Chapter 12
Local Groundwater Sustainability Policies and Global Spillovers

Iman Haqiqi, Laura Bowling, Sadia Jame, Uris Lantz C. Baldos, Jing Liu, and Thomas W. Hertel

1 Introduction

The Simplified International Model of agricultural Prices, Land useLand use, and the Environment-Gridded version (SIMPLE-G) is a quantitative framework that allows for global-to-local-to-global (GLG) analysis of a broad range of environmental research topics. The philosophy of this multiscale, GLG approach is to evaluate the long-run implications of global changes for local environmental stresses and assess the implications of local responses for global conditions through global and regional agricultural markets. In this chapter, we focus on groundwater sustainability as an application of this GLG framework. First, we look at the local changes in water withdrawal caused by global drivers using a decomposition approach, which allows us to quantify the separate contribution of global economic changes to growth in irrigation water withdrawal in the United States. We look at population and income

This chapter is a slightly revised version of a paper originally published as Haqiqi, Iman, Laura Bowling, Sadia Jame, Uris Baldos, Jing Liu, and Thomas Hertel. 2023. Global drivers of local water stresses and global responses to local water policies in the United States. *Environmental Research Letters* 18: 065007. https://doi.org/10.1088/1748-9326/acd269.

Data availability statement: The files needed to replicate this application are available at https://gtap.agecon.purdue.edu/simple-g/.

I. Haqiqi (✉) · U. L. C. Baldos · J. Liu · T. W. Hertel
Center for Global Trade Analysis, Department of Agricultural Economics, Purdue University, West Lafayette, IN, USA
e-mail: ihaqiqi@purdue.edu

L. Bowling
Department of Agronomy, Purdue University, West Lafayette, IN, USA

S. Jame
Department of Agricultural and Biological Engineering, Purdue University, West Lafayette, IN, USA

as demand-side drivers and growth in total factor productivity (TFP, the ratio of outputs to inputs) as a supply-side driver. Second, we quantify changes in crop production and agricultural water and land use around the world in response to local groundwater sustainability policies in the United States. This research highlights the trade-offs between economic production and sustainability as well as the unintended consequences of local sustainability policies, including the associated spillover effects. This multiscale framework can better inform local environmental policy by providing a holistic evaluation.

1.1 Motivation

Groundwater is a crucial resource for many communities worldwide, supplying water for drinking, irrigation, and industry. In recent decades, rapid economic growth has intensified the impacts of human systems on natural ecosystems, with degraded land and water resources creating a global crisis (Jury and Vaux Jr. 2007; Hanjra and Ejaz Qureshi 2010; Srinivasan et al. 2012; Famiglietti 2014). Rapid groundwater depletion is among the most challenging of these environmental problems (Changming et al. 2001; Qureshi et al. 2010; Karami et al. 2012; Konikow 2013; Voss et al. 2013; Castle et al. 2014; Liesch and Ohmer 2016; Dalin et al. 2017; Nabavi 2018). The pressure on farmers to produce more output has led to unsustainable use of water resources in many locations (Seckler et al. 1999; Faunt 2009; McGuire 2017; Reitz et al. 2017; Russo and Lall 2017; Rodell et al. 2018).

Groundwater overexploitation can result in several challenges, including increased pumping costs because of declining groundwater levels. This can lead to a rise in costs for farmers, households, and businesses. Moreover, overexploitation can cause water quality degradation due to saltwater intrusion and other issues, which can increase the costs of treating water for drinking and other purposes. Land subsidence, which is another problem that can arise due to groundwater overexploitation, can cause damage to infrastructure and reduce property values and the availability of water for ecosystems, leading to a loss of ecosystem services (e.g., water filtration and flood control). In extreme cases, if groundwater wells become completely depleted, it may be necessary to explore options such as relocating individuals, farmers, and businesses or implementing water transfer initiatives.

In addition, reductions in groundwater storage threaten many regions' ability to meet future water needs (Cook et al. 2015). Groundwater demand is likely to be particularly strong in the coming decades, as irrigation becomes a more important element of agricultural adaptation in a warming climate (Perry et al. 2009; Schlenker and Roberts 2009; Nepal and Shrestha 2015; Pathak et al. 2017; Haqiqi 2018; Haqiqi et al. 2019, 2021).

1.2 Other Studies

Given the importance of future groundwater sustainability, a variety of different approaches have been employed to study the interactions between hydrological and human systems. The growing literature in sociohydrology involves using increasingly complex models to address challenges involving water quality, groundwater depletion, flood risk, drought, and conflicting demands (Ertsen et al. 2014; Ghosh et al. 2014; Van Emmerik et al. 2014; Fernald et al. 2015; Blair and Buytaert 2016; Giuliani et al. 2016; Di Baldassarre et al. 2019). The literature on telecoupling and virtual water trade has also explored the connections between human systems and land–water systems (Dalin et al. 2012; Bruckner et al. 2015; Wichelns 2015; Chaudhary and Kastner 2016; Hertel 2018; D'Odorico et al. 2019). However, the significance of global drivers of local sustainability stresses (e.g., changes in groundwater withdrawals) has not previously been quantified in a multiscale, economic equilibrium framework.

1.3 SIMPLE-G Contribution: Spillovers

Effective water sustainability policies require an understanding of the critical interactions between environmental and human systems. Decisions and policies in the human system can affect the environment, but changes in the environment can also affect people and their decisions. Solutions and strategies ignoring these responses may fail. For example, improving irrigation efficiency can result in increased irrigation demand motivated by lower costs (Perry et al. 2017; Grafton and Wheeler 2018; Perry 2019; Pérez-Blanco et al. 2020). Groundwater policies and changes in the groundwater table can have spillover effects, either positive or negative, and can impact a diverse group of stakeholders. It is crucial to consider the potential spillover effects of these policies and changes in the groundwater table when making decisions related to groundwater resource management.

While food security issues are typically studied at the global and country levels, resource sustainability stresses are usually felt at the local level (e.g., abandoned villages, dying wells, drying lakes and subbasins, and land subsidence). The overarching goal of this chapter is to measure how economic changes in one region can affect environmental sustainability in another region of the world as well as how the implementation of local sustainability policies can export environmental stresses to other regions. Such connections in the global food system can have important implications for global environmental justice as rising incomes in some regions support increasingly rich and specialized diets, resulting in significant environmental costs elsewhere (e.g., Harrison 2011; Marston et al. 2015).

This study also contributes to the literature on international trade in virtual water and virtual land by quantifying the impacts of population changes on remote water and land resources (Hoekstra and Mekonnen 2012; Hoekstra and Wiedmann 2014; Ramankutty et al. 2018).

The model of this study is built on previous economic studies with valuable insights on commodity trade (Armington 1969), food security (Hertel and Baldos 2016), water and food (Liu et al. 2014), water and trade (Liu et al. 2017), land allocation (Ahmed et al. 2008), irrigation demand (Haqiqi and Hertel 2016), value of water (Haqiqi 2023), water rights (Jame and Bowling 2020), water stress (Roath 2013), and groundwater (Befus et al. 2017). This study extends the previous modeling work of Baldos et al. (2020) by incorporating a new land allocation framework to improve the modeling of land-use changes at each location (Zhao et al. 2020).

1.4 Limitations

To focus on the decomposition of the impacts of global changes, we do not consider changes in water and heat stress due to climate change. We are aware that this simplification is likely to result in the underestimation of future water stress in many regions by midcentury. However, this simplification allows us to focus more sharply on the main messages of the study, namely, the importance of global–local–global linkages in the context of land and water sustainability.

2 SIMPLE-G Version Employed for This Application

In this application, we focus on the groundwater module of SIMPLE-G (Fig. 12.1) to analyze economic decisions about groundwater withdrawals for irrigated cropland and its responses to global and local changes. These withdrawals are defined as the amount of water extracted from groundwater resources; this amount is usually greater than crop groundwater consumption, with the difference recharging groundwater or running off into streams.

2.1 Determinants of Groundwater Withdrawal

The variables affecting withdrawal decisions vary and include surface water availability and costs, groundwater availability and costs, irrigation extent, irrigation rent gaps, irrigation infrastructure (equipment and technology), crop prices, and crop production technology. When modeling groundwater withdrawals, hydrological models usually consider crop water requirements, irrigation extent, and irrigation efficiency while ignoring economic decisions.

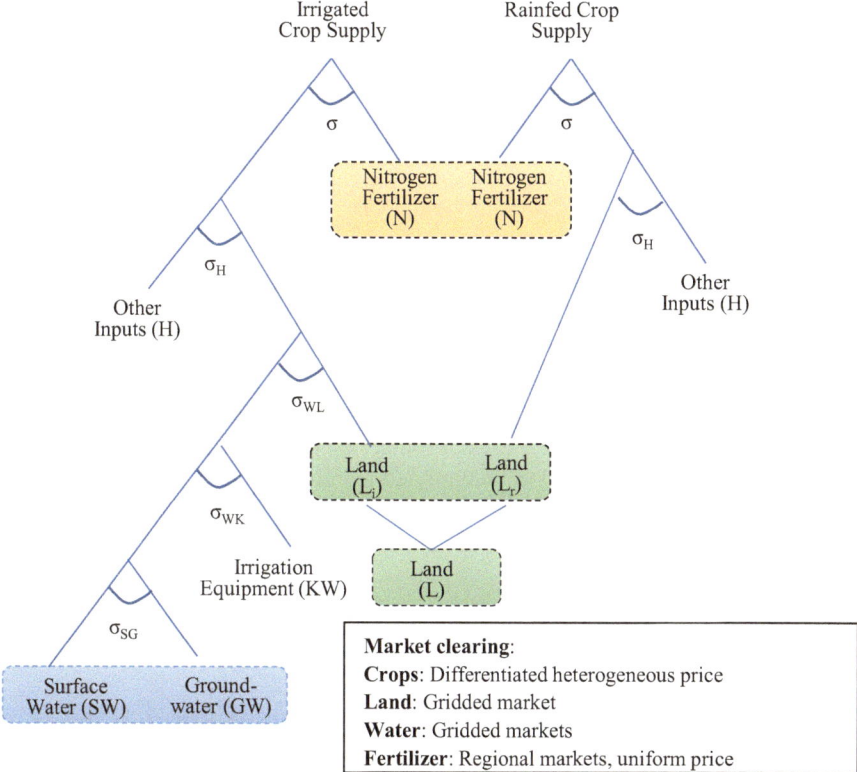

Fig. 12.1 Structure of crop production at each grid cell
Shocks and policy variables are defined for surface water, groundwater, irrigation equipment, land, nitrogen fertilizer, and overall crop production

In SIMPLE-G, both the economic and hydrologic components of water resources are considered. While hydroclimatic sources are exogenous drivers of the changes, the economic forces are endogenously modeled. In this framework, the availability and cost of surface water play determining roles, as farmers tend to withdraw less groundwater if surface water is readily available and less expensive. The cost of pumping groundwater is another factor influenced by the depth of the aquifer, the type of pumping equipment used, and energy costs. The extent of irrigation is endogenously determined in SIMPLE-G and affects the economic demand for water, while irrigation rent gaps and irrigation yield gaps can incentivize farmers to expand irrigated areas. In SIMPLE-G, using efficient irrigation equipment and technology can help reduce groundwater withdrawals. Crop prices and production technology are other variables that can impact groundwater withdrawals. Government policies, climate change, population growth, and economic development can also affect groundwater withdrawals in the model. Groundwater management regulations are one example of government policy that can influence groundwater withdrawals; others include energy, agriculture, or trade policy. Climate change affects groundwater resources in various ways. It can result in changes in rainfall patterns that affect hydrological supply (i.e., availability), and changes in temperature will alter agronomic crop water demand, which can result in increased groundwater withdrawals and decreased surface water availability. Population growth and economic development also increase demand for water resources through increases in food demand.

2.2   Decomposition

The significance of these factors varies based on the specific circumstances of each region. Therefore, it is essential to consider all these factors when developing policies for sustainable groundwater resource management. A key feature of our analysis is decomposition, which we provide with respect to the exogenous drivers of sustainability stresses. We follow the numerical integration approach developed by Harrison et al. (2000) to decompose the contribution of each exogenous variable to changes in output variables. Exogenous variables include global population growth, economic developments, technological progress, weather, and policies.

2.3   Land Allocation

In this application, we incorporate the quantity-preserving land allocation framework to improve the modeling of land-use changes at each location (Zhao et al. 2020), as described in Model 4 (see Chap. 5). This framework is particularly appropriate because we expect to see cropland converted from irrigated to rainfed uses, and vice versa when demand for groundwater is altered. The new approach ensures that we preserve the physical area as opposed to the economic volume.

2.4   Product Differentiation

Additionally, we employ product differentiation or the subregional relocation module to improve the modeling of changes in patterns of agricultural activity across grid cells and crop production zones defined by the USDA Farm Resource Regions (FRR), treating crop commodities from each FRR as differentiated commodities. This allows us to capture the tendency of crops to shift, first and foremost, within the same FRR (e.g., the Fruitful Rim).

2.5 SIMPLE-G Database Version

This application utilizes a comprehensive gridded dataset based on the reference year of 2010. To ensure consistency with SIMPLE-G, we aggregate all crops into one composite category by weighting crop outputs by relative prices. The resulting database contains detailed information regarding agricultural production and input costs for approximately 75,000 grid cells throughout the United States. This approach allows us to capture a greater level of detail regarding agricultural production and input costs and account for regional and local variations. Compared to national- or state-level models, this gridded dataset provides a more accurate and comprehensive understanding of agricultural production and input cost distributions in the United States, which is essential for effective policy-making and resource allocation.

3 Experimental Design

Two experiments are considered here. First, we integrate relevant factors anticipated to shape the future of land and water use, such as evolving population dynamics, shifting income, and technical growth (Baldos and Hertel, 2013). Second, we introduce a specific policy intervention that serves as the focal point of our sustainability analysis, allowing us to assess the impacts on agriculture.

3.1 Global Change Scenarios

Table 12.1 presents key elements of our global change scenario. In this scenario, we consider changes in population and income, as in Baldos and Hertel (2014), based on the business-as-usual shared socioeconomic pathway 2 (SSP2) from 2010 to 2050. Regarding productivity, we assume that the historical rates of productivity growth persist to midcentury (Fuglie 2012) and apply uniform cumulative TFP growth in the processed food sector. Here, TFP growth shows how much more production is possible using the same volume of inputs.

The purpose of this exercise is not to provide comprehensive future projections but rather to disentangle the significance of global drivers of local stresses within a GLG framework. The change in income, population, and TFP provides a regionally heterogeneous pattern of demand and supply shocks. Given different values of behavioral parameters, the outcome of these shocks can be quite complicated. Thus, we perform a decomposition to evaluate the significance of each driver considering GLG linkages.

Table 12.1 Projected percentage changes in population, income, and productivity, 2010–2050

Region	Population	Income per capita	Crop TFP	Livestock TFP
Eastern Europe	−12.7	239.5	17.6	50.0
North Africa	44.0	224.7	42.2	17.7
Sub-Saharan Africa	139.4	401.0	17.9	17.7
South America	31.1	176.3	132.8	157.0
Australia	33.8	70.7	64.2	17.7
Europe	0.0	66.3	78.6	17.7
South Asia	40.8	640.6	83.3	96.0
Central America	41.2	154.6	115.4	157.0
South Africa	16.0	239.5	26.8	17.7
Southeast Asia	32.1	363.6	47.0	157.0
Canada	25.8	56.4	71.1	17.7
United States	25.0	58.6	71.1	17.7
China	−6.3	606.7	121.7	157.0
Middle East	65.2	102.6	41.5	17.7
Japan + Korea	−14.5	97.6	75.0	17.7
Central Asia	52.3	394.2	25.2	50.0

Productivity is measured by growth in TFP

Sources: Percentage changes in population and income are obtained from Baldos and Hertel (2014) aggregated to 16 regions from country-level information based on SSP2 (O'Neill et al. 2014). The changes in productivity are calculated based on Fuglie (2012). The cumulative rate of productivity growth in the processed food sector is 42% worldwide

3.2 Local Sustainability Scenarios

In addition to the global-to-local analysis, we also evaluate the global impacts of a sustainable groundwater policy for the United States (i.e., the local-to-global linkages). Here, the sustainability policy is defined as restricting groundwater withdrawals to the rate of groundwater recharge in unsustainable grid cells across the United States. The data on extraction and recharge rates used in this study are obtained from the US Geological Survey (Reitz et al. 2017). We first constructed a ratio of groundwater extraction to recharge at each location circa 2010. As the year 2010 was a relatively wet year with abnormally high recharge rates, the 2010 ratio likely underestimates the long-run sustainability stress. Therefore, we use a 5-year average over 2007–2012 to calculate the ratio of groundwater extraction relative to local groundwater recharge by 5 arcmin grid cells circa 2010. We base the shocks to the groundwater supply module on this ratio (Haqiqi, 2023).

The appropriate experimental design for any given study will benefit from multidisciplinary collaboration and discussions. To design a groundwater sustainability policy, a multidisciplinary scientific collaboration mechanism should include—at minimum—economists, agronomists, and hydrologists. Economists can provide insights into the economic aspects and likely impacts of groundwater conservation policy, while hydrologists can offer insights into the required intensity of sustainability restrictions considering groundwater aquifers' physical characteristics, the hydrological impacts of changes in withdrawals, and the potential for recharge. Agronomists can assess the impact of restricting water availability throughout the growing season. Policymakers and scientists from other disciplines can provide additional insights into the applicability or acceptability of proposed conservation scenarios.

The goal of this application is to provide insights into the unintended consequences of possible groundwater conservation policies. The SIMPLE-G framework can provide a more comprehensive and informed understanding of the spillover effects of local groundwater sustainability solutions. To construct a comprehensive scenario, it is necessary to engage with a wider range of stakeholders. This approach can help to ensure that the policy design process is more spatially inclusive and that the ensuing adverse impacts do not exceed the policy's benefits.

4 Results

This section presents the findings of our investigation into the potential impacts of the two scenarios, utilizing the SIMPLE-G-US model solved in GEMPACK as our modeling framework. Focusing on two key metrics—agricultural production and groundwater withdrawals—we analyze plausible future conditions to assess their impact on water resource sustainability. The results reveal how production levels and groundwater use might shift based on policies and assumptions about population growth, economic development, and technological change, offering valuable insights for water management strategies.

4.1 Global to Local Analysis of Groundwater Demands

The global change scenario includes changes in income per capita, population, and TFP along a business-as-usual pathway from 2010 to 2050 for which key drivers are shown in Table 12.1.

4.1.1 Global Decomposition

The results of this scenario suggest that the changes in income and population, taken on their own, boost equilibrium crop production and input use (i.e., land, water, and fertilizer). However, global TFP plays a critical role in offsetting a major portion of the pressure from income and population growth on agricultural inputs. For example, changes in population and income alone would increase global cropland by around 29.91% (13.27% + 16.64%), while TFP is projected to reduce global cropland by 22.99%, leading to a far more modest 7.14% final increase in cropland in this scenario. This reveals the significance of TFP as a tool for reducing stress on global land and water resources (recall the discussion of TFP in Chap. 10).

4.1.2   Regional Decomposition

Figure 12.2a shows the major drivers of production change by region. It also decomposes the projected contribution of each driver to this aggregate outcome. In the United States, production is projected to rise by nearly 50% or roughly 0.4 billion corn-equivalent metric tons, valued at US$97 billion (in 2010 prices), over this four-decade period. The drivers of US crop output growth are relatively evenly divided between both domestic and foreign (red bar) increases in population, rising per capita incomes around the world (green bar), and improved productivity in global crop production (blue bar). In Eastern Europe and China—regions with little projected population growth (red bar)—income is the main driver of crop production (green bar). Income growth is also a key driver of output growth in South Asia, while population growth is the most important food demand driver in Africa. Note that global productivity growth has a negative impact on output in Central Asia, South Africa, Sub-Saharan Africa, and Eastern Europe. This is mainly caused by a loss in competitiveness due to low productivity growth in these regions compared with the rest of the world.

Although productivity growth leads to higher yields and therefore moderates the demand for land, the impact of population and income growth on land use is dominant in all regions except for Europe. Sub-Saharan Africa (+120 Mha) and South Asia (+36 Mha) are projected to experience the largest increases in cropland due to strong domestic demand growth in those two regions. Our analysis indicates that growth in income and population outside the United States is far more important in driving US crop production (contributing to a 22.3% increase) than income and population growth within the United States (contributing just a 6.6% increase). This is due to higher income growth rates in developing and emerging economies and higher rates of population growth in Africa and other low-income regions. For details on the regional decomposition of the global drivers of changes in water, land, and fertilizer inputs, see Haqiqi et al. (2023).

Figure 12.2b illustrates the drivers of change in US groundwater withdrawals in more detail. The greatest impacts on US groundwater are from population increases in the United States, South Asia, and Sub-Saharan Africa as well as income growth in Sub-Saharan Africa, China, and South Asia. US TFP growth—makes US agriculture more competitive—also contributes to increased US groundwater withdrawals. However, global TFP growth offsets the impacts of increased income and population.

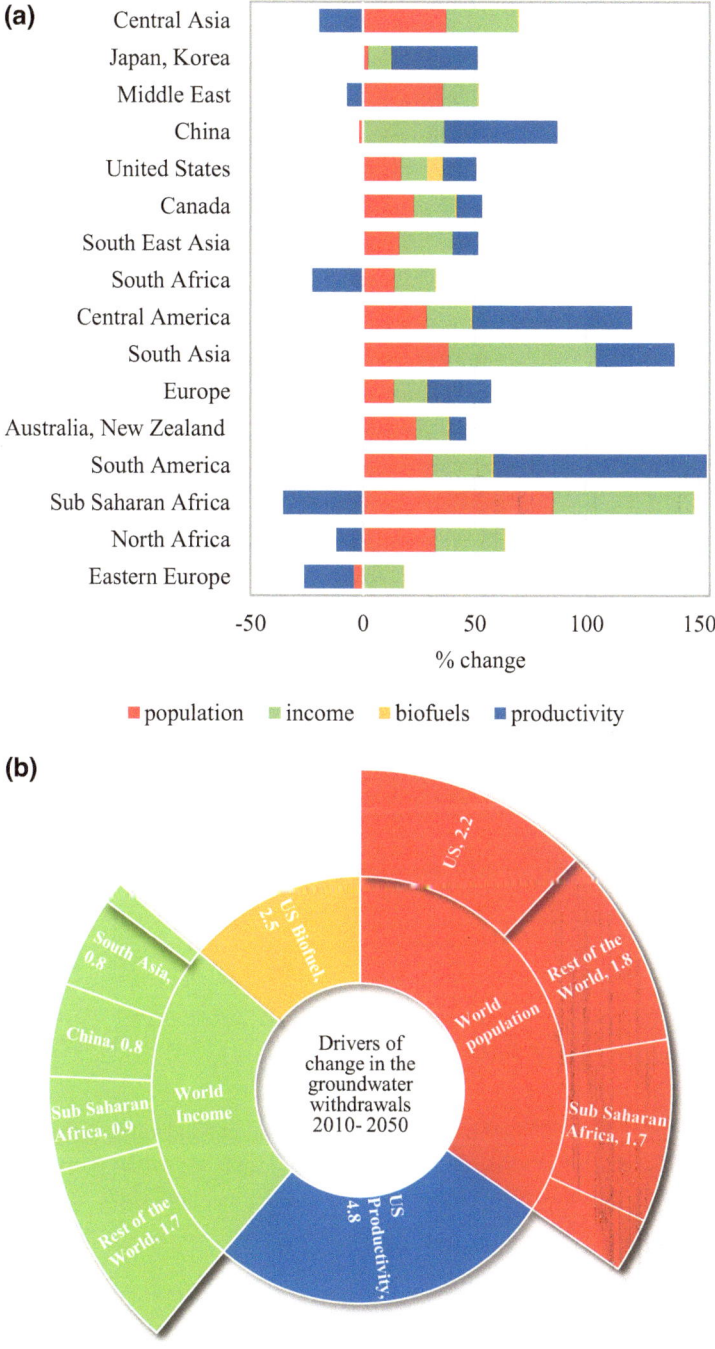

Fig. 12.2 Change in (**a**) crop production by region and (**b**) US groundwater withdrawals, 2010–2050

(**a**) Decomposition of changes in regional crop production from 2010 to 2050 due to the combined effect of changes in per capita income, population, and technology, around the world, as well as

4.1.3 Local Decomposition

Figure 12.3 illustrates the pattern of changes in groundwater withdrawals and irrigated areas across the United States due to a population increase of one million people in each world region. These changes are calculated by dividing the computed contribution of population change in each global region by the population change headcount in that region. For example, if the model calculates that an increase of Y people in region r has caused an X hectare change in irrigated area in subregion z, then the average impact of a change in population in region r on irrigated area in subregion z is equal to X_z/Y_r. We perform similar calculations for cubic meters of water. The general finding is that normalized population changes in Canada and Europe will have greater contributions to changes in water withdrawal and irrigation contraction—four times greater than those in China—for a given size of population change. Note that the population change in any region has a direct relationship with water and irrigation in the United States. Thus, a decline in population, in China, for example, means less stress on land and water resources in the United States. Another general finding is that most of the changes occur in the Fruitful Rim and Basin and Range in the Western United States. Finally, a one-million-person increase in world population can increase irrigated area in the United States by 100–400 ha depending on the source of population growth.

There are two major observations at the grid-cell level. First, the most rapid irrigation expansion (in percentage terms) is expected to occur in the Eastern United States. This finding is in line with current observations showing rapid irrigation expansion in the East and declines in the West during the 1997–2017 period (USDA 2019; Xie et al. 2021). Second, foreign demand drivers of groundwater withdrawals are more important than US demand drivers. SIMPLE-G can also be employed to provide economic insights into the likely impacts of changes and unintended consequences of policies for water and land resources—a topic to which we next turn.

4.2 Local-to-Global Analysis of US Groundwater Sustainability Policies

We implement the sustainability policy by reducing groundwater withdrawals to the rate of recharge (for details, see Haqiqi et al. 2023). To achieve this, we apply a shifter to the groundwater supply curve. This is accomplished by shocking the

---

Fig. 12.2 (continued) growth in biofuels demand. (**b**) Decomposition of drivers of changes in US groundwater withdrawals from 2010 to 2050 as computed by SIMPLE-G-US-Allcrops, based on SSP2 (Shared Socioeconomic Pathways, middle of the road) in the absence of climate change. Improvement in global productivity can completely offset the impacts of increased population and income, conditional to sufficient investments. Also, the rebound effect causes an increase in groundwater withdrawals

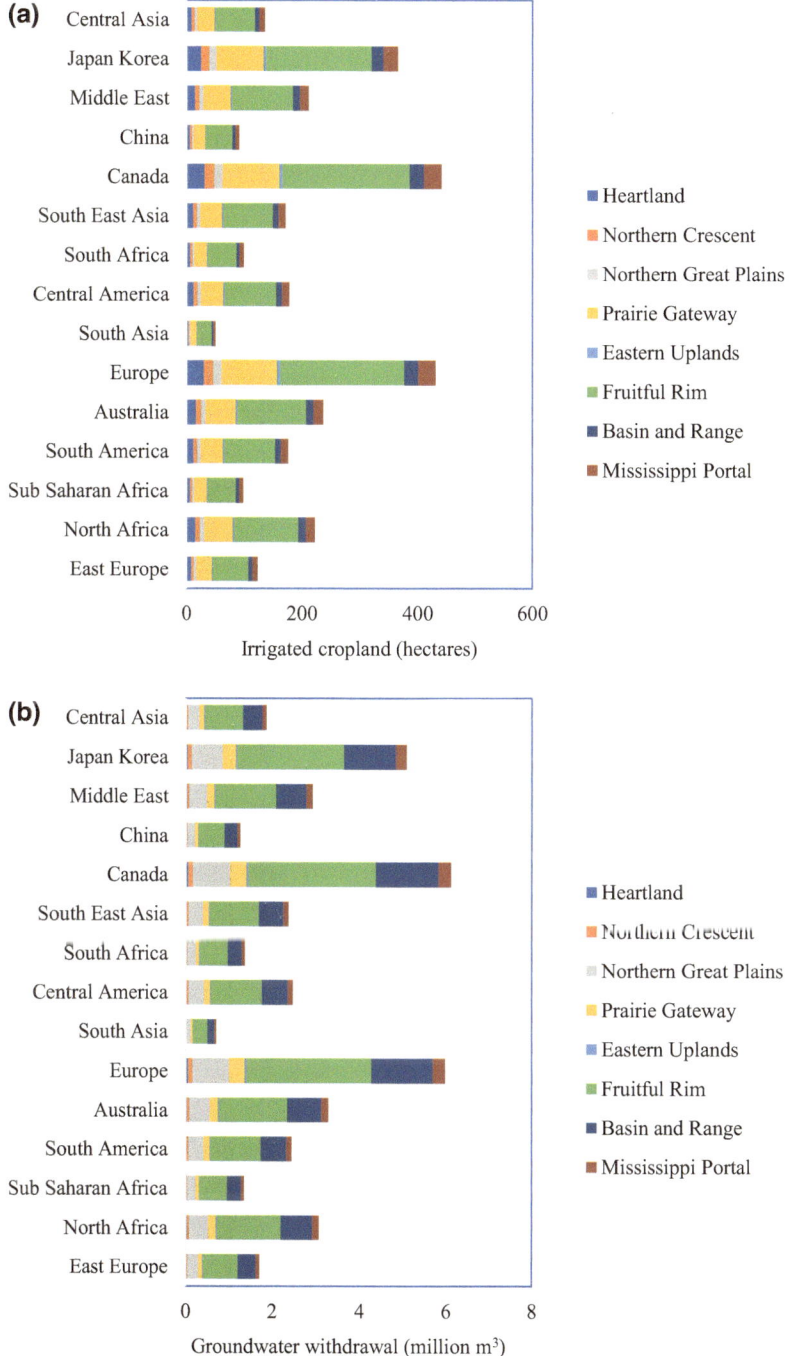

Fig. 12.3 Change in (**a**) US irrigated land and (**b**) US groundwater withdrawals in response to a population increase of one million, by region

Uniform population increases/decreases in Europe, Canada, Japan, and Korea have larger impacts on water and land resources in the United States than population increases/decreases elsewhere. A decline in the population of these regions can cause a bigger decline in cropland in the United States. In general, impacts on the western regions (Fruitful Rim, Basin, and Range) are largest appropriate slack variable in the model code. The size of this shock indicates the amount of reduction required from 2010 levels to ensure that the rate of withdrawal does not exceed the recharge rates. If a grid cell is extracting groundwater at a rate higher than the recharge rate, it must reduce withdrawals. To ease the implementation of these shocks, we truncated them at −90%, so we do not wholly eliminate groundwater use in any grid cells.

Out of the approximately 75,000 grid cells, this policy will have a direct impact on 10%, all of which are located in the Western United States. These grid cells face a significant reduction in groundwater availability, ranging from −1% to −90%. More than half of these cells will experience a reduction of more than 80%. The strict policy will cause some agricultural activities to relocate to other parts of the country that have abundant water resources and that are already major agricultural producers. However, these regions may not be able to accommodate all of the agricultural activities displaced from the Western United States, leading to increased production overseas.

4.2.1 Local Impacts and Spillovers

At the local level, the first response to this policy is a reduction in groundwater withdrawals. Figure 12.4a shows the resulting change in groundwater withdrawals by grid cell. Green areas require groundwater withdrawal reductions of more than 50%. This change induces an increase in groundwater withdrawals elsewhere in the United States. The greatest increase in groundwater withdrawals occurs in the southern and eastern parts of the Fruitful Rim region, which includes Texas and Florida, which have appropriate agroecological conditions for producing crops similar to those produced in California (the targeted regions).

The groundwater restriction scenario is expected to increase the shadow price (marginal value of water to the producer) of water by more than 50% in the targeted regions. This raises the average crop production cost by 9–15%. Figure 12.4b shows the projected change in the equilibrium price (value) of groundwater in the United States. For the targeted regions, this increasing scarcity value is mainly due to a lower supply of water. However, for other parts of the United States, the increase in the groundwater shadow price is due to changes in the demand for irrigation.

Another potential local response to this groundwater sustainability scenario is to increase the use of surface water, as this approach offers a potential substitute—albeit an imperfect one—for groundwater. It is not considered a perfect substitute: The regulations and the extraction methods are different, and farmers may not have the same control over the timing of water availability. Figure 12.5a shows the estimated change in US gridded surface water withdrawal in the wake of the groundwater sustainability policy. Depending on water availability and institutions, surface water withdrawals might increase from 5% to more than 25% in nontargeted parts of the United States. The surface water withdrawal is projected to decline in the restricted grid cells due to large reductions in the irrigated area (Fig. 12.5b). While the global response is small, the base over which this applies is large, so if the US

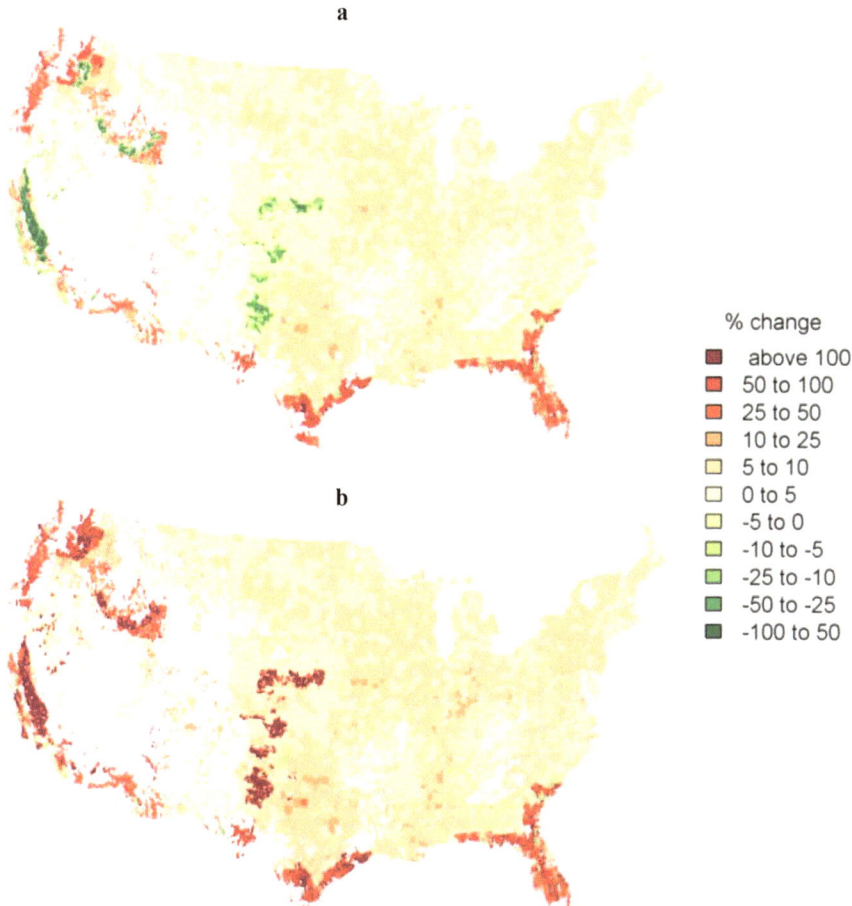

Fig. 12.4 Percentage change in (**a**) groundwater withdrawals and (**b**) value (shadow price) of groundwater to producers by 5 arcmin grid cells in response to a policy restricting groundwater withdrawals to the level of average annual recharge
Areas in *white* are not cultivated

cropland area declines by 1 million hectares, the global cropland area increases by 2.3–2.5 million hectares in response to US groundwater restrictions. This reflects the extremely high yields on US irrigated croplands which must be replaced by production elsewhere to meet global food demands, which are only slightly reduced in the face of higher prices.

4.2.2   Global Impacts

Figure 12.6a reports the estimated percentage change in global cropland area—less than 1% in each region—due to US groundwater sustainability restrictions. This

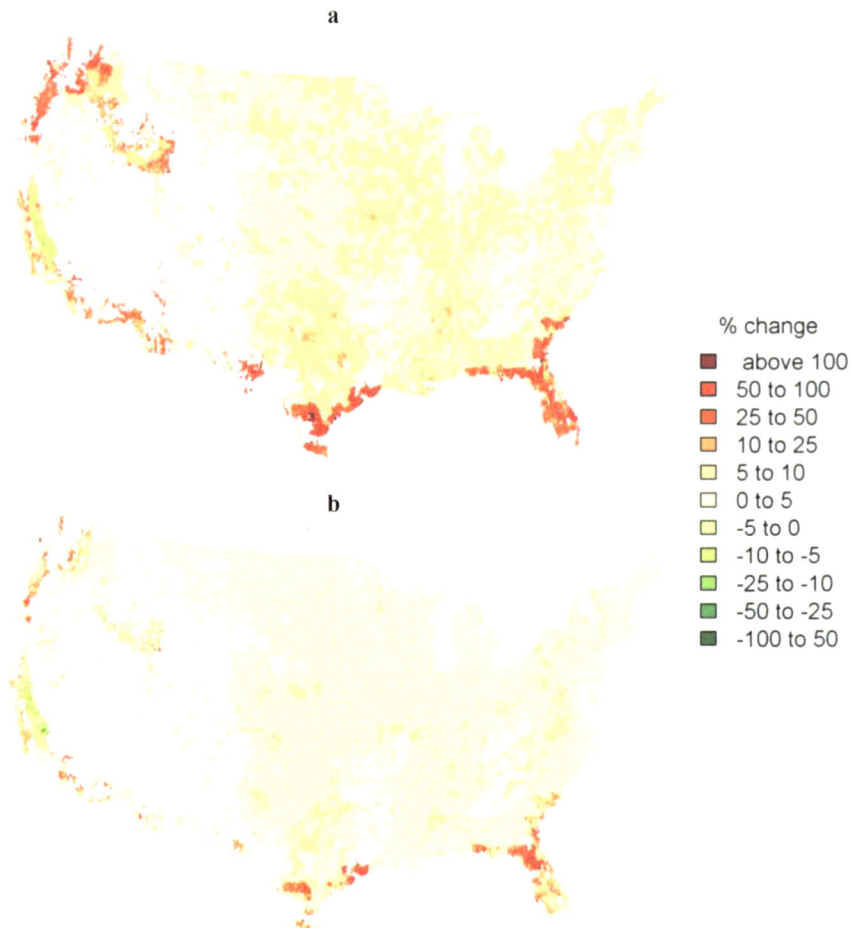

Fig. 12.5 Percentage change in (**a**) irrigation surface water withdrawals and (**b**) total cropland area by 5 arcmin grid cells in response to groundwater restrictions
Areas in *white* are not cultivated

experiment shows that local sustainability solutions may be viewed less favorably in a global context. The unintended consequences include possible deforestation (increase in cropland) and water quality issues (higher fertilizer applications) in nontargeted areas. This shows why we need to take a global approach to evaluating water policies to consider significant feedback from the human system to natural resource use when designing effective and efficient sustainability policies (Biswas 2008).

Table 12.2 depicts the impacts of US groundwater sustainability restrictions on the world and the United States. It is projected that this local groundwater sustainability policy may lead to a 0.1% increase in global cropland area (1.7 million hectares) and a 0.3% increase in fertilizer application worldwide. In the United

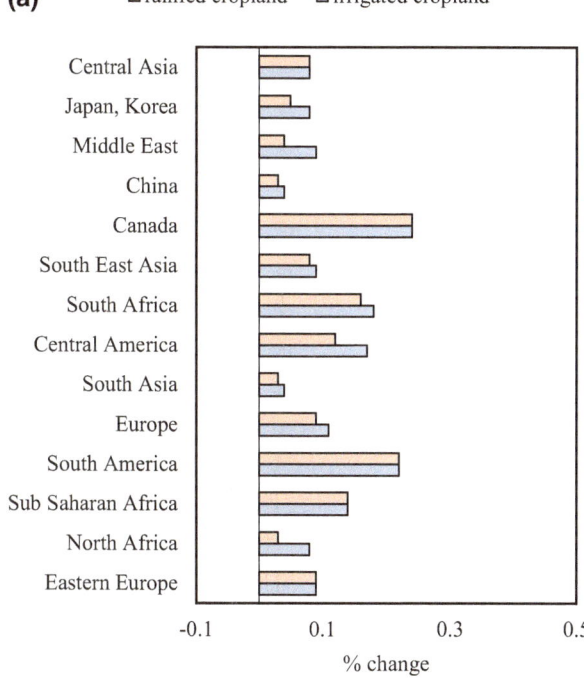

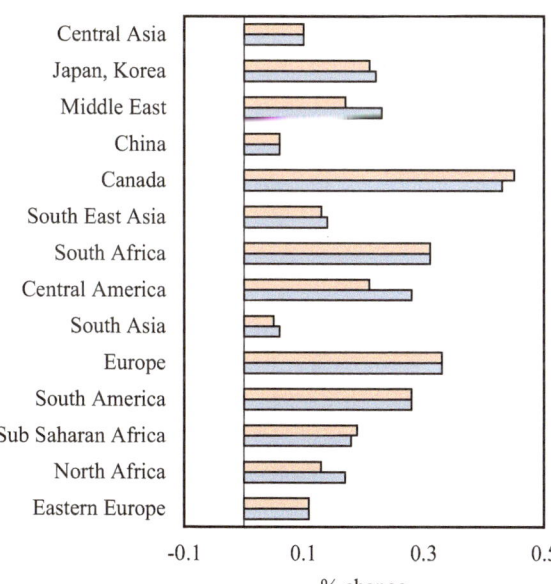

Fig. 12.6 Percentage change in (**a**) global production and (**b**) land use in response to a US groundwater sustainability constraint in 2050

Table 12.2 Projected changes in cropland area, water withdrawal, fertilizer applications, and crop productions in response to US groundwater restriction

	World			US		
	Irrigated	Rainfed	Total	Irrigated	Rainfed	Total
%						
Crop production (%)	−1.03	0.61	−0.06	−16.39	5.37	−2.98
Cropland area (%)	−0.35	0.23	0.11	−4.68	1.18	0.13
Surface water withdrawal (%)	1.36	na	1.36	4.19	na	4.19
Groundwater withdrawal (%)	−9.25	na	−9.25*	−61.70	na	−61.70
Fertilizer application (%)	−0.32	0.39	0.25	−11.01	4.50	1.46
Δ						
Crop production (m ton)	−46.07	39.32	−6.77	−50.91	28.97	−21.93*
Cropland area (m ha)	−1.12	2.82	1.70	−1.33	1.54	0.21

(* See application files for more details)

States, the irrigated area declines by 4.7%, leading to a 16.4% reduction in irrigated production (50.9 million tons corn-equivalent). The responses within the United States are estimated to increase rainfed areas (extensive margins) and fertilizer application in rainfed areas (intensive margins). This is mainly due to a 9.4% increase in local crop prices in Fruitful Rim, which will motivate other farmers to increase their production and new farmers to enter the market. A model closure with low local production flexibility could increase the non-US cropland area by up to 20 million hectares. For detailed analysis, see the chapter application files for *simpleg_leon* scenario.

5   Discussion

There have been several attempts to project future water withdrawals based on gridded models (Herbert and Döll 2019), county-level analysis (Roy et al. 2012), Hydrologic Unit Codes (Brown et al. 2019), or at global scales (Boretti and Rosa 2019). However, most current studies have ignored the role of technological progress (measured here as TFP). We show that productivity improvements in crop production, the livestock sector, and processed food activities can offset the pressure on water resources over the coming decades. However, investments in research and development (R&D) are required to achieve the suggested productivity improvements (see Chap. 10 as well as Baldos et al. 2018; Baldos and Hertel 2018). Continued improvements in productivity—which have followed investments in R&D over the past—are essential to avoid increasing stress on water resources. Although improvements in irrigation efficiency are expected to increase water use (Perry et al. 2017; Grafton and Wheeler 2018; Perry 2019), growth in productivity can reduce the stress on water resources by offsetting demand pressures. Given the uncertainty in TFP calculations, future studies should explore the consequences of alternative R&D and TFP trajectories (Fuglie et al. 2022).

Over the last few decades, policy discussions around groundwater sustainability have evolved from a legal and economic definition of safe yield (i.e., the amount of water that can be withdrawn before it is no longer economically feasible) to a more socially and environmentally based definition of sustainability that can provide a long-term balance between withdrawals and impacts (Alley et al. 1999; Alley and Leake 2004). Even a withdrawal amount that is replenished each year can lead to a decrease in natural discharge and cause harm to ecosystems that rely on groundwater contributions; in many locations, the sustainable yield is substantially less than the natural recharge rate (Sophocleous 2000). The sustainability scenario undertaken here, therefore, provides a conservative estimate of the economic impacts of a sustainable groundwater policy. In some cases, restricting groundwater withdrawals to estimated annual rates of recharge allows for extraction far above sustainable yields.

Aeschbach-Hertig and Gleeson (2012) and Piemontese et al. (2020) call for a more comprehensive socioeconomic strategy and more attention to the significance of local population and equity in sustainability studies. This study contributes to the growing body of evidence on the importance of the global–local–global approach for sustainable agriculture (Hertel et al. 2019, 2023; Haqiqi et al. 2022; Ray et al. 2023). While global costs of sustainability might be small due to economic responses and reallocation of resources, the local benefits of sustainability and local costs of ignoring it are significant.

Gleeson et al. (2012) argue that groundwater sustainability is a values-driven process involving social, economic, and environmental factors that cannot be defined in terms of a given withdrawal target and that sustainability planning for groundwater must utilize a long time horizon (50–100 years); involve local, adaptive input; and utilize models capable of backcasting from desired outcomes to policy options that can support them. This work utilizes a publicly available, multiscale modeling tool (SIMPLE-G) that can be incorporated into regional community planning efforts to examine the socioeconomic impacts of various management strategies over long time horizons. However, we would argue that, in addition to a local, physically based analysis to inform policy options, given linkages to the global food system, community-level planning for groundwater sustainability options must also include a broader perspective, considering the role of global economic drivers.

Another benefit of the multiscale modeling framework is its ability to capture the interactions of different countries' policies and their local impacts. Disaggregating production in the United States while keeping other regions ungridded in the background permits us to consider the market responses of US trade partners, even as we explore US domestic policies in detail. These global market developments are potentially important when they are implemented by major trade partners and food-producing countries. One outstanding example is China, which has been aggressively pursuing national policies to achieve sustainable agricultural development during the past decade. The National Sustainable Agricultural Development Plan (2015–2030), released in 2015, continues to emphasize the contributions of agricultural science and technology (more than 60% to total output growth), cropland protection (no less than 120 million hectares of cropland), and efficient irrigation (capped national irrigation water use at 700 billion m³ by 2030 and an irrigation efficiency index no lower than 0.60). It is important to be able to factor such major developments into the analysis of US agriculture and sustainability policies at the national and local levels.

This study shows that SIMPLE-G is a useful tool that can provide insights into the spatial distribution of solutions to environmental problems on cultivated land. This information can be used by policymakers to identify the areas most affected by environmental problems, allowing them to focus their efforts on resolving these problems in the most efficient way possible. It is also helpful when evaluating the effectiveness of environmental policies and understanding the unintended consequences that may arise. As shown in this chapter, as well as in other applications in this book, the SIMPLE-G framework can be used to assess the market-mediated impacts of environmental policies by analyzing the bidirectional feedback between distant locations through crop markets, labor markets, and fertilizer markets.

6   Conclusion

With limited technological progress, increasing stresses on water and land resources are projected to emerge because of anticipated growth in demand for food. Specifically, groundwater withdrawals in some locations are expected to increase by more than 50%. The underlying drivers of these stresses are global in nature, with demand growth in South Asia and China alone accounting for roughly one-quarter of US irrigated cropland expansion. However, strong productivity growth can offset the pressure from global changes in population and growth. Unfortunately, groundwater pumping for irrigation and other uses now exceeds annual recharge rates by more than 10 times in parts of the Central Valley of California, the High Plains Aquifer, and the Snake River Basin of the United States. This situation is expected to further deteriorate in the absence of regulation.

We find that local water policies can have unanticipated spillover effects, thereby exacerbating water stresses in other parts of the world. Furthermore, any attempt to restrict water for irrigation will result in the reallocation of cropping activity to other parts of the country and the world. Given the increase in water and land use in other parts of the world in the wake of the US sustainability policy (around 20 million hectares), the global net environmental benefit of such a policy may be limited or even negative due to spillover effects. Cropland expansion in other parts of the world can cause more biodiversity loss, deforestation, and water pollution. Addressing this sustainability challenge will require global coordination of resource governance as well as institutional reforms. The global–local–global framework here offers a practical means for analyzing the consequences of such policy proposals as they emerge.

The relationships among groundwater policy, land use, and its spillover effects reveal challenges and opportunities. SIMPLE-G has illuminated the pathways through which local water conservation resonates across borders, potentially amplifying environmental pressures elsewhere. The questions it provokes—about labor market dynamics, global food security, and the cascading impacts of sustainability strategies—remain central threads in the ongoing dialog about managing our precious, shared aquifer resources. Further research and policy evaluations are crucial for ensuring groundwater sustainability not just within localized contexts but also as a keystone for the health of our planet and its diverse inhabitants.

Acknowledgments and Competing Interests The authors acknowledge support from the National Science Foundation HDR award # 2118329: "NSF Institute for Geospatial Understanding through an Integrative Discovery Environment (I-GUIDE)," the US Department of Energy, Office of Science, Biological and Environmental Research Program, Earth and Environmental Systems Modeling, Multi-Sector Dynamics Contract #DE-SC0016162; the United States Department of Agriculture AFRI grant #2019-67023-29679, "Economic Foundations of Long Run Agricultural Sustainability," and the National Science Foundation INFEWS award #1855937, "Identifying Sustainability Solutions Through Global-Local-Global Analysis of a Coupled Water-Agriculture-Bioenergy System."

The findings and conclusions presented in this chapter are those of the authors and should not be construed to represent any official determination or policy of the US Department of Agriculture (USDA), the US government, the Department of Energy (DOE), or the National Science Foundation (NSF). Furthermore, we declare that there is no conflict of interest related to this work.

References

Aeschbach-Hertig, Werner, and Tom Gleeson. 2012. Regional strategies for the accelerating global problem of groundwater depletion. *Nature Geoscience* 5: 853–861. https://doi.org/10.1038/ngeo1617.

Ahmed, Syud Amer, Thomas W. Hertel, and Ruben Lubowski. 2008. *Calibration of a land cover supply function using transition probabilities*. Research Memorandum 14. Department of Agricultural Economics, Purdue University, West Lafayette: Global Trade Analysis Project (GTAP). https://doi.org/10.21642/GTAP.RM14.

Alley, William M., and Stanley A. Leake. 2004. The journey from safe yield to sustainability. *Groundwater* 42: 12–16. https://doi.org/10.1111/j.1745-6584.2004.tb02446.x.

Alley, William M., Thomas E. Reilly, and O. Lehn Franke. 1999. *Sustainability of ground-water resources*, Circular 1186. Denver: US Geological Survey.

Armington, Paul S. 1969. A theory of demand for products distinguished by place of production. *Staff Papers* 16: 159–178. https://doi.org/10.2307/3866403.

Baldassarre, Di, Murugesu Sivapalan Giuliano, Maria Rusca, Christophe Cudennec, Margaret Garcia, Heidi Kreibich, Megan Konar, et al. 2019. Sociohydrology: Scientific challenges in addressing the sustainable development goals. *Water Resources Research* 55: 6327–6355. https://doi.org/10.1029/2018WR023901.

Baldos, Uris Lantz C., Iman Haqiqi, Thomas W. Hertel, Mark Horridge, and J. Liu. 2020. SIMPLE-G: A multiscale framework for integration of economic and biophysical determinants of sustainability. *Environmental Modelling & Software* 133: 104805. https://doi.org/10.1016/j.envsoft.2020.104805.

Baldos, U.L.C., Viens, F.G., Hertel, T.W. and Fuglie, K.O., 2019. R&D spending, knowledge capital, and agricultural productivity growth: A Bayesian approach. *American Journal of Agricultural Economics, 101*(1): 291–310. https://doi.org/10.1093/ajae/aay039

Baldos, Uris Lantz C., Hertel, Thomas W. 2014. Global food security in 2050: the role of agricultural productivity and climate change. *Australian Journal of Agricultural and Resource*

Economics, *58*(4): 554–570. https://doi.org/10.1111/ajar.2014.58.issue-4. https://doi.org/10.1111/1467-8489.12048.

Baldos, Uris Lantz C., and Thomas W. Hertel. 2013. Looking back to move forward on model validation: Insights from a global model of agricultural land use. *Environmental Research Letters* 8: 034024. https://doi.org/10.1088/1748-9326/8/3/034024.

Baldos, Uris, and Thomas Hertel. 2018. Productivity growth is key to achieving long run agricultural sustainability. *Purdue Policy Research Institute (PPRI) Policy Briefs* 4: 8.

Befus, Kevin M., Scott Jasechko, Elco Luijendijk, Tom Gleeson, and M. Bayani Cardenas. 2017. The rapid yet uneven turnover of Earth's groundwater. *Geophysical Research Letters* 44: 5511–5520. https://doi.org/10.1002/2017GL073322.

Biswas, Asit K. 2008. Integrated water resources management: Is it working? *International Journal of Water Resources Development* 24: 5–22. https://doi.org/10.1080/07900620701871718.

Blair, P., and W. Buytaert. 2016. Socio-hydrological modelling: A review asking "why, what and how?". *Hydrology and Earth System Sciences* 20: 443–478. https://doi.org/10.5194/hess-20-443-2016.

Boretti, Alberto, and Lorenzo Rosa. 2019. Reassessing the projections of the World Water Development Report. *npj Clean Water* 2: 15. https://doi.org/10.1038/s41545-019-0039-9.

Brown, Thomas C., Vinod Mahat, and Jorge A. Ramirez. 2019. Adaptation to future water shortages in the United States caused by population growth and climate change. *Earth's Future* 7: 219–234. https://doi.org/10.1029/2018EF001091.

Bruckner, Martin, Günther Fischer, Sylvia Tramberend, and Stefan Giljum. 2015. Measuring telecouplings in the global land system: A review and comparative evaluation of land footprint accounting methods. *Ecological Economics* 114: 11–21. https://doi.org/10.1016/j.ecolecon.2015.03.008.

Castle, Stephanie L., Brian F. Thomas, John T. Reager, Matthew Rodell, Sean C. Swenson, and James S. Famiglietti. 2014. Groundwater depletion during drought threatens future water security of the Colorado River Basin. *Geophysical Research Letters* 41: 5904–5911. https://doi.org/10.1002/2014GL061055.

Changming, Liu, Yu Jingjie, and Eloise Kendy. 2001. Groundwater exploitation and its impact on the environment in the North China Plain. *Water International* 26: 265–272. https://doi.org/10.1080/02508060108686913.

Chaudhary, Abhishek, and Thomas Kastner. 2016. Land use biodiversity impacts embodied in international food trade. *Global Environmental Change* 38: 195–204. https://doi.org/10.1016/j.gloenvcha.2016.03.013.

Cook, Benjamin I., Toby R. Ault, and Jason E. Smerdon. 2015. Unprecedented 21st century drought risk in the American Southwest and Central Plains. *Science Advances* 1. American Association for the Advancement of Science: e1400082. https://doi.org/10.1126/sciadv.1400082.

D'Odorico, Paolo, Joel Carr, Carole Dalin, Jampel Dell'Angelo, Megan Konar, Francesco Laio, Luca Ridolfi, et al. 2019. Global virtual water trade and the hydrological cycle: Patterns, drivers, and socio-environmental impacts. *Environmental Research Letters* 14: 053001. https://doi.org/10.1088/1748-9326/ab05f4.

Dalin, Carole, Megan Konar, Naota Hanasaki, Andrea Rinaldo, and Ignacio Rodriguez-Iturbe. 2012. Evolution of the global virtual water trade network. *Proceedings of the National Academy of Sciences* 109: 5989–5994. https://doi.org/10.1073/pnas.1203176109.

Dalin, Carole, Yoshihide Wada, Thomas Kastner, and Michael J. Puma. 2017. Groundwater depletion embedded in international food trade. *Nature* 543: 700–704. https://doi.org/10.1038/nature21403.

Ertsen, M.W., J.T. Murphy, L.E. Purdue, and T. Zhu. 2014. A journey of a thousand miles begins with one small step – Human agency, hydrological processes and time in socio-hydrology. *Hydrology and Earth System Sciences* 18: 1369–1382. https://doi.org/10.5194/hess-18-1369-2014.

Famiglietti, J.S. 2014. The global groundwater crisis. *Nature Climate Change* 4: 945–948. https://doi.org/10.1038/nclimate2425.

Faunt, Claudia C., ed. 2009. *Groundwater availability of the Central Valley Aquifer, California,* Professional Paper 1766. Reston: US Geological Survey. https://doi.org/10.5066/F79S1PX3.

Fernald, A., S. Guldan, K. Boykin, A. Cibils, M. Gonzales, B. Hurd, S. Lopez, et al. 2015. Linked hydrologic and social systems that support resilience of traditional irrigation communities. *Hydrology and Earth System Sciences* 19: 293–307. https://doi.org/10.5194/hess-19-293-2015.

Fuglie, Keith O. 2012. Productivity growth and technology capital in the global agricultural economy. In *Productivity growth in agriculture: an international perspective,* ed. K.O. Fuglie, S.L. Wang, and V.E. Ball, 335–368. CABI Books. Wallingford: CAB International. https://doi.org/10.1079/9781845939212.0335.

Fuglie, Keith, Srabashi Ray, Uris Lantz C. Baldos, and Thomas W. Hertel. 2022. The R&D cost of climate mitigation in agriculture. *Applied Economic Perspectives and Policy* 44: 1955–1974. https://doi.org/10.1002/aepp.13245.

Ghosh, Sanchari, Kelly M. Cobourn, and Levan Elbakidze. 2014. Water banking, conjunctive administration, and drought: The interaction of water markets and prior appropriation in southeastern Idaho. *Water Resources Research* 50: 6927–6949. https://doi.org/10.1002/2014WR015572.

Giuliani, M., Y. Li, A. Castelletti, and C. Gandolfi. 2016. A coupled human-natural systems analysis of irrigated agriculture under changing climate. *Water Resources Research* 52: 6928–6947. https://doi.org/10.1002/2016WR019363.

Gleeson, Tom, Yoshihide Wada, Marc F.P. Bierkens, and Ludovicus P.H. Van Beek. 2012. Water balance of global aquifers revealed by groundwater footprint. *Nature* 488: 197–200. https://doi.org/10.1038/nature11295.

Grafton, R. Quentin, and Sarah Ann Wheeler. 2018. Economics of water recovery in the Murray-Darling Basin, Australia. *Annual Review of Resource Economics* 10: 487–510. https://doi.org/10.1146/annurev-resource-100517-023039.

Haqiqi, I. 2023. A Gridded Dataset for Groundwater Sustainability Restriction Policy Scenarios for the Contiguous US. GLASSNET Geospatial Data for Sustainability. MyGeoHUB. https://doi.org/10.13019/AHZR-4843

Hanjra, Munir A., and M. Ejaz Qureshi. 2010. Global water crisis and future food security in an era of climate change. *Food Policy* 35: 365–377. https://doi.org/10.1016/j.foodpol.2010.05.006.

Haqiqi, Iman. 2018. The impacts of climate change on yields of irrigated and rainfed crops: Length, depth, and correlation of damages. In *Annual Meeting.* Washington, DC: Agricultural and Applied Economics Association. https://doi.org/10.22004/ag.econ.274417.

———. 2023. *The value of water in US agriculture: Integrating spatially and temporally heterogeneous hydroclimatic and economic data (version 1.0).* MyGeoHUB. https://doi.org/10.13019/9MXE-T280.

Haqiqi, Iman, and Thomas W. Hertel. 2016. Decomposing irrigation water use changes in equilibrium models. In *Annual Meeting.* Boston: Agricultural and Applied Economics Association.

Haqiqi, Iman, Danielle S. Grogan, Thomas W. Hertel, and Wolfram Schlenker. 2019. Predicting crop yields using soil moisture and heat: An extension to Schlenker and Roberts (2009). In *Annual Meeting.* Atlanta: Agricultural and Applied Economics Association. https://doi.org/10.22004/ag.econ.291093.

———. 2021. Quantifying the impacts of compound extremes on agriculture. *Hydrology and Earth System Sciences* 25: 551–564. https://doi.org/10.5194/hess-25-551-2021.

Haqiqi, Iman, Chris J. Perry, and Thomas W. Hertel. 2022. When the virtual water runs out: Local and global responses to addressing unsustainable groundwater consumption. *Water International* 47: 1060–1084. https://doi.org/10.1080/02508060.2023.2131272.

Haqiqi, Iman, Laura Bowling, Sadia Jame, Uris Baldos, Jing Liu, and Thomas Hertel. 2023. Global drivers of local water stresses and global responses to local water policies in the United States. *Environmental Research Letters* 18: 065007. https://doi.org/10.1088/1748-9326/acd269.

Harrison, Jill Lindsey. 2011. *Pesticide Drift and the Pursuit of Environmental Justice*. Cambridge, MA: MIT Press. https://doi.org/10.7551/mitpress/9780262015981.001.0001.

Harrison, W. Jill, J. Mark Horridge, and K.R. Pearson. 2000. Decomposing simulation results with respect to exogenous shocks. *Computational Economics* 15: 227–249. https://doi.org/10.1023/A:1008739609685.

Herbert, Claudia, and Petra Döll. 2019. Global assessment of current and future groundwater stress with a focus on transboundary aquifers. *Water Resources Research* 55: 4760–4784. https://doi.org/10.1029/2018WR023321.

Hertel, Thomas W. 2018. Economic perspectives on land use change and leakage. *Environmental Research Letters* 13: 075012. https://doi.org/10.1088/1748-9326/aad2a4.

Hertel, Thomas W., and Uris Lantz C. Baldos. 2016. Attaining food and environmental security in an era of globalization. *Global Environmental Change* 41: 195–205. https://doi.org/10.1016/j.gloenvcha.2016.10.006.

Hertel, Thomas W., Thales A.P. West, Jan Börner, and Nelson B. Villoria. 2019. A review of global-local-global linkages in economic land-use/cover change models. *Environmental Research Letters* 14: 053003. https://doi.org/10.1088/1748-9326/ab0d33.

Hertel, Thomas W., Elena Irwin, Stephen Polasky, and Navin Ramankutty. 2023. Focus on global–local–global analysis of sustainability. *Environmental Research Letters* 18: 100201. https://doi.org/10.1088/1748-9326/acf8da.

Hoekstra, Arjen Y., and Mesfin M. Mekonnen. 2012. The water footprint of humanity. *Proceedings of the National Academy of Sciences* 109: 3232–3237. https://doi.org/10.1073/pnas.1109936109.

Hoekstra, Arjen Y., and Thomas O. Wiedmann. 2014. Humanity's unsustainable environmental footprint. *Science* 344: 1114–1117. https://doi.org/10.1126/science.1248365.

Jame, Sadia A., and Laura C. Bowling. 2020. Groundwater doctrine and water withdrawals in the United States. *Water Resources Management* 34: 4037–4052. https://doi.org/10.1007/s11269-020-02642-0.

Jury, William A., and Henry J. Vaux. 2007. The emerging global water crisis: Managing scarcity and conflict between water users. *Advances in Agronomy* 95: 1–76. https://doi.org/10.1016/S0065-2113(07)95001-4.

Karami, Ayatollah, Abdoulkarim Esmaeili, and Bahadin Najafi. 2012. Assessing effects of alternative food subsidy reform in Iran. *Journal of Policy Modeling* 34: 788–799. https://doi.org/10.1016/j.jpolmod.2011.08.002.

Konikow, Leonard F. 2013. *Groundwater depletion in the United States (1900–2008)*, 2013–5079. Reston: US Geological Survey. https://doi.org/10.3133/sir20135079.

Liesch, Tanja, and Marc Ohmer. 2016. Comparison of GRACE data and groundwater levels for the assessment of groundwater depletion in Jordan. *Hydrogeology Journal* 24: 1547–1563. https://doi.org/10.1007/s10040-016-1416-9.

Liu, Jing, Thomas W. Hertel, Farzad Taheripour, Tingju Zhu, and Claudia Ringler. 2014. International trade buffers the impact of future irrigation shortfalls. *Global Environmental Change* 29: 22–31. https://doi.org/10.1016/j.gloenvcha.2014.07.010.

Liu, Jing, Thomas W. Hertel, Richard B. Lammers, Alexander Prusevich, Uris Lantz C. Baldos, Danielle S. Grogan, and Steve Frolking. 2017. Achieving sustainable irrigation water withdrawals: Global impacts on food security and land use. *Environmental Research Letters* 12: 104009. https://doi.org/10.1088/1748-9326/aa88db.

Marston, Landon, Megan Konar, Ximing Cai, and Tara J. Troy. 2015. Virtual groundwater transfers from overexploited aquifers in the United States. *Proceedings of the National Academy of Sciences* 112: 8561–8566. https://doi.org/10.1073/pnas.1500457112.

McGuire, Virginia L. 2017. *Water-level and recoverable water in storage changes, High Plains aquifer, predevelopment to 2015 and 2013–15*, 2017–5040. Reston: US Geological Survey. https://doi.org/10.3133/sir20175040.

Nabavi, Ehsan. 2018. Failed policies, falling aquifers: Unpacking groundwater overabstraction in Iran. *Water Alternatives* 11: 699–724.

Nepal, Santosh, and Arun Bhakta Shrestha. 2015. Impact of climate change on the hydrological regime of the Indus, Ganges and Brahmaputra River Basins: A review of the literature.

International Journal of Water Resources Development 31: 201–218. https://doi.org/10.1080/07900627.2015.1030494.

O'Neill, Brian C., Elmar Kriegler, Keywan Riahi, Kristie L. Ebi, Stephane Hallegatte, Timothy R. Carter, Ritu Mathur, and Detlef P. van Vuuren. 2014. A new scenario framework for climate change research: The concept of Shared Socioeconomic Pathways. *Climatic Change* 122: 387–400. https://doi.org/10.1007/s10584-013-0905-2.

Pathak, Pratik, Ajay Kalra, and Sajjad Ahmad. 2017. Temperature and precipitation changes in the Midwestern United States: Implications for water management. *International Journal of Water Resources Development* 33: 1003–1019. https://doi.org/10.1080/07900627.2016.1238343.

Pérez-Blanco, C. Dionisio, Arthur Hrast-Essenfelder, and Chris Perry. 2020. Irrigation technology and water conservation: A review of the theory and evidence. *Review of Environmental Economics and Policy* 14: 216–239. https://doi.org/10.1093/reep/reaa004.

Perry, Chris. 2019. Will irrigation technology, pricing, or quotas ensure sustainable water use? In *The Oxford handbook of food, water and society*, ed. Tony Allan, Brendan Bromwich, Martin Keulertz, and Anthony Colman, 76–96. Oxford University Press. https://doi.org/10.1093/oxfordhb/9780190669799.013.2.

Perry, Chris, Pasquale Steduto, Richard G. Allen, and Charles M. Burt. 2009. Increasing productivity in irrigated agriculture: Agronomic constraints and hydrological realities. *Agricultural Water Management* 96: 1517–1524. https://doi.org/10.1016/j.agwat.2009.05.005.

Perry, Chris, Pasquale Steduto, and Fawzi Karajeh. 2017. *Does improved irrigation technology save water? A review of the evidence.* Cairo: Food and Agriculture Organization of the United Nations.

Piemontese, Luigi, Giulio Castelli, Ingo Fetzer, Jennie Barron, Hanspeter Liniger, Nicole Harari, Elena Bresci, and Fernando Jaramillo. 2020. Estimating the global potential of water harvesting from successful case studies. *Global Environmental Change* 63: 102121. https://doi.org/10.1016/j.gloenvcha.2020.102121.

Qureshi, Asad Sarwar, Mushtaq A. Gill, and Asrar Sarwar. 2010. Sustainable groundwater management in Pakistan: Challenges and opportunities. *Irrigation and Drainage* 59: 107–116. https://doi.org/10.1002/ird.455.

Ramankutty, Navin, Zia Mehrabi, Katharina Waha, Larissa Jarvis, Claire Kremen, Mario Herrero, and Loren H. Rieseberg. 2018. Trends in global agricultural land use: Implications for environmental health and food security. *Annual Review of Plant Biology* 69: 789–815. https://doi.org/10.1146/annurev-arplant-042817-040256.

Ray, Srabashi, Iman Haqiqi, Alexandra E. Hill, J. Edward Taylor, and Thomas W. Hertel. 2023. Labor markets: A critical link between global-local shocks and their impact on agriculture. *Environmental Research Letters* 18: 035007. https://doi.org/10.1088/1748-9326/acb1c9.

Reitz, Meredith, Ward E. Sanford, Gabriel Senay, and J. Cazenas. 2017. Annual estimates of recharge, quick-flow runoff, and ET for the contiguous U.S. using empirical regression equations. *Journal of the American Water Resources Association* 53: 961983. https://doi.org/10.1111/1752-1688.12546.

Roath, Jennifer. 2013. *An evaluation of spatial variability of water stress index across the United States: Implications of supply and demand in the East vs the West.* Thesis, West Lafayette: Purdue University.

Rodell, M., J.S. Famiglietti, D.N. Wiese, J.T. Reager, H.K. Beaudoing, F.W. Landerer, and M.-H. Lo. 2018. Emerging trends in global freshwater availability. *Nature* 557: 651–659. https://doi.org/10.1038/s41586-018-0123-1.

Roy, Sujoy B., Limin Chen, Evan H. Girvetz, Edwin P. Maurer, William B. Mills, and Thomas M. Grieb. 2012. Projecting water withdrawal and supply for future decades in the U.S. under climate change scenarios. *Environmental Science & Technology* 46: 2545–2556. https://doi.org/10.1021/es2030774.

Russo, Tess A., and Upmanu Lall. 2017. Depletion and response of deep groundwater to climate-induced pumping variability. *Nature Geoscience* 10: 105–108. https://doi.org/10.1038/ngeo2883.

Schlenker, Wolfram, and Michael J. Roberts. 2009. Nonlinear temperature effects indicate severe damages to U.S. crop yields under climate change. *Proceedings of the National Academy of Sciences* 106: 15594–15598. https://doi.org/10.1073/pnas.0906865106.

Seckler, David, Randolph Barker, and Upali Amarasinghe. 1999. Water scarcity in the twenty-first century. *International Journal of Water Resources Development* 15: 29–42. https://doi.org/10.1080/07900629948916.

Sophocleous, M. 2000. From safe yield to sustainable development of water resources—the Kansas experience. *Journal of Hydrology* 235: 27–43. https://doi.org/10.1016/S0022-1694(00)00263-8.

Srinivasan, V., E.F. Lambin, S.M. Gorelick, B.H. Thompson, and S. Rozelle. 2012. The nature and causes of the global water crisis: Syndromes from a meta-analysis of coupled human-water studies. *Water Resources Research* 48. https://doi.org/10.1029/2011WR011087.

US Department of Agriculture. 2019. *2017 Census of Agriculture*. Washington, DC: USDA-NASS. https://www.nass.usda.gov/AgCensus/.

van Emmerik, T.H.M., Z. Li, M. Sivapalan, S. Pande, J. Kandasamy, H.H.G. Savenije, A. Chanan, and S. Vigneswaran. 2014. Socio-hydrologic modeling to understand and mediate the competition for water between agriculture development and environmental health: Murrumbidgee River basin, Australia. *Hydrology and Earth System Sciences* 18: 4239–4259. https://doi.org/10.5194/hess-18-4239-2014.

Voss, Katalyn A., James S. Famiglietti, MinHui Lo, Caroline de Linage, Matthew Rodell, and Sean C. Swenson. 2013. Groundwater depletion in the Middle East from GRACE with implications for transboundary water management in the Tigris-Euphrates-Western Iran region. *Water Resources Research* 49: 904–914. https://doi.org/10.1002/wrcr.20078.

Wichelns, Dennis. 2015. Virtual water and water footprints do not provide helpful insight regarding international trade or water scarcity. *Ecological Indicators* 52: 277–283. https://doi.org/10.1016/j.ecolind.2014.12.013.

Xie, Yanhua, Holly K. Gibbs, and Tyler J. Lark. 2021. Landsat-based Irrigation Dataset (LANID): 30 m resolution maps of irrigation distribution, frequency, and change for the US, 1997–2017. *Earth System Science Data* 13: 5689–5710. https://doi.org/10.5194/essd-13-5689-2021.

Zhao, Xin, Dominique Y. van der Mensbrugghe, Roman M. Keeney, and Wallace E. Tyner. 2020. Improving the way land use change is handled in economic models. *Economic Modelling* 84: 13–26. https://doi.org/10.1016/j.econmod.2019.03.003.

Chapter 13
The Role of Labor Markets in Determining the Efficacy and Distributional Impact of Sustainability Policies

Srabashi Ray, Iman Haqiqi, Alexandra E. Hill, J. Edward Taylor, and Thomas W. Hertel

1 Introduction

Labor is a critical input in agriculture and an important determinant of rural household and community welfare in the wake of policies directed at the farm sector. In 2017, labor accounted for more than 30% of production costs in the agricultural sector globally and close to 60% in Sub-Saharan Africa (Aguiar et al. 2019). In the United States, average labor cost as a share of gross farm income (2017–2019) ranges from 25% to 35% for the production of fruits, nuts, vegetables, and other specialty crops (US Department of Agriculture (USDA) 2021). Thus, agriculture brings not only land and water resources but also human resources into the food system.

This chapter is a slightly revised version of a paper originally published as Ray, Srabashi, Iman Haqiqi, Alexandra E. Hill, J. Edward Taylor, and Thomas W. Hertel. 2023. Labor markets: A critical link between global-local shocks and their impact on agriculture. *Environmental Research Letters* 18: 035007. https://doi.org/10.1088/1748-9326/acb1c9.

Data availability statement: The files needed to replicate this application are available at https://gtap.agecon.purdue.edu/simple-g/.

S. Ray (✉) · I. Haqiqi · T. W. Hertel
Center for Global Trade Analysis, Department of Agricultural Economics, Purdue University, West Lafayette, IN, USA
e-mail: ray152@purdue.edu

A. E. Hill
Department of Agricultural and Resource Economics, University of California, Berkeley, CA, USA

J. E. Taylor
Department of Agricultural and Resource Economics, University of California, Davis, CA, USA

© The Author(s) 2025
I. Haqiqi, T. W. Hertel (eds.), *SIMPLE-G*,
https://doi.org/10.1007/978-3-031-68054-0_13

199

Agricultural labour markets operate at a "meso" level that can shape the impacts of conservation policies (e.g., groundwater conservation) at the local level and feed back into national-level changes.[1] Several global modeling frameworks resolve processes with high spatial resolution (i.e., individual grid cells) (Lotze-Campen et al. 2008; Valin et al. 2013; Shin et al. 2016; Baldos et al. 2020). However, the meso level is also important when analyzing sustainability challenges (Johnson et al. 2023).

In this chapter, we highlight the critical role of labor markets in governing the effectiveness and distributional impacts of global food price shocks and local sustainability policies targeting land and water use in agriculture. We begin by reviewing the empirical literature on agricultural labor markets. We then offer insights from economic theory regarding the link between the functioning of labor markets and agricultural outcomes. Finally, we build these components into the SIMPLE-G model of US agriculture developed in Chap. 12 and demonstrate how labor market responses alter the impacts of global price shocks and local sustainability policies on agricultural production, employment, and land use.

Agricultural labor markets in the United States are complex, involving different types of labor with unique characteristics. In the United States, field crops such as maize, soy, and wheat are largely grown in the Midwest and Great Plains, where mechanization has permitted many farms to operate with small labor forces consisting primarily of family labor, supplemented by hired labor from local communities (USDA 2022). In contrast, in the Fruitful Rim, where the bulk of US fruit, vegetable, and horticultural (FVH) farms operate, immigrant labor—particularly from Mexico—is the main source of hired labor. The labor requirements for the FVH sector are highly seasonal. Follow-the-crop migration has traditionally facilitated agricultural production by redistributing agricultural workers across localities and seasons. However, farmworker mobility has decreased significantly over time (Fan et al. 2015), while the overall supply of farm workers from Mexico has declined (Charlton and Taylor 2016). Farms have therefore become more reliant on workers who have settled in nearby localities, and they are more likely to compete with nearby nonfarm businesses for scarce labor. These trends highlight the importance of representing labor markets in a global–local–global framework.

The growing reluctance of workers to relocate to new regions of the country means that wage differentials can emerge and persist across regions. This was evident in nonagricultural labor markets after China's accession to the World Trade Organization (WTO) and rapid growth in Chinese exports over the past two decades (Autor et al. 2016, 2021). Trade economists analyzing the impacts of WTO accession have paid little attention to labor markets (Bhattasali et al. 2004). As a result, few anticipated the slow adjustment of the US manufacturing sector and the depressed regional labor markets that emerged (Autor et al. 2021).

---

[1] The meso level refers to an intermediate scale arising between the aggregated global (or macro) scale and the local (or micro) scale.

We bring these recent labor market insights into the SIMPLE-G framework by explicitly accounting for agricultural workers' mobility. We illustrate the importance of labor market responsiveness by assessing the impacts of the recent commodity price boom and local conservation policies that withdraw overexploited resources from farming. The emphasis is on how labor market rigidities shape the responses to global and local shocks. While the focus here is on the United States, these issues are also relevant in developing countries, where agriculture is more labor intensive and labor market rigidities are prevalent (Campos and Nugent 2018; Konte et al. 2022).

2   Agricultural Labor Markets in the United States

Labor is a key input for planting, weeding, harvesting, and postharvest activities in agriculture. People's willingness to perform farm work at prevailing wages and to engage in follow-the-crop migration determines the availability of labor at particular geographic locations and seasons and thus are key determinants of agricultural production. Environmental stressors and economic conditions affect agricultural labor markets. For example, Jessoe et al. (2018) and Feng et al. (2012) show that environmental shocks on crop yields (i.e., rising temperatures and declining precipitation) cause out-migration from rural areas and decrease farm employment. Fan et al. (2016) show that agricultural wages rise during recessions due to a leftward shift in labor supply from decreased migration and an inelastic farm labor demand (because people must eat). This contrasts with labor market conditions in other industries with high proportions of immigrant workers (e.g., construction), in which both supply and demand are likely to decline during recessions. In the wake of new conservation policies, the optimal behavior of agricultural producers might be to contract and relocate production. However, producers' ability to make these changes depends on workers' mobility.

Historically, the US agricultural workforce has been characterized by a large number of migrant workers willing to travel long distances for employment. Today, fewer individuals are willing to work in agriculture, and those in the labor pool are more settled (Fan et al. 2015; Martin 2017). A variety of factors have contributed to the decline in farm labor supply. These include increases in immigration enforcement (Kostandini et al. 2014; Charlton and Taylor 2016), growing employment opportunities in nonfarm sectors (Richards and Patterson 1998; Martin 2017), and relative changes in economic conditions in the United States and Mexico (Taylor et al. 2012). These factors have contributed to the overall decline in labor supply from Mexico and a reduced willingness to work in US agriculture.

Migrant agricultural workers follow seasonal paths of crop production across large geographic regions. These paths are typically circular: Workers begin in the south early in the year, move northward, and then head south again, following the crop and weather patterns that dictate harvest times (Taylor 1937; Arnedo et al.

2011).[2] In recent years, however, agricultural workers have become less mobile (Fan et al. 2015). More people report being settled in a particular location and working nearby (Martin 2017). The specific causes of this transition to a more settled workforce are not immediately obvious, but the trend correlates with increases in immigration enforcement, which limits historic cross-border migration patterns and increases in the number of workers with families in the United States. It is also linked to the aging of the farm workforce (a consequence of decreasing immigration of young workers from Mexico); growing employment opportunities in nonfarm sectors, which reduce the need to move for employment (Martin 2017); and possibly to welfare policies (Green et al. 2003).

Combined, the decline in agricultural labor supply and reduced intra-US migration have led to a relative scarcity of agricultural labor and widespread reports of worker shortages (Hertz and Zahniser 2013; Bronars 2015; Richards 2018). The implications for US agricultural production are vast. In the short run, agricultural producers seeking to move or expand their production area might be unable to do so if their operations are not located near where many workers live. Moreover, to attract a sufficient workforce in the long run, job opportunities (e.g., wages, hours, and desired skills) must be enticing enough for workers to remain settled in the area.

In sum, employers face stickier labor markets and can no longer rely on having an elastic migrant workforce that arrives at the farm gate when and where they desire. Models that assume that the agricultural labor supply is highly (or perfectly) elastic are likely to present a biased picture of local and global agricultural and environmental impacts of policy interventions.

3   Insights About Labor Market Rigidities from Economic Theory

In Part II, we developed the two-input theoretical framework underlying the SIMPLE framework. Eqs. 3.5, 3.6, 3.7, 3.8, and 3.9 in Box 3.1 show the changes in input use, input prices, and overall production due to any exogenous shock. We can use these equations to understand how the impacts of a policy can vary under different assumptions about agricultural labor markets.

Agricultural labor is subsumed under human inputs in the two-input framework. When agricultural labor markets are assumed to be perfectly elastic, then $v_H \rightarrow \infty$. Under this condition, we can rewrite the changes in natural resource use (Eq. 13.1), human input use (Eq. 13.2), and total production (Eq. 13.3) as follows:

---

[2] In one historic route, West Coast agricultural laborers began in California's Imperial Valley, next to the Mexican border, harvesting truck crops in the winter, moved to the Los Angeles or San Bernardino County areas in the spring, then spent most of the summer and fall in central California near Bakersfield and Fresno (Taylor and Rowell 1938).

$$q_R = p^* \frac{v_R}{\theta_R} - \phi_R \qquad (13.1)$$

$$q_H = v_R \frac{p^*}{\theta_R} + \sigma \frac{p^*}{\theta_R} - \phi_R \qquad (13.2)$$

$$q = p^* \frac{v_R}{\theta_R} + p^* \frac{\sigma \theta_H}{\theta_R} - \phi_R \qquad (13.3)$$

A comparison of these sets of equations shows that if we ignore labor markets, we are foregoing significant richness in our analysis. If we reintroduce labor market rigidities to Eq. 13.3 (i.e., the change in total production), we observe that both the extensive (v_R) and intensive ($\theta_H \sigma$) margin responses are modified by the terms Γ_R and Γ_σ in Eq. 3.9 in Box 3.1. Both the modifying terms are less than 1, indicating that labor market rigidities diminish the grid cell's responsiveness to commodity price changes. The "stickier" the labor market is, the smaller Γ_R and Γ_σ are, leading to a diminished production response. Intuitively, it is more expensive to hire workers in the relatively tight labor market required to manage the resources under production.

Further examination of the terms Γ_R and Γ_σ reveals the crucial role that the labor supply elasticity, v_H, plays in governing local responses to a commodity price boom, particularly the size of this elasticity relative to the supply response of natural inputs, v_R. Given the geographic immobility of natural resources, we expect v_R to be lower than the labor supply response; therefore, $v_H > v_R$. The greater the difference in the supply elasticities of the two inputs, the larger the intensification component of the supply response. If the factor supply elasticities are equal (i.e., $v_H = v_R$), there is no incentive to intensify production in response to the commodity price hike. In this case, both input demands will rise proportionately with output.

When commodity prices increase, the returns to both inputs increase. We expect land and water resources to capture a greater share of the increase in crop prices (i.e., $p_H < p_R$) because producers are willing to pay a higher price for the relatively scarce input. Thus, the relative price elasticity of supply for labor versus natural resources determines the distribution of gains from a higher crop price between the two inputs. In the extreme case of perfect labor mobility—as assumed in most integrated assessment models (IAMs)—labor supply is perfectly elastic, and wages are constant despite increased production. In this case, all of the gains accrue to landowners.

In the case of the groundwater policy shock, as with the commodity market boom, the presence of inelastic labor markets mediates the production impact of this resource conservation policy. Eq. 13.3 shows that the maximum impact of the policy would be realized under perfect labor mobility ($\Gamma_H \to 1$ as $v_H \to \infty$). We saw that the production impact of the crop price boom is smaller when labor market rigidities are introduced. However, in the case of a groundwater conservation policy, the ramifications of ignoring imperfect labor mobility are ambiguous. The direct impact of the policy shock on production is negative (i.e., land and water are forced out of production in targeted grid cells), but the spillover effects are positive (i.e., labor moves from targeted to untargeted cells). Both these effects are overestimated when we ignore labor market rigidities. Therefore, whether ignoring labor market rigidities causes the aggregate effect of the groundwater policy to be over- or underestimated varies by grid cell, depending on the grid-level parameters.

From the theoretical model, it is clear that we need to explicitly model agricultural labor markets to accurately capture the policy impacts, distributional consequences, and—importantly—the effectiveness of the policy in terms of the resources conserved. In the SIMPLE-G-CZ model, we address this problem by considering a range of long-run labor supply elasticity estimates that are supported by the literature (Hill et al. 2021). Shifting migration patterns are also pivotal elements in agricultural labor markets' role in mediating farms' ability to respond to stressors. We expect the decline in the mobility of the agricultural workforce to alter the optimal producer response to crop price shocks and groundwater sustainability policies. The extent to which producers expand or relocate production in response to these shocks will be mediated by the availability of workers at specific locations. Thus, models that assume perfect labor mobility are likely to be biased. Informed by studies of agricultural labor-supply elasticity (Hill et al. 2021), we geographically restrict labor mobility to areas, or laborsheds, in which modern farmworkers are willing to work and compare the results with those under an assumption of perfect labor mobility.[3]

4   SIMPLE-G Version Employed for This Study

In this chapter, we use the SIMPLE-G-CZ (Ray et al. 2023) model (Fig. 13.1), which builds on the foundation of the SIMPLE-G framework outlined in Part III of this book and incorporates rigidities in agricultural labor markets. Specifically, the model in Fig. 13.1 elaborates on the version of SIMPLE-G used in the previous chapter of this book (Chap. 12) by disaggregating the agricultural labor input and introducing geographically limited "laborsheds."

There are multiple empirical and conceptual challenges to defining the geographic extent of agricultural labor markets. The three most widely used geographic delineations of US labor markets are the Office of Management and Budget's core-based statistical areas (CBSAs), the USDA Economic Research Service's commuting zones (CZs), and the Bureau of Economic Analysis's economic areas (EAs). Fowler and Jensen (2020) compare these methods on the extent to which labor markets represent core EAs, the degree of economic connectivity within areas, and the degree to which individuals live and work within the same areas. They conclude that no existing labor market delineation is adequate; geographic delineations must be tailored to the problem at hand.

---

[3] One can think of agricultural labor markets as "laborsheds" in much the same way as hydrologists think about watersheds. Watersheds are shaped by geological features; laborsheds are shaped by farm workers' willingness to move from one place to another in response to changing labor demands.

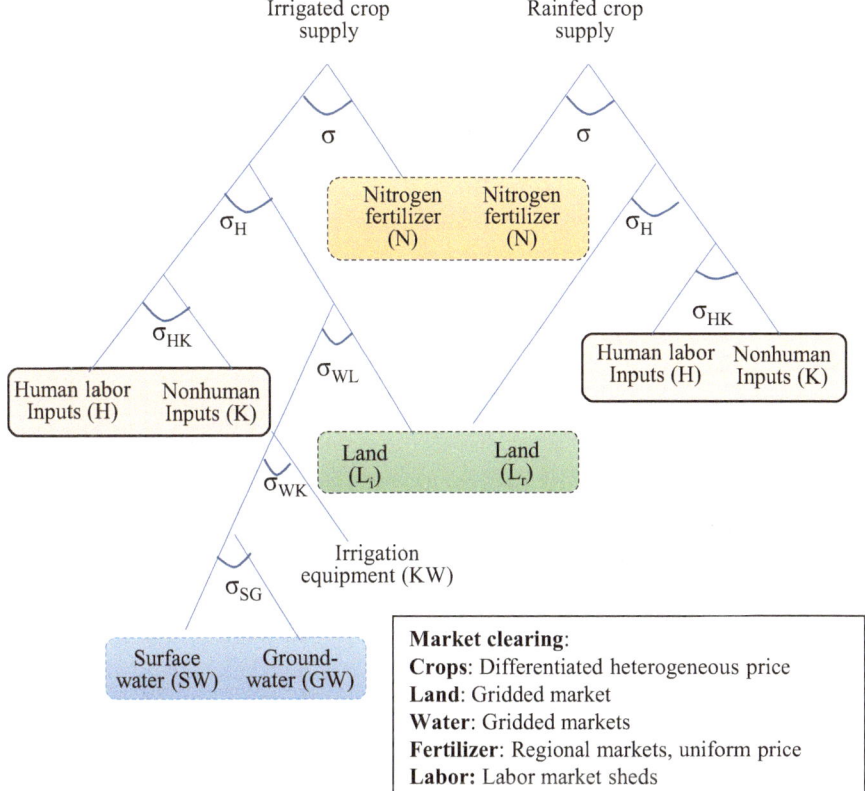

Fig. 13.1   Structure of crop production in SIMPLE-G-CZ

The SIMPLE-G-CZ model is based on SIMPLE-G-US (Baldos et al. 2020). In SIMPLE-G-CZ, "other Inputs" are split into human labor inputs (H) and nonhuman inputs (K), which comprise capital and other production inputs (e.g., seeds and chemicals). Human labor refers to all types of labor, including hired, migrant, and family labor. The SIMPLE-G-CZ model introduces spatially explicit markets for the labor input using the concept of commuting zones (CZs) (Fowler et al. 2016). Each CZ has a labor supply curve parameterized based on the extensive margin labor supply elasticity reported by Hill et al. (2021). Labor demand is determined by profit-maximizing producers in each grid cell. The allocation of labor within a CZ is determined by a quantity-preserving CET function, which determines the equilibrium wages at local laborsheds (i.e., commuting zones). The elasticity of substitution parameters is shown by σ, indexed by the relevant inputs

The SIMPLE-G-CZ model employed in this chapter builds on the literature on CZs (Fowler et al. 2016; Fowler and Jensen 2020). The SIMPLE-G-CZ model incorporates labor market rigidities in the SIMPLE-G framework using estimates of agricultural labor supply elasticity from Hill et al. (2021). In the SIMPLE-G-CZ model, labor markets are clear and wages are determined at the CZ level. We also incorporate within-CZ frictions across grid cells.

We use CZs as the geographical areas over which we delineate the pool of available workers because, among existing labor market delineations, they are most consistent with our research objectives. The CZ methodology, which was first developed by Tolbert and Sizer (1996), is based on central place theory, which assigns counties to nodes based on a hierarchical cluster algorithm that groups counties with strong commuting ties (Carpenter et al. 2022). CZs are widely used in population and labor economic analyses of areas sharing a common labor market as an alternative to counties (which are problematic because they are largely arbitrary political units) and metropolitan statistical areas or CBSAs (which by definition *exclude* nonmetropolitan places and do not span the entire United States). Agricultural workers often cross county and state lines, and the CZ approach is well suited to address this mobility.

5   Experimental Design: Global Change Scenarios

Our experiments consider two cases that allow us to highlight the significance of labor markets in a multiscale, integrated socioeconomic–biophysical framework. The first case examines a global commodity price boom. The second case considers a local sustainable groundwater policy. We simulate the impacts of these two shocks on both local and national production, wages, and employment. For each case, we consider two labor market scenarios: (1) perfect labor mobility across the United States, in keeping with traditional IAMs, and (2) restricted mobility, as discussed in Sect. 2.

5.1   Global Commodity Boom

Agricultural commodity prices are notably volatile, and crop producers must continually form expectations about prices prior to harvest. Commodity prices began to rise sharply at the onset of the Ukraine conflict in early 2022. This was particularly notable for wheat and oilseeds, for which Ukraine and Russia account for a large share of global exports. Due to substitutability in use as well as in production, crop commodities tend to move in tandem, and the prices of other crops also rose during this period. While they have subsequently declined, agricultural commodity prices remain above the level forecast prior to the conflict, and the future for agricultural prices is uncertain. Similar shocks to the global food system occurred in the 2006–2013 period when a combination of rapid growth in biofuel demand, low commodity stocks, and adverse weather events resulted in a series of commodity price spikes (Abbott et al. 2008). Of course, the price that matters for producer profits is the one prevailing at the time of sale, but this price is unknown at the time of planting, hence the importance of expectations. For longer-term decision-making, prices in subsequent years are also relevant. For example, the price of maize prevailing in the spring of 2022 was 19% higher than it had been in the spring of 2021 (World Bank 2022). Price increases vary by crop, and it is unclear whether farmers understood these increases to signal persistently higher prices. For this reason, we study the effect of a commodity price boom using a stylized 10% price hike.

5.2 Sustainable Groundwater Policy in the Western United States

In this chapter, we implement the same groundwater sustainability shock discussed in Chap. 12. Water scarcity is one of the major challenges facing states in the Western United States, where there is widespread unsustainable use of groundwater. As in Chap. 12, we implement a groundwater sustainability policy that restricts groundwater extraction rates to average recharge rates in each grid cell. This policy ensures that the groundwater table does not decline further over the long run. Viewed from a local perspective, the implications of this policy are dramatic, implying up to a 90% reduction in groundwater pumping in some locations. In total, the policy imposes a 66.7% reduction in total groundwater withdrawals in the United States (Haqiqi et al. 2023). Restrictions on groundwater extraction have local and regional economic impacts, and labor markets may play a critical role in transmitting impacts across grid cells. We quantify the consequences of such policies for agricultural production, employment, and land use at local and regional scales.

6 Results

In the cases of both policy shocks, we assume that the perturbation to the system is permanent, as the model is designed to elicit long-run equilibrium responses to shocks. Thus, producers assume that the groundwater sustainability policy will not be reversed, despite the political pressures that inevitably emerge following such policies. In the case of the commodity price boom, we assume that there are structural features causing prices to be persistently higher relative to baseline expectations so that farmers adjust their land, water, and labor usage accordingly. If the commodity price boom is perceived as temporary (as is typical for commodity booms), then the supply response will be smaller than that shown here.

6.1 Impacts of an Agricultural Commodity Price Boom

The price boom has an expansionary effect on national crop output and on all underlying inputs used (Table 13.1). However, the extent of the expansionary effect is starkly different depending on agricultural labor mobility. When labor is perfectly

Table 13.1 National change (%) in crop production, employment, wages, land use, groundwater use, and crop price in the United States

Percentage change in	Perfect labor mobility (SIMPLE-G model)	Restricted labor markets (SIMPLE-G-CZ)
Crop price boom		
Crop production	35.3	9.2
Employment	12.7	4.8
Wages	0	18.6
Land use	1.9	1.3
Groundwater use	9.1	6.4
Sustainable groundwater policy		
Crop production	−2.9	−2.7
Employment	−3.7	0.9
Wages	0	−0.26
Land use	0.3	0.2
Groundwater use	−41.9	−41.5
Crop price	2.2	2.1

mobile, the 10% crop price increase leads to a 35.3% increase in total crop production and a 12.7% increase in employment in the United States. The local-level increase in employment ranges from 12% to 25% across the West and Midwestern United States and rises to more than 25% in the South and Southeastern United States (Fig. 13.2). This supply response to a 10% global price shock is implausibly large, considering estimates in the literature (Lee and Helmberger 1985; Keeney and Hertel 2009). In contrast, total crop production and employment increase by 9.2% and 4.8%, respectively, when labor market rigidities are included in the model. Local-level increases in employment range from 2% to 12% (Fig. 13.2).

Under the imperfect labor mobility scenario, land use expands along the margins of the Corn Belt as well as parts of the Southern and Eastern United States, where agriculture is less reliant on scarce groundwater than in the Western United States (Fig. 13.2f). This is consistent with recent observations in the wake of the biofuel boom (Lark et al. 2015). In contrast, there is almost no change in cropland use in the Western United States.

Under restricted labor mobility, the smaller increase in land use in response to the commodity price hike is associated with a smaller increase in groundwater use: 6.4%, compared with 9.1% under perfect labor mobility (Fig. 13.3). Differences in land use between the two labor market scenarios are small compared with differences in production and employment because—based on empirical estimates—both models limit the mobility of land across uses (Villoria and Liu 2018).

Agricultural labor markets mediate the regional impacts of the global crop price boom on agricultural workers. With perfect labor mobility, agricultural wages do not change, whereas wage increases are evident given more realistic, sticky labor market assumptions. In the presence of sticky labor markets, national agricultural wages increase by 18.6%. However, the CZs in the Corn Belt that respond to the global

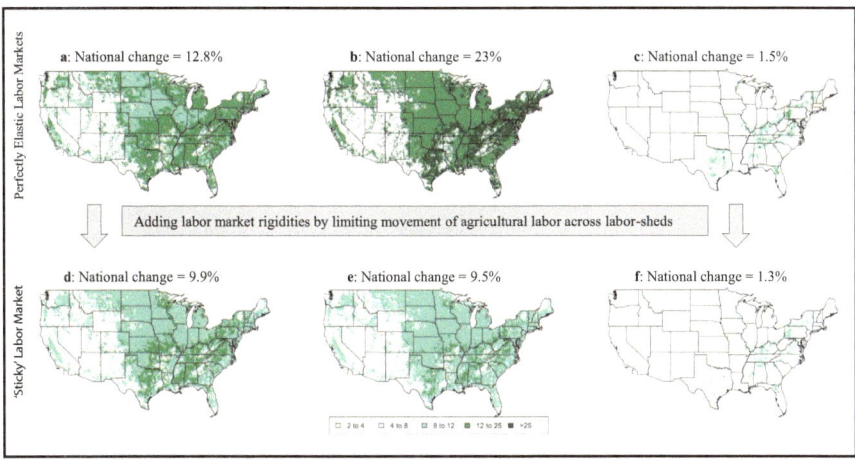

Fig. 13.2 Grid-cell level response to a permanent 10% price hike as a percentage change in (**a**) production, (**b**) employment, and (**c**) land use under a perfectly elastic labor market contrasted with responses in (**d**) production, (**e**) employment, and (**f**) land use under restricted labor mobility

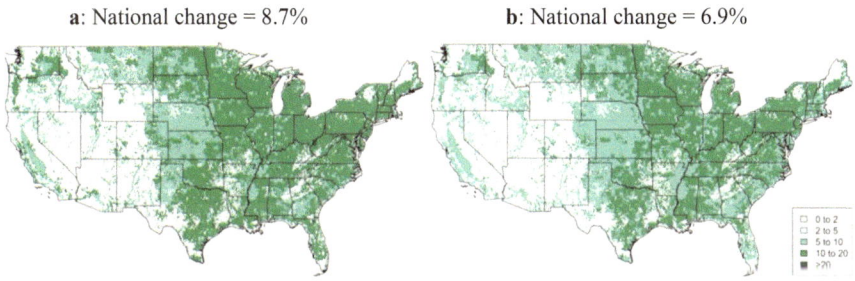

Fig. 13.3 Changes in groundwater use due to a global price shock in (**a**) perfectly elastic and (**b**) "sticky" labor markets

price shock with the strongest increases in crop production experience a wage increase of 20–30% (Fig. 13.4). The structure of agricultural labor markets plays a significant role in determining the impacts of global shocks on local-level farmworkers.

6.2   Impacts of a Sustainable Groundwater Conservation Policy

The sustainable groundwater policy affects primarily the Western United States, where irrigated agriculture relies on severely overexploited groundwater resources. The aggregate impacts of the policy in terms of reduction in crop production

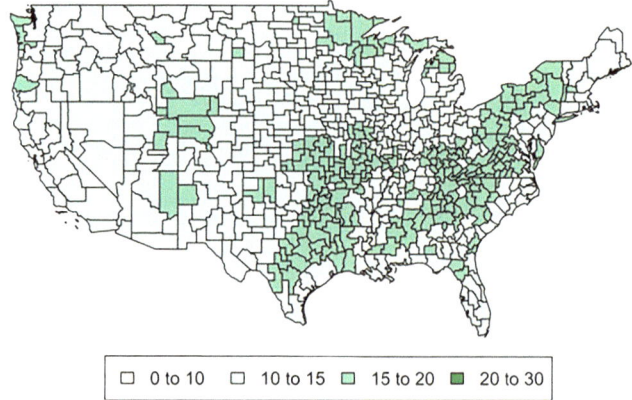

| ☐ 0 to 10 | ☐ 10 to 15 | ☐ 15 to 20 | ■ 20 to 30 |

Fig. 13.4 Impact of crop price shock on wages at the commuting zone level (restricted labor mobility model)
Note that the change in wages is concentrated in the Corn Belt, with a relatively smaller impact in the western United States, where agriculture is relatively more reliant on overexploited groundwater

(2.7–2.9%) and the change in land use (an increase of less than 1%) are comparable across the two assumptions about the structure of agricultural labor markets. However, if we ignore labor market rigidities, the national-level reduction in employment is overestimated: 3.7% compared with 0.9% (Table 13.1).

Due to the reduction in national crop production following the groundwater conservation policy, US crop prices increase by roughly 2% under both policies. This increase affects all producers, leading to spillover effects in grids that are not directly targeted by the policy. These spillover effects could contribute to the overexploitation of groundwater resources in the nontargeted regions. (We do not explore this overexploitation here.) The national impact of this policy on groundwater use under both labor market scenarios is a 42% reduction in national groundwater use. While the simulation reduces groundwater withdrawals by 66.7% in the directly affected cells, the increase in national crop prices stimulates an increase in groundwater use in locations not targeted by the policy. Thus, the net reduction in groundwater use, accounting for the spillover effects, is lower: just 42%.

As in the case of the global price shock, the local impacts of a groundwater conservation policy can vary significantly from what appears to be small impacts at the national level. In the directly affected regions of the Central Valley of California, crop production falls by up to 50% and employment decreases significantly (Fig. 13.5). The magnitude of the policy impact on employment depends on the structure of the local agricultural labor markets. Under perfect labor mobility, the fall in employment closely matches the corresponding fall in grid-level production. However, under restricted labor mobility, employment falls by up to 15% in California's Central Valley (Fig. 13.5).

The spillover effects, on both production and employment, in the indirectly affected grid cells are also overestimated under the assumption of perfect labor

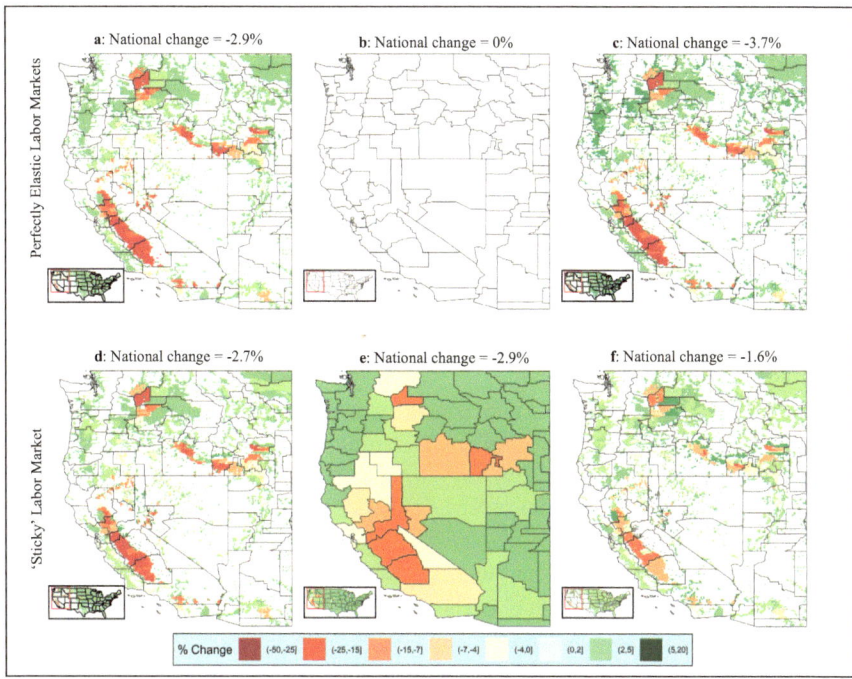

Fig. 13.5 Impact of US sustainable groundwater policy shock as a percentage change in (**a**) crop production, (**b**) employment, and (**c**) wages in the western United States under perfect mobility versus impact on (**d**) crop production, (**e**) employment, and (**f**) wages under restricted labor mobility Figure **b** is blank because there is no change in wages in this scenario. Borders within the wage change maps show the commuting zone delineations across the western United States

mobility. Grids that are not directly targeted by the groundwater policy in a CZ expand production due to the higher crop prices. These increases in production also lead to an increase in employment. When we consider labor market rigidities, the spillover effects on employment generate up to a 5% increase in employment in most grids (Fig. 13.5). This effect is estimated to reach 20% under perfect labor mobility. The spillover effects of a conservation policy give farmworkers the opportunity to find employment in neighboring grid cells. If farmworkers' mobility is restricted, the spillover effects are also limited.

By assumption, the perfect labor mobility case completely overlooks the potential wage impacts (Fig. 13.5b) of conservation policies. Figure 13.5e shows that agricultural wages decrease by 7–50% in the Central Valley and parts of the Snake River Basin, where the groundwater conservation policy is targeted. Farmworkers who lose their jobs when unsustainable use of natural resources is restricted are likely to absorb wage cuts if alternative opportunities for employment are limited.

7 Discussion

This chapter documents an important first step toward integrating labor markets into multiscale sustainability analyses. We detail conceptually why agricultural labor markets are increasingly likely to influence agriculture's ability to respond to a variety of factors, including changes in global commodity prices and local conservation policies. In recent years, US agricultural workers have become less mobile, contributing to geographically smaller and more contained laborsheds. Our simulations demonstrate how assumptions about labor market mobility shape the impacts of market and policy changes. Labor markets can be crucial determinants of how agricultural production changes over time and across space in response to economy-wide, sectoral, or regional shocks.

Incorporating labor market responses can improve the accuracy and utility of multiscale models of the economy and environment (e.g., SIMPLE-G). However, challenges remain, and future research should be directed toward addressing them. A top priority is to develop better estimates of the underlying parameters used in the model. Given the importance of factor supply responses within this framework, estimating the agricultural labor supply response within and across laborsheds is a priority for future research. There are few reliable agricultural labor-supply elasticity estimates (Hill et al. 2021), and land and groundwater supply elasticities can also be improved. Work is currently underway to estimate cropland supply elasticities at the grid cell level and use those elasticities to validate the model's predictions over a historical period (Villoria et al. 2022). This will be an important advance.

An important adaptation to groundwater restrictions involves changes in crop mix that imply changes in the labor intensity of production. Although incorporating dozens of crops into a gridded model of agriculture would be appealing, the absence of necessary data and parameters would pose major challenges to such an endeavor. However, the proposed framework approximates the interplay between multiple crop types with different water needs by estimating an aggregate relationship between the water intensity of all crops grown in a given region and water availability. Crop mix also plays a role on the demand side of this analysis, where we have assumed a single, national crop price. Chap. 5 and other chapters of this book develop an approach that allows for the differentiation of the composite crop by the USDA production region. This is also an important advance.

There are potential limitations to using CZs as geographical restrictions on labor mobility. The CZ methodology is based on commuting patterns for the general US population, the majority of whom do not regularly move for employment throughout the year. Most workers commute from their homes to nearby workplaces (e.g., living in a suburb and commuting to a city center). The CZ methodology is not designed to depict the movements of follow-the-crop migrants. While this type of worker movement has become less common in recent years, follow-the-crop migrants still represent an important subset of the agricultural workforce and are crucial for ensuring a sufficient labor supply when and where it is needed. One emerging solution to the decline in migrant workers is the H-2A visa program for temporary agricultural workers. H-2A visa holders are beholden to specific agricultural employers for a duration that is agreed upon prior to the workers' arrival. As these workers cannot move from employer to employer, we omit them from our current analysis; in effect, their labor market area encompasses the home country and a single US employer (usually a farmer or farmer association, but in some cases a labor contractor). All of these constraints point to the need for future work to build on our framework and explore the implications of richer delineations of agricultural labor markets.

In summary, we believe that there is great potential for the explicit modeling of agricultural labor markets to contribute to our understanding of the impacts of global food price shocks as well as current debates over the agriculture–environment interface and the equity implications of conservation policies. The results of this study show that the assumption of a less than perfectly elastic labor supply can significantly alter conclusions from multiscale sustainability modeling regarding the effectiveness and distributional consequences of global shocks as well as local conservation policies. We see this work as a first step toward understanding how labor markets mediate between local and global economies, thereby enabling the refinement of integrated assessment models with richer representations of labor markets.

Acknowledgments and Competing Interests The authors acknowledge support from USDA-AFRI grants #NIFA-2022-67023-36403, "Labor Markets and the Impacts of Environmental Stresses and Conservation Policies on US Agriculture," and #2019-67023-29679, "Economic Foundations of Long Run Agricultural Sustainability," and the National Science Foundation INFEWS award #1855937, "Identifying Sustainability Solutions through Global-Local-Global Analysis of a Coupled Water-Agriculture-Bioenergy System."

The findings and conclusions presented in this chapter are those of the authors and should not be construed to represent any official determination or policy of the US Department of Agriculture (USDA), or the National Science Foundation (NSF). Furthermore, we declare that there is no conflict of interest related to this work.

References

Abbott, Philip C., Christopher Hurt, and Wallace E. Tyner. 2008. *What's driving food prices?* Issue report 37951. Farm Foundation. https://doi.org/10.22004/ag.econ.37951.

Aguiar, Angel, Maksym Chepeliev, Erwin L. Corong, Robert McDougall, and Dominique van der Mensbrugghe. 2019. The GTAP data base: Version 10. *Journal of Global Economic Analysis* 4: 1–27. https://doi.org/10.21642/JGEA.040101AF.

Arnedo, A, S Rose, and M Borges. 2011. Mapping migration. *Mexican Migration and Apple Mosaic*. https://blogs.dickinson.edu/latinomosaic/history-of-the-apple/. Accessed 4 Feb 2024.

Autor, David H., David Dorn, and Gordon H. Hanson. 2016. The China shock: Learning from labor-market adjustment to large changes in trade. *Annual Review of Economics* 8: 205–240. https://doi.org/10.1146/annurev-economics-080315-015041.

Autor, David, David Dorn, and Gordon Hanson. 2021. On the persistence of the China shock. *Brookings Papers on Economic Activity* Fall: 381–447.

Baldos, Uris Lantz C., Iman Haqiqi, Thomas W. Hertel, Mark Horridge, and J. Liu. 2020. SIMPLE-G: A multiscale framework for integration of economic and biophysical determinants of sustainability. *Environmental Modelling & Software* 133: 104805. https://doi.org/10.1016/j.envsoft.2020.104805.

Bhattasali, Deepak, Shantong Li, and William J. Martin. 2004. *China and the WTO: Accession, policy reform, and poverty reduction strategies.* Washington, DC: World Bank. https://doi.org/10.1596/0-8213-5667-4.

Bronars, Stephen G. 2015. *A vanishing breed: How the decline in U.S. farm laborers over the last decade has hurt the U.S. economy and slowed production on American farms.* New York: Partnership for a New American Economy.

Campos, Nauro F., and Jeffrey B. Nugent. 2018. The dynamics of the regulation of labour in developing and developed countries since 1960. In *The political economy of structural reforms in Europe*, ed. Nauro F. Campos, Paul De Grauwe, and Yuemei Ji, 75–88. Oxford University Press. https://doi.org/10.1093/oso/9780198821878.003.0003.

Carpenter, Craig Wesley, Michael C. Lotspeich-Yadao, and Charles M. Tolbert. 2022. When to use commuting zones? An empirical description of spatial autocorrelation in U.S. counties versus commuting zones. *PLoS One* 17: e0270303. https://doi.org/10.1371/journal.pone.0270303.

Charlton, Diane, and J. Edward Taylor. 2016. A declining farm workforce: Analysis of panel data from rural Mexico. *American Journal of Agricultural Economics* 98: 1158–1180. https://doi.org/10.1093/ajae/aaw018.

Fan, Maoyong, Susan Gabbard, Anita Alves Pena, and Jeffrey M. Perloff. 2015. Why do fewer agricultural workers migrate now? *American Journal of Agricultural Economics* 97: 665–679. https://doi.org/10.1093/ajae/aau115.

Fan, Maoyong, Anita Alves Pena, and Jeffrey M. Perloff. 2016. Effects of the great recession on the U.S. agricultural labor market. *American Journal of Agricultural Economics* 98: 1146–1157. https://doi.org/10.1093/ajae/aaw023.

Feng, Shuaizhang, Michael Oppenheimer, and Wolfram Schlenker. 2012. *Climate change, crop yields, and internal migration in the United States*, w17734. Cambridge, MA: National Bureau of Economic Research. https://doi.org/10.3386/w17734.

Fowler, Christopher S., and Leif Jensen. 2020. Bridging the gap between geographic concept and the data we have: The case of labor markets in the USA. *Environment and Planning A: Economy and Space* 52: 1395–1414. https://doi.org/10.1177/0308518X20906154.

Fowler, Christopher S., Danielle C. Rhubart, and Leif Jensen. 2016. Reassessing and revising commuting zones for 2010: History, assessment, and updates for U.S. "labor-sheds" 1990–2010. *Population Research and Policy Review* 35: 263–286. https://doi.org/10.1007/s11113-016-9386-0.

Green, Richard D., Philip L. Martin, and J. Edward Taylor, ed. 2003. Welfare reform in agricultural California. *Journal of Agricultural and Resource Economics* 28: 169–183. https://doi.org/10.22004/ag.econ.30715.

Haqiqi, Iman, Laura Bowling, Sadia Jame, Uris Baldos, Jing Liu, and Thomas Hertel. 2023. Global drivers of local water stresses and global responses to local water policies in the United States. *Environmental Research Letters* 18: 065007. https://doi.org/10.1088/1748-9326/acd269.

Hertz, Tom, and Steven Zahniser. 2013. Is there a farm labor shortage? *American Journal of Agricultural Economics* 95: 476–481. https://doi.org/10.1093/ajae/aas090.

Hill, Alexandra E., Izaac Ornelas, and J. Edward Taylor. 2021. Agricultural labor supply. *Annual Review of Resource Economics* 13: 39–64. https://doi.org/10.1146/annurev-resource-101620-080426.

Jessoe, Katrina, Dale T. Manning, and J. Edward Taylor. 2018. Climate change and labour allocation in rural Mexico: Evidence from annual fluctuations in weather. *The Economic Journal* 128: 230–261. https://doi.org/10.1111/ecoj.12448.

Johnson, Justin Andrew, Molly E. Brown, Erwin Corong, Jan Philipp Dietrich, Roslyn C. Henry, Patrick José von Jeetze, David Leclère, Alexander Popp, Sumil K. Thakrar, and David R. Williams. 2023. The meso scale as a frontier in interdisciplinary modeling of sustainability from local to global scales. *Environmental Research Letters* 18: 025007. https://doi.org/10. 1088/1748-9326/acb503.

Keeney, Roman, and Thomas W. Hertel. 2009. The indirect land use impacts of United States biofuel policies: The importance of acreage, yield, and bilateral trade responses. *American Journal of Agricultural Economics* 91: 895–909. https://doi.org/10.1111/j.1467-8276.2009. 01308.x.

Konte, Maty, Wilfried A. Kouamé, and Emmanuel B. Mensah. 2022. Structural reforms and labor productivity growth in developing countries: Intra or inter-reallocation channel? *The World Bank Economic Review* 36: 646–669. https://doi.org/10.1093/wber/lhac002.

Kostandini, Genti, Elton Mykerezi, and Cesar Escalante. 2014. The impact of immigration enforcement on the U.S. farming sector. *American Journal of Agricultural Economics* 96: 172–192. https://doi.org/10.1093/ajae/aat081.

Lark, Tyler J., J. Meghan Salmon, and Holly K. Gibbs. 2015. Cropland expansion outpaces agricultural and biofuel policies in the United States. *Environmental Research Letters* 10: 044003. https://doi.org/10.1088/1748-9326/10/4/044003.

Lee, David R., and Peter G. Helmberger. 1985. Estimating supply response in the presence of farm programs. *American Journal of Agricultural Economics* 67: 193–203. https://doi.org/10.2307/ 1240670.

Lotze-Campen, Hermann, Christoph Müller, Alberte Bondeau, Stefanie Rost, Alexander Popp, and Wolfgang Lucht. 2008. Global food demand, productivity growth, and the scarcity of land and water resources: A spatially explicit mathematical programming approach. *Agricultural Economics* 39: 325–338. https://doi.org/10.1111/j.1574-0862.2008.00336.x.

Martin, Philip L. 2017. *Immigration and farm labor: Challenges and opportunities.* Berkeley: Giannini Foundation of Agricultural Economics.

Ray, Srabashi, Iman Haqiqi, Alexandra E. Hill, J. Edward Taylor, and Thomas W. Hertel. 2023. Labor markets: A critical link between global-local shocks and their impact on agriculture. *Environmental Research Letters* 18: 035007. https://doi.org/10.1088/1748-9326/acb1c9.

Richards, Timothy J. 2018. Immigration reform and farm labor markets. *American Journal of Agricultural Economics* 100: 1050–1071. https://doi.org/10.1093/ajae/aay027.

Richards, Timothy J., and Paul M. Patterson. 1998. Hysteresis and the shortage of agricultural labor. *American Journal of Agricultural Economics* 80: 683–695. https://doi.org/10.2307/1244056.

Shin, Jaewoo, Christoph Müller, and Joshua Elliot. 2016. *Global gridded crop model evaluation tool.* https://mygeohub.org/resources/ggcmevaluation. Accessed 26 Mar 2018.

Taylor, Paul S. 1937. Migratory farm labor in the United States. *Monthly Labor Review* 44: 537–549.

Taylor, Paul S., and Edward J. Rowell. 1938. Patterns of agricultural labor migration within California. *Monthly Labor Review* 47: 980–990.

Taylor, J. Edward, Diane Charlton, and Antonio Yúnez-Naude. 2012. The end of farm labor abundance. *Applied Economic Perspectives and Policy* 34: 587–598. https://doi.org/10.1093/ aepp/pps036.

Tolbert, Charles M., and Molly Sizer. 1996. *U.S. commuting zones and labor market areas: A 1990 update,* Staff Paper 9614. Washington, DC: US Department of Agriculture, Economic Research Service. https://doi.org/10.22004/ag.econ.278812.

USDA-ERS. 2021. *Labor cost share of total gross revenues.* Washington, DC: US Department of Agriculture, Economic Research Service. https://www.ers.usda.gov/topics/farm-economy/ farm-labor/#laborcostshare. Accessed 4 Feb 2024.

———. 2022. *Farm labor.* Washington, DC: US Department of Agriculture, Economic Research Service. https://www.ers.usda.gov/topics/farm-economy/farm-labor.aspx. Accessed 4 Feb 2024.

Valin, Hugo, Petr Havlík, Niklas Forsell, Stefan Frank, Aline Mosnier, Daan Peters, Carlo Hemlinck, Matthis Spöttle, and Maarten van den Berg. 2013. *Description of the GLOBIOM (IIASA) model and comparison with the MIRAGE-BioF (IFPRI) model.* Laxenburg: International Institute for Applied Systems Analysis.

Villoria, Nelson B., and Jing Liu. 2018. Using continental grids to improve understanding of global land supply responses: Implications for policy-driven land use changes in the Americas. *Land Use Policy* 75: 411–419. https://doi.org/10.1016/j.landusepol.2018.04.010.

Villoria, Nelson B., Alfredo Cisneros-Pineda, Iman Haqiqi, Shourish Chakravarty, Michael Delgado, and Thomas W. Hertel, ed. 2022. Heterogeneous land supply responses in U.-S. agriculture: Exploring changes in land use from reductions in biofuel mandates. In *Annual Meeting*. Anaheim: Agricultural and Applied Economics Association. https://doi.org/10.22004/ag.econ.322315.

World Bank. 2022. *World Bank commodities price data* (The Pink Sheet). https://thedocs. worldbank.org/en/doc/5d903e848db1d1b83e0ec8f744e55570-0350012021/related/CMO-Pink-Sheet-August-2022.pdf. Accessed 4 Feb 2024.

Chapter 14
Tackling Policy Leakage and Targeting Hot Spots Could Be Key to Addressing the "Wicked" Challenge of Nutrient Pollution from Corn Production in the United States

Jing Liu, Laura Bowling, Christopher Kucharik, Sadia Jame, Uris Lantz C. Baldos, Larissa Jarvis, Navin Ramankutty, and Thomas W. Hertel

1 Introduction

Widespread and intensive agricultural activity has resulted in the loss of large amounts of nitrogen (N) from soils (Goolsby et al. 2001; Turner et al. 2007). Elevated N levels in streams and rivers cause a spectrum of challenging problems, including biodiversity loss and threatened human health (Vitousek et al. 1997). Nutrients transported through the Mississippi River Basin (MRB) have been blamed for what are referred to as "dead zones" (i.e., hypoxic or low-oxygen water) that have formed in the Gulf of Mexico (Rabalais et al. 2001; Diaz and Rosenberg 2008). The

This chapter is a slightly revised version of a paper originally published as Liu, Jing, Laura Bowling, Christopher Kucharik, Sadia Jame, Uris Lantz C. Baldos, Larissa Jarvis, Navin Ramankutty, and Thomas W. Hertel. 2023. Tackling policy leakage and targeting hotspots could be key to addressing the "wicked" challenge of nutrient pollution from corn production in the U.S. *Environmental Research Letters* 18: 105002. https://doi.org/10.1088/1748-9326/acf727.

Data availability statement: The files needed to replicate this application are available at https://gtap.agecon.purdue.edu/simple-g/.

J. Liu (✉) · U. L. C. Baldos · T. W. Hertel
Center for Global Trade Analysis, Department of Agricultural Economics, Purdue University, West Lafayette, IN, USA
e-mail: liu207@purdue.edu

L. Bowling
Department of Agronomy, Purdue University, West Lafayette, IN, USA

C. Kucharik
Department of Agronomy, University of Wisconsin, Madison, WI, USA

© The Author(s) 2025
I. Haqiqi, T. W. Hertel (eds.), *SIMPLE-G*,
https://doi.org/10.1007/978-3-031-68054-0_14

217

largest hypoxic zone measured since 1985 was 22,730 km^2 (8776 square miles) in 2017 (US EPA Hypoxia Task Force). Reducing the size of this zone to an acceptable level by 2035 will require a reduction of 48% in total nitrogen and phosphorus load (Fennel and Laurent 2018; US EPA 2023).

It is widely recognized that there is no silver bullet for resolving the "wicked" problem of nonpoint source water pollution in the Mississippi watershed (Shortle and Horan 2017; McLellan et al. 2018). To achieve the 48% nutrient reduction goal, in-field nutrient management must be combined with edge-of-field measures as well as downstream nutrient removal practices (Schilling and Wolter 2009; Iowa Nutrient Reduction Strategy 2013; McLellan et al. 2015, 2018). While agronomic and environmental management techniques to control and remove lost N have advanced, there is limited evidence that existing policies effectively facilitate the adoption of these techniques (Shortle et al. 2012; McLellan et al. 2015; Roy et al. 2021). Programs to promote improved water quality in the United States have been found to be largely inefficient, as the incremental cost of water quality protection has exceeded the incremental benefits (Olmstead 2010; Laukkanen and Nauges 2014; Savage and Ribaudo 2016). This low efficiency is often attributed to a failure to identify the proper value of N effluent mitigation (Shortle et al. 2012; Biffi et al. 2021; Fleming et al. 2022). The uniform value assumed in the current policy design does not reflect the spatially varying marginal cost of mitigating water quality damages (Shortle and Horan 2017). Quantifying this cost is challenging in practice because nonpoint source pollution is often not measurable. Without knowing the site-specific biophysical and ecological characteristics of N loss, economic instruments cannot be efficiently deployed.

This chapter introduces a special version of the SIMPLE-G model, called SIM-PLE-G-US-CS, that overcomes these problems by estimating and embedding key biophysical relationships in an economic model. Using this integrated multiscale analytic tool, we compare the effectiveness of various policies in reducing nitrate loading in the MRB and the spatial patterns of mitigation.

S. Jame
Department of Agricultural and Biological Engineering, Purdue University, West Lafayette, IN, USA

L. Jarvis
McGill Sustainability Systems Initiative, McGill University, QC, Canada

N. Ramankutty
School of Public Policy and Global Affairs, University of British Columbia, Vancouver, BC, Canada

2   SIMPLE-G Version Employed for This Study

The version of the model that we created for this application is called SIMPLE-G-US-CS, where CS stands for "corn–soy." Instead of aggregating production across all crops, this model focuses on the two dominant crops in the Midwest: corn and soybeans. Corn is a very N-intensive crop that is often rotated with soybeans. According to the 2017 Census of Agriculture (USDA NASS), these two crops account for 77% of the annual harvested area in the MRB watershed, and 42% of basin-wide N fertilizer use was attributed to the production of corn. The corn–soy model advances the general SIMPLE-G framework by introducing grid-cell-level N loss and crop yield response parameters estimated from the Agro-IBIS model (Kucharik 2003; Donner and Kucharik 2008; Kucharik et al. 2013), as shown in Fig. 14.1. Specifically, the yield responses to N simulated by Agro-IBIS are used to compute the elasticity of substitution (σ) between N fertilizer and augmented land by grid cell and irrigation type. The N loss processes simulated by Agro-IBIS translate the economic equilibrium level of N application into N loss. More information about the two models, validation, and coupling of the two can be found in the supplemental information of Liu et al. (2023).

3   Experimental Design

Our experimental design considers four strategies individually and in combination to study the impacts of different conservation options, as highlighted in Fig. 14.2. The first strategy is an N loss tax that increases the cost of N fertilizer application in proportion to the estimated N loss rate for a given practice in that grid cell. N loss refers to the N fertilizer nutrient that is applied but not taken up by the crop and subsequently leaves the root zone. The final cost is determined by the nationally uniform N fertilizer price (US$/kg of N applied)[1] and the product of a nationwide tax rate of US$1/kg of N loss and the N loss intensity (kg of N loss/kg of N application), which varies by location and practice. After being adjusted by the N loss intensity, the tax imposes the highest penalty on the heavy polluters, whose profit margin will be affected directly by the tax and indirectly by the adverse yield impacts of less N fertilizer application.

Unlike the N loss tax, which reduces N use and nitrate loss via higher input costs, our second strategy achieves the same goal by applying less fertilizer but using it more efficiently. We select two relatively easy—and therefore more likely to be

---

[1] The N loss tax is a hypothetical nationwide policy that is expected to yield comparable mitigation outcomes as the other nationwide policy—split N application. The base tax rate is set uniformly nationwide for the sake of practical necessity. It is further adjusted by the N loss intensity to create the site-specific N loss tax rate such that the final rate is higher for the locations with higher N loss intensity. See Section B of the SI in Liu et al. (2023) for more information.

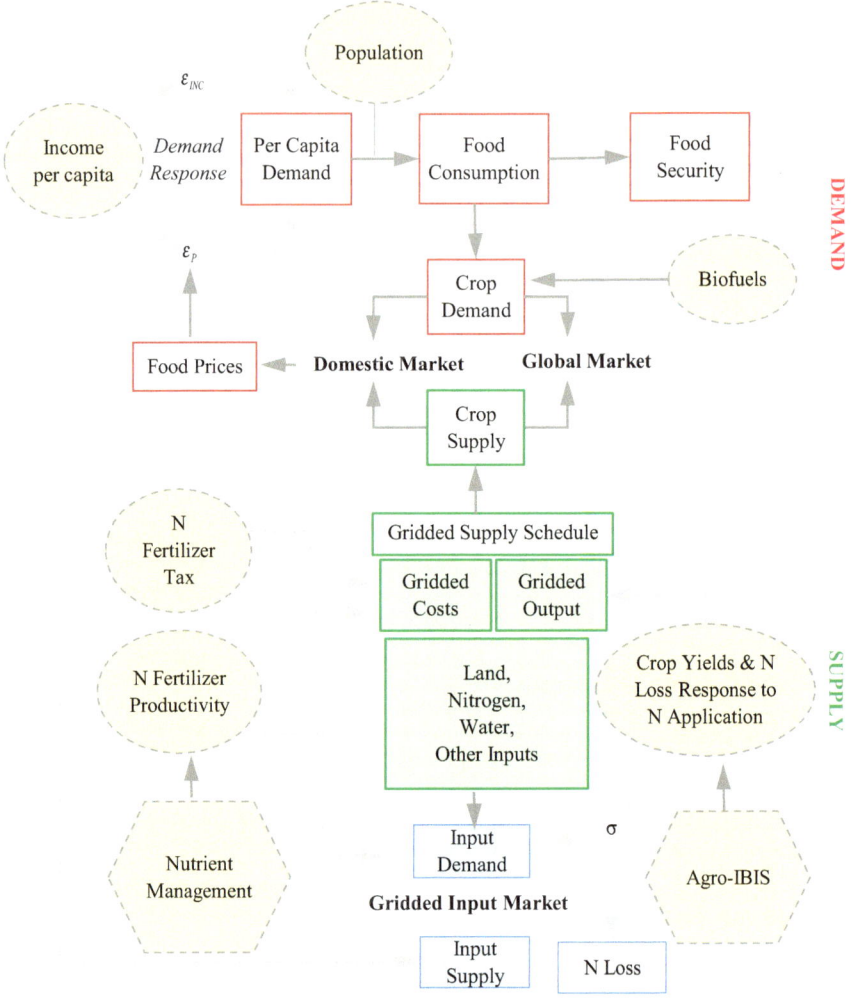

Fig. 14.1 Schematic of the SIMPLE-G-US-CS model
SIMPLE-G-US-CS is a modification of the standard SIMPLE-G model (Fig. 4.1). Key modifica-
tions include the interaction between Agro-IBIS and SIMPLE-G and accounting for nitrogen
(N) loss resulting from N input use

adopted—practices: split N and side-dressing.[2] According to the Iowa Nutrient
Reduction Strategy (2013), moving from fall to spring pre-planting application
and side-dressing application reduce nitrate-N loss by an average of 6% and 4%,
respectively. These improved practices are implemented in the model by increasing

---

[2] Split N means that growers make two or more N fertilizer applications during the growing season
rather than supplying all of the crop's N requirements with a single treatment prior to or at planting.
Side-dressing refers to applying fertilizers in a shallow furrow or band along the side of row crops.

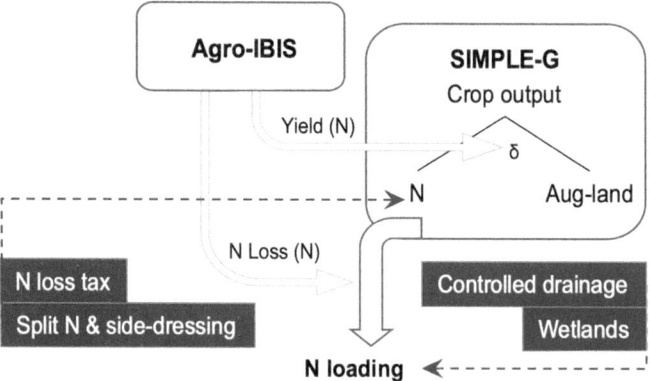

Fig. 14.2 Connections between Agro-IBIS and SIMPLE-G-US-CS and nitrogen (N) loss mitigation policies

Yield (N) and N loss (N) are univariate transfer functions (with respect to N) through which biophysical characteristics are embedded into the economic model. The N loss tax and split N affect N application rates and therefore N loading through the transfer functions. Controlled drainage and wetland restoration affect nitrate loads mainly through post-application nitrate removal

the productivity of N fertilizer to reduce N loss by 10% while keeping the baseline crop output unchanged.

The other two mitigation strategies focus on locally feasible nutrient management practices: controlled drainage[3] and wetland restoration.[4] Both practices yield spatially varying N loss removal rates that are determined by local conditions (e.g., water runoff, subsurface-drained area, and soil and vegetation characteristics). These strategies do not affect N fertilizer application directly but remove pollutants after application before they enter a stream. Additional information for each conservation effort can be found in the supplemental information of Liu et al. (2023).

4 Results

We implement the four policies individually as well as together in the SIMPLE-G-US-CS and explore the results, both at the grid cell level and at more aggregate levels, including states and over the entire Mississippi Basin.

---

[3] Controlled drainage uses a water control structure to adjust the depth of the subsurface drainage outlet in order to control water in the field.

[4] Wetlands in this paper refer to constructed integrated systems that use the natural functions of vegetation, soil, and microorganisms as well as the environment to improve water quality.

Table 14.1 Nitrogen (N) loss reduction outcomes, impacts on crop output and price, and mitigation efficiency across management strategies

Policy	N application Million tons	Crop output Million tons	Crop price US$/ton	N loss Million tons	Efficiency US$/ kg N	kg N/ ha
N loss tax	−0.65 (−5.91%)	−9.98 (−2.08%)	4.16 (+1.84%)	−0.34 (−9.02%)	10.09	5.30
Split N	−0.75 (−6.80%)	0.00 (0.00%)	0.00 (0.00%)	−0.42 (−11.18%)	3.57	6.57
Controlled drainage	−0.02 (−0.17%)	−0.88 (−0.18%)	0.36 (+0.16%)	−0.46 (−12.15%)	0.79	31.70
Wetlands	−0.05 (−0.47%)	−2.38 (−0.55%)	1.18 (+0.48%)	−0.58 (−15.41%)	1.81	27.26
Tax + Split N	−1.35 (−12.19%)	−8.99 (−1.87%)	3.74 (+1.65%)	−0.71 (−18.86%)	6.94	11.1
Tax + Split N + Controlled drainage	−1.28 (−11.63%)	−4.78 (−1.00%)	1.97 (+0.87%)	−1.04 (−27.66%)	5.08	16.3
Tax + Split N + Wetlands	−1.36 (−12.30%)	−9.64 (−2.01%)	4.01 (+1.78%)	−1.17 (−31.00%)	5.12	18.2

Mitigation efficiency, measured in US$/kg, indicates the economic efficiency in terms of the direct cost incurred to reduce 1 unit of N loss. Note that this measure abstracts from the potential uses of the tax revenues. The efficiency, measured in kg/ha, indicates the biophysical efficiency regarding the potential for N loss removal per cropland area

4.1 Combining N Loss Tax, Split N Fertilizer Application, and Wetland Restoration Has the Potential to Reduce N Loss from Corn Production by over 30%

Among the four strategies explored, wetland restoration appears to be the most effective single strategy, reducing N loss from corn production by 15%, followed by controlled drainage (12%) (Table 14.1). When combined, wetland restoration—along with the N loss tax and split N application—can raise the reduction potential to 31%. An N loss tax of US$1/kg of N loss boosts the average cost of N fertilizer to corn farms by 28.9% and reduces N fertilizer use by 6% and total N loss by 9%.[5] National crop yields are barely affected by the rate reduction, falling slightly from 7.48 to 7.39 corn-equivalent tons per hectare. However, the local effects are more significant. Figure S2 in Liu et al. (2023) shows that the potential crop yield declines

---

[5]Results of different tax rates from US$0.1–1/kg of N loss are reported in Section G of the supplemental information of Liu et al. (2023). The N loss charge (in US$/kg of N application) is computed as the product of a charge rate (in US$/kg of N loss) and nitrate-N loss rate (kg of N loss/ kg of N application). For example, if a farm loses 30% of the N fertilizer applied, the actual cost of applying 1 kg of N fertilizer increases from the base price of US$1/kg to US$1.3/kg, which includes the US$0.3/kg N loss charge. The 28.9% simply represents the aggregated N loss rate at the national level.

most around the edges of the MRB, where there is a higher dependence on supplementary fertilization is higher. Farms located in the Great Plains are least affected by the tax because of their relatively lower N loss intensity.

At the aggregate level, postapplication treatments such as controlled drainage and wetland construction result in much larger N loss reductions (31 and 27 kg of N/ha, respectively) than split N application and the N loss tax (7 and 5 kg of N/ha, respectively).[6] Removing 1 kg of N loss costs US$1.80 when accomplished via wetlands and US$0.80 when using controlled drainage.[7] It is more costly (US$3.60) to mitigate N loss through the adoption of side-dressing and split N application and even more costly (US$10) by imposing a pollution tax, which is calculated by dividing the total N loss tax collected through N fertilizer sales by the amount of N loss reduced. It is important to note that this economic accounting differs from the implementation costs associated with other practices and must be interpreted with caution should readers wish to compare costs across practices. In addition, the tax revenue can be recycled to support pollution abatement, which could lower the actual cost of the policy. The outcomes of alternative tax recycling schemes (e.g., cutting the existing tax on capital income or subsidizing additional programs to further enhance the mitigation effect) have been more extensively studied in the context of carbon taxation (Timilsina 2018). This work has yet to be done for nutrient management. More information about how each cost is calculated can be found in Liu et al. (2023, supplement Section F).

Crop output falls in almost all cases, albeit modestly, due to the higher input costs associated with the rising N fertilizer price, infrastructure installation and maintenance, or forgone cropland (relevant to wetlands only). The composite corn–soy price increases by no more than 2%, regardless of the scenario, given the modest change in crop output. The sum of the individual scenarios' output and price effects is greater than when they are implemented in concert, indicating the presence of complementarities among the policies.

4.2 The Most Effective Nitrate Loss Mitigation Policy Varies by Location

While total mitigation across the four individual strategies is comparable, the spatial pattern of the N loss reductions varies remarkably (Fig. 14.3). The amount of mitigation is relatively consistent across the US Corn Belt for N loss tax and split

---

[6]These results are generally comparable to those recorded in the CEAP regional reports, although a straight comparison between the two may not be reasonable given the difference of the actions considered in each study.

[7]This number accounts only the cost for the control system but not the installation of the subsurface-drains itself due to lack of information. The cost of the latter varies depending on the spacing and depth of the drainage pipes.

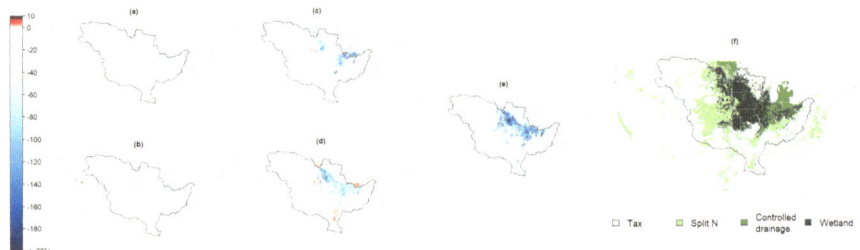

Fig. 14.3 Changes in nitrogen (N) loss under (**a**) an N loss tax, (**b**) split N application, (**c**) controlled drainage, (**d**) wetland restoration, and (**e**) combined strategies of tax, split N, and wetland restoration

Units are tons of N loss per 5 arcmin grid cells. A negative value indicates N loss reduction. Figure **f** shows the most effective single strategy at each grid cell. The maps include only the grid cells where corn and soybeans are grown in the United States

N application, as these are not tied to specific locations. This finding stands in marked contrast to the patterns associated with strategies that are contingent on local conditions. Controlled drainage is only possible in locations where subsurface drainage is installed. Wetland restoration in our analysis is limited to locations where hydric soils and subsurface drainage are present. Results show that N loss mitigation per grid cell is much higher in the heart of the Corn Belt, where controlled drainage and wetland restoration are more prevalent. Compared to N loss tax and split N application, these two strategies also lead to much higher N removal rates.

The gridded results reported in Fig. 14.3a–d allow us to identify the single practice among the four that exhibits the largest N loss reduction (in terms of tons N per grid cell) at each location (Fig. 14.3f). Controlled drainage and wetland restoration dominate the Corn Belt as the most effective practices, except at the western edge, where split N is more effective. Outside of the Corn Belt, the N loss tax stands out as the most effective strategy, especially in the Eastern United States, under the current setting of the experiments (e.g., tax rate, the extent of N fertilizer productivity being increased, and the spatial extent of controlled drainage and restorable wetlands). This stems from a combination of factors, including relatively high N loss intensity and marginal productivity of N applications, as well as the reduced prevalence of subsurface drainage and restorable wetlands in this region.

4.3 Pairing Nationwide Strategies with Site-Specific Conservation Practices Can Remedy Counterproductive Policy Spillovers

Conservation systems such as controlled drainage and wetland restoration incur an additional cost of US$10–US$20 per acre. Despite being a small share of the US $450/acre nonland cost of producing corn (e.g., in Central Illinois circa 2010,

Schnitkey et al. 2021), it could still reduce profitability and curb output on adopting farms. By removing land from production, wetland restoration could be more costly, although some lands are intentionally retired due to their low productivity. Considering both factors, output on adopting farms and demand for N fertilizer are likely to fall. When aggregated to the national level, the local effect could boost the corn price while curbing the price of N fertilizer due to the weakened demand for fertilizer. In the long run, the elevated corn price will induce production expansion and additional N application elsewhere.

Figure 14.3c, d clearly show this spatial spillover effect: N loss around the fringes of the Corn Belt rises in response to higher corn prices. Because there is little subsurface drainage in these fringe areas (Valayamkunnath et al. 2020), less of the increased N loss will contribute directly to the hypoxia problem in the Gulf of Mexico, but it could result in groundwater contamination.

To quantify the accumulated spillover effects, we decompose the overall change in N loss into two components: mitigation (a decrease in N loss) and spillovers (an increase in N loss). Although the magnitude of spillover effect is relatively small compared to the mitigation potential, the additional N applications driven by spillovers are more environmentally harmful. This extra N application leads to higher N loss intensity regardless of the measurement method. For example, on average, 41% of the additional N applied to the untreated cropland area is lost, compared to a 33% N loss rate on the same land before wetland restoration is introduced. Both N losses per hectare of cropland and per ton of crop output increase significantly. However, these spillovers are sharply reduced when policies are combined (Fig. 14.3e), as the uniform coverage greatly limits market-mediated leakage.

4.4   Targeting N Loss Hot Spots Would Make Conservation Efforts More Efficient and Cost-Effective

Substantial reductions in N loss are spatially concentrated in areas with extensive corn acreage, intensive N fertilizer use, and/or effective conservation practices. We find that, across the four practices studied, less than 10% of the total 48,317 grid cells contribute 50% of the mitigation, with only a small reduction in crop output (5% or less) (Fig. 14.4). These top-mitigating grid cells shown in Fig. S11 of Liu et al. (2023) account for 39.4% of the corn–soy area and 38.8% of US corn–soy output. They also use 42.5% of N fertilizer and produce 46.7% of N loss in US corn–soy production. Implementing a combined strategy of tax, split N, and wetland restoration to reduce N loss by 30% would cost US$6 billion annually, or approximately US$38/acre/year. Focusing on this 10% of the grid cells that contribute half of the 30% N loss reduction would reduce costs to US$2.4 billion/year, while removing the other half of the 30% would cost more (US$3.6 billion/year) due to lower N removal efficiency. At higher adoption rates, reducing an additional unit of N costs more—

and N removal per hectare also declines—because more "expensive" locations are included, where either the N loss intensity is low, the marginal product of N is high, or both.

Since environmental policies are typically set at the state or federal level rather than by individual grid cells, we also report state-level mitigation potentials in Fig. 14.4. Collectively, these nine selected states produce 80% of US corn and soy output and use 83% of the N fertilizer applied to corn production. These states also account for 80–85% of total N loss reduction under the tax and split N strategies, and almost all the reduction under the controlled drainage and wetland scenarios. Controlled drainage is especially effective for Iowa, Illinois, Indiana, Minnesota, and Ohio, where subsurface drainage is widely used (Valayamkunnath et al. 2020). Wetland restoration is also effective across most of the Corn Belt. However, due to spillover effects, these policies may increase N loss in states lacking controlled drainage and wetland systems. Not surprisingly, policy combinations can outperform individual policies, and this difference is particularly pronounced in Iowa, Illinois, and Minnesota.

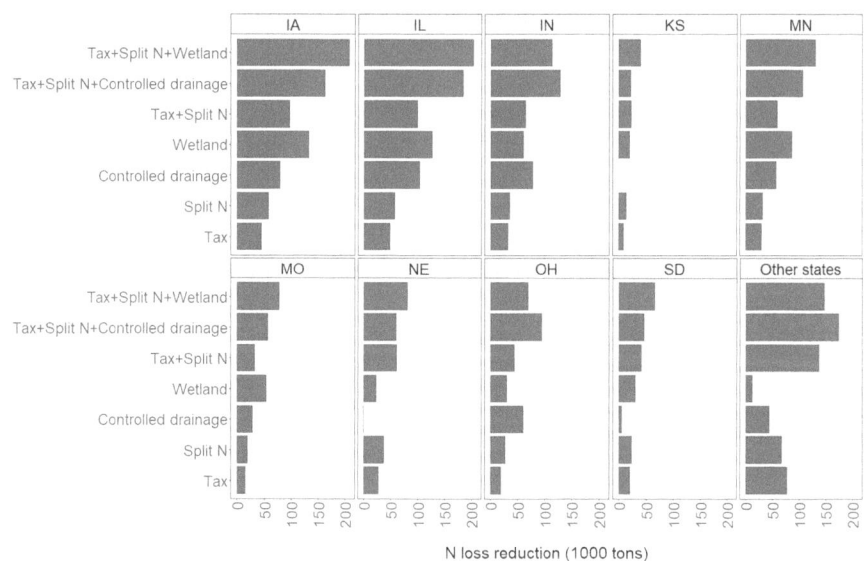

Fig. 14.4 N loss reduction by (**a**) state and mitigation strategy and (**b**) accumulated percentage In Fig. **b**, mitigation at the grid-cell level is first sorted in descending order and then accumulated. Therefore, the order of the grid cells varies by policy. The *light gray, dotted horizontal line* indicates a 50% reduction in total nitrate load

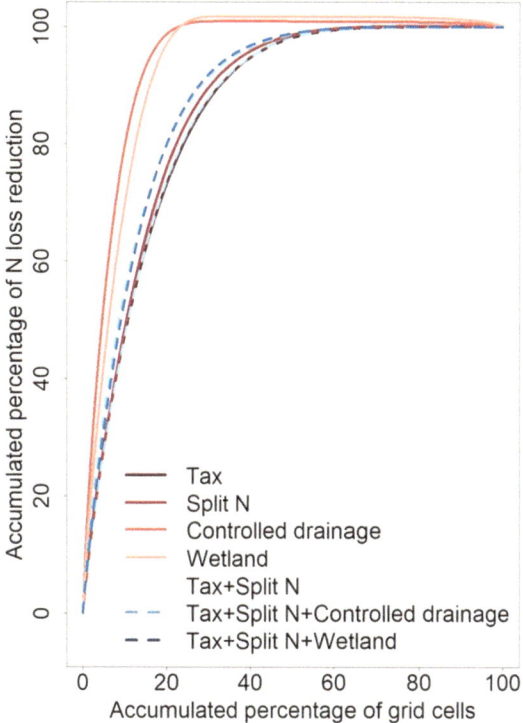

Fig. 14.4 (continued)

5   Discussion

There are several limitations to this study that warrant further investigation. First, we focus only on N loss through water and do not consider nitrous oxide emissions. Therefore, we have not comprehensively evaluated the effectiveness of these conservation policies. A recent study suggests that policies targeting water quality also provide substantial co-benefits by reducing nitrous oxide emissions (Weng et al. 2024). Second, our estimation of the area feasible for controlled drainage and restorable wetlands focus on regions with high potential but does not cover the entire continental United States due to limited data at the time of analysis. Alternative data sources, such as AgTile-US (Valayamkunnath et al. 2020) and Potentially Restorable Wetlands on Agricultural Land provided by EPA EnviroAtlas, could help estimate feasibility in future studies. Third, the N loss reported in our study differs from the amount of nitrate that reaches the Gulf of Mexico, and we do not consider N legacy (Van Meter et al. 2016, 2018; Basu et al. 2022). Both will depend on local hydrological and biogeochemical processes, which will affect the amount of nitrate ultimately reaching the Gulf (Masuda et al. 2021). In future work, it will be valuable to integrate our multiscale analytical framework with hydroecological models to track nutrient movement through ecosystems.

Our study contributes several advancements to the literature. The wider impact of local decisions, or the spillover effect, is the most intriguing result we would like to emphasize. When some—but not all—farms are targeted, the comparative advantage of farms is altered, and the change is transmitted by prices in input and output markets across different scales, leading to unintended displacement of crop production and pollution. Similar "leakage effects" associated with spatially targeted environmental policy intervention have been well recognized in the deployment of climate (Hertel and Tyner 2013; Dou et al. 2018) and air pollution (Fang et al. 2019) policies, but the relevant literature is sparse in the context of water pollution except for a few economic studies (Turner et al. 1999; Xu et al. 2022). A growing concern about governance is that the hidden external cost outside of the target area could offset direct gains (Rajagopal and Zilberman 2013; Dou et al. 2018). However, the spillover phenomenon in policy-making remains poorly understood (Bastos Lima et al. 2019). The coupling of SIMPLE-G-US-CS with Agro-IBIS helps unravel this mechanism by explicitly characterizing production technologies and biophysical characteristics across locations. We find that the spillover effects depend on the cost burden of conservation on farmers. The higher the farmers' burden, the larger the ensuing output reduction and market-mediated spillovers. The magnitude of the spillover depends on the specifics of the policy implementation, including forgone production value, copayments required of farmers, and adoption rates. In our case, leakage is still strongly outweighed by the mitigation efforts but could hinder conservation goals and raise equity and efficiency concerns.

This caveat, however, should not deter the targeting of policy interventions that have been linked to efficiency gains and extensively recommended by the literature (Babcock et al. 1997; van der Horst 2007). Our model's ability to identify areas with high mitigation efficiency provides an empirical foundation for shaping policies that enhance the effectiveness of these interventions (van der Horst 2007). The finding that the leakage effect can be mitigated by combining nationwide and regional strategies offers fresh insights for future policy design. Additionally, our agroecosystem-supported economic model can be used to explore differentiated taxes or subsidies that discourage excessive fertilizer application and compensate farmers for behaviorial changes. While numerous studies have confirmed the effectiveness of conservation practices in improving water quality, much less evidence supports the idea that existing conservation programs succeed in enrolling low-cost adopters or achieve wide adoption of these practices. One possible reason why undifferentiated policies fail is the mismatch between the policy-authorized payment and farmers' expectations, particularly when there are potential negative impacts on yields. And this concern is not unfounded. Roy et al. (2021) show that N application rates in many Midwest counties remain below the N input break point, beyond which crop yield plateaus or declines. Our grid-cell-based analysis also finds locations where nutrient deficiencies could limit crop yields. Assessing this concern requires a thorough understanding of site-specific N balance, the uncertainties associated with climate, production technologies, and the prices of crops and inputs (especially N fertilizer).

6 Conclusion

By integrating a version of SIMPLE-G with the agroecosystem model Agro-IBIS, we evaluate the effectiveness of four conservation strategies—N loss tax, split N application, controlled drainage, and wetland restoration, both individually and in combination—to manage nitrate-N loss from U.S. corn production. Collectively, these practices could reduce N loss from US corn production by 30%, at an estimated annual cost of US$6 billion. Several studies (Rabotyagov et al. 2010, 2014; Tallis et al. 2019; Xu et al. 2022) have reported similar levels of expenditure to achieve comparable mitigation goals.[8] The mitigation effect of each practice varies significantly across regions, highlighting the importance of spatial targeting for both the selection of practices and locations to improve the cost-effectiveness. This aligns with findings from a growing body of research on this topic (Kurkalova 2015; Lintern et al. 2020; Hansen et al. 2021). A successful transition from research to policy and practice requires innovative policy design. Future policies should also consider regulating fertilizer products as a complement to voluntary, farmer-oriented conservation programs. For example, Kanter and Searchinger (2018) propose a municipal minimum sales share of enhanced efficiency fertilizers, akin to how Corporate Average Fuel Economy Standards regulate auto manufacturers to improve fuel efficiency rather than trying to regulate millions of individual drivers. Our modeling framework can be extended to evaluate and test these alternative policy options

Competing Interests The authors acknowledge support from USDA-AFRI grants #2016-67007-24957 and #2019-67023-29679 and NSF grants CBET INFEWS/T2-1855937, INFEWS/T1-1855996, and AccelNet OISE-2020635.

The findings and conclusions presented in this chapter are those of the authors and should not be construed to represent any official determination or policy of the US Department of Agriculture (USDA), or the National Science Foundation (NSF). Furthermore, we declare that there is no conflict of interest related to this work.

---

[8]These include the US$1.4 billion/year to reduce N loading in the Upper MRB by 30% through in-field and edge-of-field practices and retirement of land (Rabotyagov et al. 2010), US$2.6 billion/year through market and regulatory instruments to reduce N flows in the Ohio River Basin and Upper MRB by 25% (Tallis et al. 2019), US$2.7 billion/year to reduce the hypoxic zone in the Gulf of Mexico to 5000 km^2 through cropland conservation and fertilizer management practices (Rabotyagov et al. 2014), and US$6 billion/year in opportunity costs in terms of the value of forgone crop production by changing N fertilizer intensification and crop acreage in order to reduce N runoff from crop production to the Gulf by 45% (Xu et al. 2022).

References

Babcock, Bruce A., P.G. Lakshminarayan, JunJie Wu, and David Zilberman. 1997. *Targeting tools for the purchase of environmental amenities*, Staff General research papers archive. Ames: Iowa State University, Department of Economics.

Basu, Nandita B., Kimberly J. Van Meter, Danyka K. Byrnes, Philippe Van Cappellen, Roy Brouwer, Brian H. Jacobsen, Jerker Jarsjö, et al. 2022. Managing nitrogen legacies to accelerate water quality improvement. *Nature Geoscience* 15: 97–105. https://doi.org/10.1038/s41561-021-00889-9.

Biffi, Sofia, Rebecca Traldi, Bart Crezee, Michael Beckmann, Lukas Egli, Dietrich Epp Schmidt, Nicole Motzer, et al. 2021. Aligning agri-environmental subsidies and environmental needs: A comparative analysis between the US and EU. *Environmental Research Letters* 16: 054067. https://doi.org/10.1088/1748-9326/abfa4e.

Diaz, Robert J., and Rutger Rosenberg. 2008. Spreading dead zones and consequences for marine ecosystems. *Science* 321: 926–929. https://doi.org/10.1126/science.1156401.

Donner, Simon D., and Christopher J. Kucharik. 2008. Corn-based ethanol production compromises goal of reducing nitrogen export by the Mississippi River. *Proceedings of the National Academy of Sciences* 105: 4513–4518. https://doi.org/10.1073/pnas.0708300105.

Dou, Yue, Ramon Felipe Bicudo Da Silva, Hongbo Yang, and Jianguo Liu. 2018. Spillover effect offsets the conservation effort in the Amazon. *Journal of Geographical Sciences* 28: 1715–1732. https://doi.org/10.1007/s11442-018-1539-0.

Fang, Delin, Bin Chen, Klaus Hubacek, Ruijing Ni, Lulu Chen, Kuishuang Feng, and Jintai Lin. 2019. Clean air for some: Unintended spillover effects of regional air pollution policies. *Science Advances* 5: eaav4707. https://doi.org/10.1126/sciadv.aav4707.

Fennel, Katja, and Arnaud Laurent. 2018. N and P as ultimate and proximate limiting nutrients in the northern Gulf of Mexico: Implications for hypoxia reduction strategies. *Biogeosciences 15* (10): 3121–3131.

Fleming, P.M., K. Stephenson, A.S. Collick, and Z.M. Easton. 2022. Targeting for nonpoint source pollution reduction: A synthesis of lessons learned, remaining challenges, and emerging opportunities. *Journal of Environmental Management* 308: 114649. https://doi.org/10.1016/j.jenvman.2022.114649.

Goolsby, Donald A., William A. Battaglin, Brent T. Aulenbach, and Richard P. Hooper. 2001. Nitrogen input to the Gulf of Mexico. *Journal of Environmental Quality* 30: 329–336. https://doi.org/10.2134/jeq2001.302329x.

Hansen, Amy T., Todd Campbell, Se Jong Cho, Jonathan A. Czuba, Brent J. Dalzell, Christine L. Dolph, Peter L. Hawthorne, et al. 2021. Integrated assessment modeling reveals near-channel management as cost-effective to improve water quality in agricultural watersheds. *Proceedings of the National Academy of Sciences* 118: e2024912118. https://doi.org/10.1073/pnas.2024912118.

Hertel, Thomas W., and Wallace E. Tyner. 2013. Market-mediated environmental impacts of biofuels. *Global Food Security* 2: 131–137. https://doi.org/10.1016/j.gfs.2013.05.003.

Iowa State University. 2013. *Iowa nutrient reduction strategy*. Ames: Iowa Department of Agriculture and Land Stewardship, Iowa Department of Natural Resources, and Iowa State University College of Agriculture and Life Sciences.

Kanter, David R., and Timothy D. Searchinger. 2018. A technology-forcing approach to reduce nitrogen pollution. *Nature Sustainability* 1: 544–552.

Kucharik, Christopher J. 2003. Evaluation of a process-based agro-ecosystem model (Agro-IBIS) across the U.S. Corn Belt: Simulations of the interannual variability in maize yield. *Earth Interactions* 7: 1–33. https://doi.org/10.1175/1087-3562(2003)007<0001:EOAPAM>2.0.CO;2.

Kucharik, Christopher J., Andy VanLoocke, John D. Lenters, and Melissa M. Motew. 2013. Miscanthus establishment and overwintering in the Midwest USA: A regional modeling study of crop residue management on critical minimum soil temperatures. *PLoS One* 8: e68847. https://doi.org/10.1371/journal.pone.0068847.

Kurkalova, Lyubov A. 2015. Cost-effective placement of best management practices in a watershed: Lessons learned from conservation effects assessment project. *JAWRA Journal of the American Water Resources Association* 51: 359–372. https://doi.org/10.1111/1752-1688. 12295.

Laukkanen, Marita, and Céline Nauges. 2014. Evaluating greening farm policies: A structural model for assessing agri-environmental subsidies. *Land Economics* 90: 458–481. https://doi. org/10.3368/le.90.3.458.

Lima, Bastos, G. Mairon, U. Martin Persson, and Patrick Meyfroidt. 2019. Leakage and boosting effects in environmental governance: A framework for analysis. *Environmental Research Letters* 14: 105006. https://doi.org/10.1088/1748-9326/ab4551.

Lintern, Anna, Lauren McPhillips, Brandon Winfrey, Jonathan Duncan, and Caitlin Grady. 2020. Best management practices for diffuse nutrient pollution: Wicked problems across urban and agricultural watersheds. *Environmental Science & Technology* 54: 9159–9174. https://doi.org/ 10.1021/acs.est.9b07511.

Liu, Jing, Laura Bowling, Christopher Kucharik, Sadia Jame, Lantz C. Uris, Larissa Jarvis Baldos, Navin Ramankutty, and Thomas W. Hertel. 2023. Tackling policy leakage and targeting hotspots could be key to addressing the "wicked" challenge of nutrient pollution from corn production in the U.S. *Environmental Research Letters* 18: 105002. https://doi.org/10.1088/ 1748-9326/acf727.

Masuda, Yuta J., Seth C. Harden, Pranay Ranjan, Chloe B. Wardropper, Collin Weigel, Paul J. Ferraro, Sheila M.W. Reddy, and Linda S. Prokopy. 2021. Rented farmland: A missing piece of the nutrient management puzzle in the Upper Mississippi River Basin? *Journal of Soil and Water Conservation* 76: 5A–9A. https://doi.org/10.2489/jswc.2021.1109A.

McLellan, Eileen, Dale Robertson, Keith Schilling, Mark Tomer, Jill Kostel, Doug Smith, and Kevin King. 2015. Reducing nitrogen export from the Corn Belt to the Gulf of Mexico: Agricultural strategies for remediating hypoxia. *JAWRA Journal of the American Water Resources Association* 51: 263–289. https://doi.org/10.1111/jawr.12246.

McLellan, Eileen L., Keith E. Schilling, Calvin F. Wolter, Mark D. Tomer, Sarah A. Porter, Joe A. Magner, Douglas R. Smith, and Linda S. Prokopy. 2018. Right practice, right place: A conservation planning toolbox for meeting water quality goals in the Corn Belt. *Journal of Soil and Water Conservation* 73: 29A–34A. https://doi.org/10.2489/jswc.73.2.29A.

Olmstead, Sheila M. 2010. The economics of water quality. *Review of Environmental Economics and Policy* 4: 44–62. https://doi.org/10.1093/reep/rep016.

Rabalais, Nancy N., R. Eugene Turner, and William J. Wiseman Jr. 2001. Hypoxia in the Gulf of Mexico. *Journal of Environmental Quality* 30: 320–329. https://doi.org/10.2134/jeq2001. 302320x.

Rabotyagov, Sergey, Todd Campbell, Manoj Jha, Philip W. Gassman, Jeffrey Arnold, Lyubov Kurkalova, Silvia Secchi, Hongli Feng, and Catherine L. Kling. 2010. Least-cost control of agricultural nutrient contributions to the Gulf of Mexico hypoxic zone. *Ecological Applications* 20: 1542–1555. https://doi.org/10.1890/08-0680.1.

Rabotyagov, Sergey S., Todd D. Campbell, Michael White, Jeffrey G. Arnold, M. Jay Atwood, Lee Norfleet, Catherine L. Kling, et al. 2014. Cost-effective targeting of conservation investments to reduce the northern Gulf of Mexico hypoxic zone. *Proceedings of the National Academy of Sciences* 111: 18530–18535. https://doi.org/10.1073/pnas.1405837111.

Rajagopal, Deepak, and David Zilberman. 2013. On market-mediated emissions and regulations on life cycle emissions. *Ecological Economics* 90: 77–84. https://doi.org/10.1016/j.ecolecon.2013. 03.006.

Roy, Eric D., Courtney R. Hammond Wagner, and Meredith T. Niles. 2021. Hot spots of opportunity for improved cropland nitrogen management across the United States. *Environmental Research Letters* 16: 035004. https://doi.org/10.1088/1748-9326/abd662.

Savage, Jeff, and Marc Ribaudo. 2016. Improving the efficiency of voluntary water quality conservation programs. *Land Economics* 92: 148–166. https://doi.org/10.3368/le.92.1.148.

Schilling, Keith E., and Calvin F. Wolter. 2009. Modeling nitrate-nitrogen load reduction strategies for the Des Moines River, Iowa using SWAT. *Environmental Management* 44: 671–682. https://doi.org/10.1007/s00267-009-9364-y.

Schnitkey, Gary, Carl Zulauf, Nick Paulson, and Krista Swanson. 2021. 2022 Break-even prices for corn and soybeans. *farmdoc daily* 11: 168.

Shortle, James, and Richard D. Horan. 2017. Nutrient pollution: A wicked challenge for economic instruments. *Water Economics And Policy* 03: 1650033. https://doi.org/10.1142/S2382624X16500338.

Shortle, James S., Marc Ribaudo, Richard D. Horan, and David Blandford. 2012. Reforming agricultural nonpoint pollution policy in an increasingly budget-constrained environment. *Environmental Science & Technology* 46: 1316–1325. https://doi.org/10.1021/es2020499.

Tallis, Heather, Stephen Polasky, Jessica Hellmann, Nathaniel P. Springer, Rich Biske, Dave DeGeus, Randal Dell, et al. 2019. Five financial incentives to revive the Gulf of Mexico dead zone and Mississippi basin soils. *Journal of Environmental Management* 233: 30–38. https://doi.org/10.1016/j.jenvman.2018.11.140.

Timilsina, Govinda R. 2018. *Where is the carbon tax after thirty years of research?* Policy research working papers. Washington, DC: World Bank. https://doi.org/10.1596/1813-9450-8493.

Turner, R. Kerry, Stavros Georgiou, Ing-Marie Gren, Fredric Wulff, Scott Barrett, Tore Söderqvist, Ian J. Bateman, et al. 1999. Managing nutrient fluxes and pollution in the Baltic: An interdisciplinary simulation study. *Ecological Economics* 30: 333–352. https://doi.org/10.1016/S0921-8009(99)00046-4.

Turner, B.L., Eric F. Lambin, and Anette Reenberg. 2007. The emergence of land change science for global environmental change and sustainability. *Proceedings of the National Academy of Sciences* 104: 20666–20671. https://doi.org/10.1073/pnas.0704119104.

US Environmental Protection Agency (EPA). 2023. *Mississippi River/Gulf of Mexico watershed nutrient task force.* 2023 Report to Congress. October 2023. https://www.epa.gov/system/files/documents/2023-11/10305_2023-htf-report-to-congress_508.pdf

Valayamkunnath, Prasanth, Michael Barlage, Fei Chen, David J. Gochis, and Kristie J. Franz. 2020. Mapping of 30-meter resolution tile-drained croplands using a geospatial modeling approach. *Scientific Data* 7: 257. https://doi.org/10.1038/s41597-020-00596-x.

van der Horst, Dan. 2007. Assessing the efficiency gains of improved spatial targeting of policy interventions; the example of an Agri-environmental scheme. *Journal of Environmental Management* 85: 1076–1087. https://doi.org/10.1016/j.jenvman.2006.11.034.

Van Meter, K.J., N.B. Basu, J.J. Veenstra, and C.L. Burras. 2016. The nitrogen legacy: Emerging evidence of nitrogen accumulation in anthropogenic landscapes. *Environmental Research Letters* 11: 035014. https://doi.org/10.1088/1748-9326/11/3/035014.

Van Meter, K.J., P. Van Cappellen, and N.B. Basu. 2018. Legacy nitrogen may prevent achievement of water quality goals in the Gulf of Mexico. *Science* 360: 427–430. https://doi.org/10.1126/science.aar4462.

Vitousek, Peter M., John D. Aber, Robert W. Howarth, Gene E. Likens, Pamela A. Matson, David W. Schindler, William H. Schlesinger, and David G. Tilman. 1997. Human alteration of the global nitrogen cycle: Sources and consequences. *Ecological Applications* 7: 737–750. https://doi.org/10.1890/1051-0761(1997)007[0737:HAOTGN]2.0.CO;2.

Weng, Weizhe, Kelly M. Cobourn, Armen R. Kemanian, Kevin J. Boyle, Yuning Shi, Jemma Stachelek, and Charles White. 2024. Quantifying co-benefits of water quality policies: An integrated assessment model of land and nitrogen management. *American Journal of Agricultural Economics* 106 (2): 547–572. https://doi.org/10.1111/ajae.12423.

Xu, Yuelu, Levan Elbakidze, Haw Yen, Jeffrey G. Arnold, Philip W. Gassman, Jason Hubbart, and Michael P. Strager. 2022. Integrated assessment of nitrogen runoff to the Gulf of Mexico. *Resource and Energy Economics* 67: 101279. https://doi.org/10.1016/j.reseneeco.2021.101279.

Chapter 15
The Role of Transportation Infrastructure Expansion in the Transmission of Global Crop Price Shocks to the Brazilian Agriculture

Zhan Wang

1 Introduction

The high logistics cost from farm to port is a major factor that hinders the export competitiveness of Brazilian agriculture (Tiller and Thill 2017; Fliehr et al. 2019; Mendes Dos Reis et al. 2020; Valdes 2022), especially for its inland Cerrado biome. Nonetheless, the Cerrado has experienced rapid expansion of both cropland and crop production in the last three decades (Bicudo Da Silva et al. 2020; Souza et al. 2020). For example, Meade et al. (2016) compare the crop prices at destinations and their decomposition (farm-gate crop price, inland transportation and handling cost, and ocean transportation cost) between two crop-producing sites—Sorriso in the Mato Grosso state (MT) to represent the Cerrado biome and Campo Mourão in Paraná state (PR) to represent the south/southeast region, which is better connected to global markets. Figure 15.1 illustrates these data, showing that the Cerrado biome is more disadvantaged in inland transportation than the south/southeast region, not only because the former is further from crop exporting ports but also because of its sparser transportation network, especially the railway network. As a result, although Sorriso faces similar or lower farm-gate crop prices for corn and soybean production, its inland transportation cost is about three times that of Campo Mourão. Thus, crops produced in Sorriso are more expensive for importers and less competitive in international markets (Fig. 15.1b).

Data availability statement: The files needed to replicate this application are available at https://gtap.agecon.purdue.edu/simple-g/.

Z. Wang (✉)
Center for Global Trade Analysis, Department of Agricultural Economics, Purdue University, West Lafayette, IN, USA
e-mail: zhanwang@purdue.edu

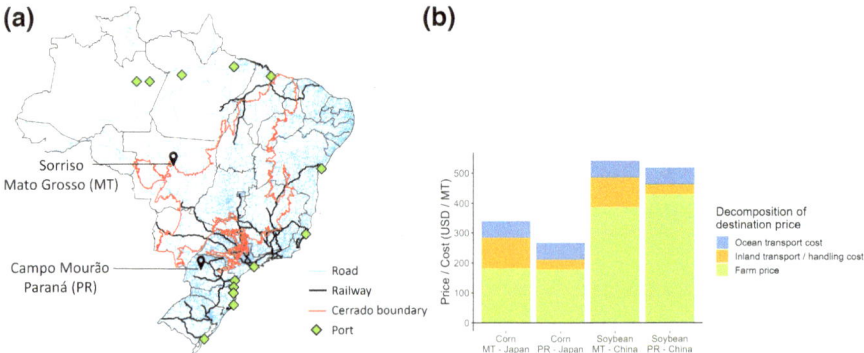

Fig. 15.1 Role of logistics costs in Brazilian crop exports. (**a**) Locations of two representative crop production sites for the Cerrado biome (*red boundary*) and south/southeast region, respectively (Sorriso in the Mato Grosso (MT) state and Campo Mourão in Paraná (PR) state) and the distribution of road, railway, and crop exporting ports in 2017. (**b**) The decomposition of destination prices from Sorriso (MT) and Campo Mourão (PR) on corn exports to Japan and soybean trade to China. (Data are obtained from Meade et al. (2016))

In 2021, Brazil launched the 2035 National Logistics Plan (PNL2035), which aims to expand domestic transportation infrastructure to improve the connectivity of inland regions to coastal regions and eventually to the global market. According to the Ministry of Infrastructure of Brazil (2022), PNL2035 will expand the railway network length by up to 90.56% upon its completion, while the expected expansion of the already developed road network would be slight (<1%). How will this large-scale infrastructure expansion plan influence the export competitiveness of crops produced from inland Brazil? To analyze the comprehensive impacts of PNL2035 on crop export competitiveness, it is necessary to capture the interactive effects between global-level crop price changes and the spatial heterogeneity of transportation cost reductions from infrastructure expansion and to consider the spillover effects among regions in Brazil. The SIMPLE-G model introduced in this book, with its global–local–global philosophy, provides an appropriate tool for this research question.

In this chapter, we use a specially designed version of SIMPLE-G to simulate the interactive impacts of global crop price change and domestic transportation infrastructure expansion on crop production, land use, and carbon balance. Brazil is the focus region (hence SIMPLE-G-Brazil), and the PNL2035 provides the policy background. Of course, the methodology described in this chapter can be generalized to other regions of the world where transportation costs are of considerable importance. First, we provide a simplified economic framework to qualitatively demonstrate the impact of a global crop price increase and domestic transportation cost reduction on both subregional and regional crop supplies. Second, we introduce SIMPLE-G-Brazil, the regional variant of SIMPLE-G developed to research gridded agricultural and environmental impacts within Brazil. In particular, we discuss the two new modules that differentiate SIMPLE-G-Brazil in this application from the fundamental SIMPLE-G framework introduced in Part III: the transportation cost and cropland supply modules. Following Wang et al. (2024), the transportation cost module connects the expansion of transportation infrastructure with the reduction in monetary transportation costs at the grid cell level, which represents the role of transportation infrastructure in local farm-gate crop prices. The cropland supply module extends the factor supply system of cropland in the basic SIMPLE-G framework to capture the impacts of conservation policy on cropland expansion potential. For example, Brazil requires that a certain share of land (depending on the biome) on each farm cannot be used as cropland to prevent deforestation from overcultivation (Metzger et al. 2019). The cropland supply module can automatically prevent cropland expansion on all grid cells that have reached or exceeded this policy constraint. Additionally, the cropland supply module allows us to conduct a sensitivity analysis with respect to the cropland supply elasticity parameters. Third, we describe the experimental design for quantitatively simulating the impacts of a global crop price change and transportation cost reduction. Finally, we report and discuss simulation results covering three aspects: the global-to-local transmission of crop prices, the corresponding response of crop production and cropland use, and, consequently, the implications for terrestrial carbon fluxes.

2   Economic Framework

In this section, we conceptually demonstrate how international crop price changes and domestic transportation cost reductions interact and influence crop production, illustrated in Fig. 15.2. In this framework, Brazil consists of two regions: coastal and inland. For simplicity, we assume that the coastal region has already fully benefited

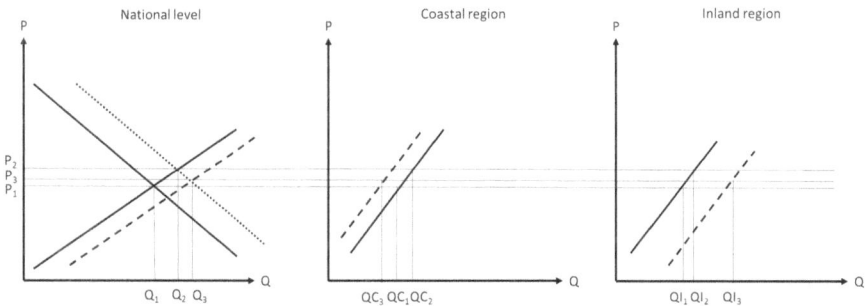

Fig. 15.2 Economic framework of national and subnational responses to a world price shock (*dotted line*) and transportation infrastructure expansion (*dashed line*). *Solid lines* show the supply–demand equilibrium without external shocks. An increase in global demand for Brazilian production shifts the national demand curve outward, resulting in an increase in the border price that producers receive for crops in both regions. They expand production to QC_2 and QI_2, respectively. The introduction of new transportation infrastructure boosts farm-gate returns in the inland region and acts as a supply shifter, resulting in output level QI_3. With the expansion of inland production, mobile input costs may increase; thus, some coastal production is displaced, as shown by the backward supply shift, resulting in production at QC_3

from existing transportation infrastructure; thus, expansion of the domestic transportation network will only reduce transportation costs in the inland region. Without external shocks, the national crop supply and crop demand (including both domestic and export demand) interact, resulting in equilibrium crop price P_1 and crop quantity Q_1. At the subnational level, the coastal and inland regions face the same free-on-board (FOB) price and produce crops QC_1 and QI_1, respectively, where $Q_1 = QC_1 + QI_1$.[1]

First, we assume an increase in Brazilian crop export demand from the global market, which is represented as the outward shifting of the demand curve (solid to dotted line) on the national level. Both the new equilibrium price, P_2, and crop quantity, Q_2, increase from the previous equilibrium. On the subnational level, without further shocks to local supply, a higher FOB price, P_2, causes local crop production (QC_2, QI_2) to increase in both regions.

In addition to the international demand shock, we consider the reduction in transportation costs for the inland region. As the farm-gate crop price equals the difference between the FOB crop price and transportation cost, a decrease in transportation costs raises the farm-gate price received by producers from the inland region and the average farm-gate price received at the national level, which is equivalent to a subsidy in production. As a result, the supply curves in both the inland region and at the national level shift rightward (solid to dashed lines). Although reductions in transportation costs do not directly influence farm-gate crop prices in the coastal region, both coastal and inland regions use factors supplied at the national level in crop production. Therefore, the reduction in transportation costs in the inland region makes it more competitive in attracting mobile factors, which reduces the availability of production factors for the coastal region and increases the cost of crop production there, resulting in an upward shift in the supply curve. The interactive effects of the increase in crop export demand and reduction in inland transportation costs cause a smaller magnitude of increase in national crop price P_3 ($P_1 < P_3 < P_2$) and further growth in national crop production Q_3 ($Q_1 < Q_2 < Q_3$). At the subnational level, we expect the reduction in transportation costs to further promote local crop production in the inland region ($QI_1 < QI_2 < QI_3$), but its spillover effect reduces crop production in the coastal region ($QC_3 < QC_1 < QC_2$).

The analysis with the simplified two-region framework provides us with the economic intuition with respect to the interactive impacts of a global crop price shock and a reduction in domestic transportation costs as well as a series of expected outcomes to be further tested as hypotheses. However, this simplified framework cannot capture the more complex agricultural production systems or rich spatial information on a fine-scale level; further analysis with the computable multiscale economic model SIMPLE-G-Brazil is needed.

---

[1]For simplicity, here we do not consider the transport cost between farmers and domestic consumers and instead assume that all domestic consumers face the same crop price as the FOB price. Given that the most densely populated regions in Brazil are located along its coast (IBGE 2010), this assumption is reasonable for our application.

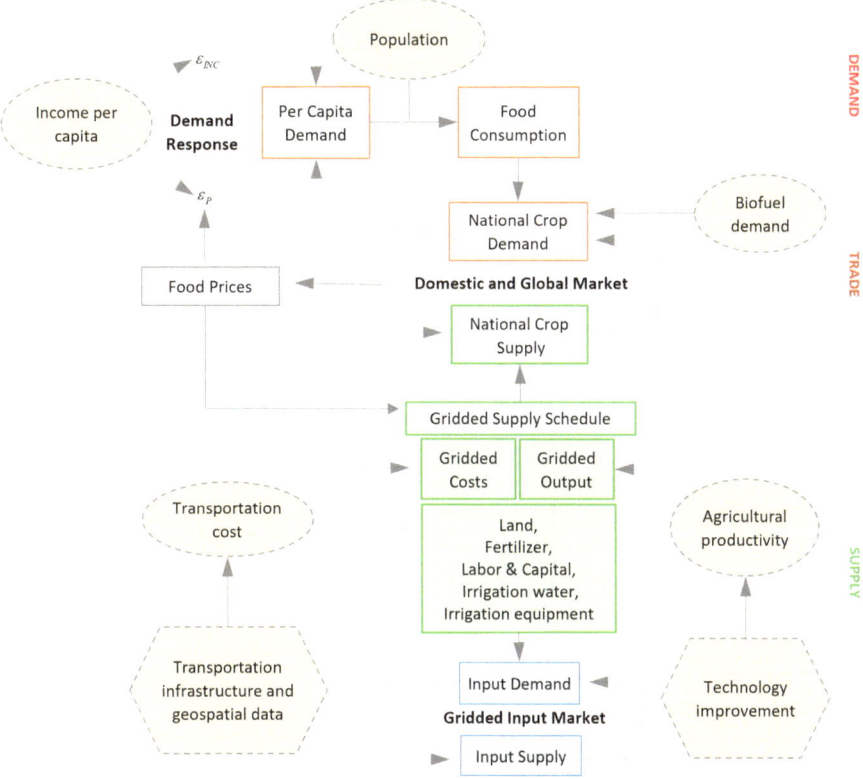

Fig. 15.3 Overview of the SIMPLE-G-Brazil model for the transportation infrastructure application. (Adapted from Wang et al. (2024))

3   SIMPLE-G-Brazil for the Transportation Application

This section takes us under the hood of the SIMPLE-G-Brazil model, showcasing key data sources and introducing two major upgrades: a nuanced land supply module to capture finer details of land use restrictions and a new transportation cost module that adds another layer of realism to the model.

3.1   Model and Data

SIMPLE-G-Brazil is constructed from the basic version of SIMPLE-G described in Part III. SIMPLE-G-Brazil divides the world into 17 regions and further disaggregates Brazil's cropland into 50,598 grid cells, each with a spatial resolution of 5 arcminutes (around 7000–8500 hectares in area, depending on latitude). Figure 15.3 summarizes the model structure of SIMPLE-G-Brazil for the transportation application. We incorporate a transportation cost module that links the Brazilian (port) crop price obtained from the national crop supply and demand equilibrium with the gridded farm-gate crop price. Furthermore, we modify the supply system of gridded cropland, which allows us to conduct a systematic sensitivity analysis (SSA) with respect to uncertainties from cropland supply parameter estimation and to control for conservation policies.

We selected 2017 as the baseline year for the model databases, and this model has been validated with historical observational data from 2000 to 2017 (for detailed information on model validation, please refer to Chap. 9). This study leverages a diverse mosaic of datasets to build a comprehensive picture of the Brazilian agricultural landscape. At the spatial level, detailed cropland area maps from *MapBiomas* (2020) are combined with yield data from Portmann et al. (2010) and further refined with microregional adjustments from the Brazilian Institute of Geography and Statistics (IBGE) (Prado Siqueira 2022). Irrigation status is estimated based on municipality-level irrigation ratios from the IBGE (2019) agricultural census. The transportation network is defined using data from the Ministry of Infrastructure of Brazil (2022), while crop exporting port locations are pinpointed via Victoria et al. (2021). Carbon stock factors from Novaes et al. (2017) are adjusted with tillage status data from Fuentes-Llanillo et al. (2021) to provide a nuanced picture of carbon storage, and vehicle carbon emission factors are drawn from Sims et al. (2014). On the aggregated level, the study draws upon FAOSTAT (FAO 2021) data for cropland area, crop output and price and World Bank open data (World Bank 2020) for population and per capita gross domestic product (GDP). Finally, GTAP database v.10 (Aguiar et al. 2019) provides insights into crop demand, while input cost shares are from Hertel and Baldos (2016). This rich tapestry of data sources fuels the intricate analyses presented in this chapter, enabling us to explore the complex interactions within the Brazilian agricultural system. Note that Table 15.1 summarizes the data sources for SIMPLE-G-Brazil and its transportation application. The simulation results from SIMPLE-G-Brazil are obtained using the GEMPACK economic modeling software (v.12) (Horridge et al. 2018).

3.2   *Transportation Cost Module*

To capture the role of transportation cost as the price wedge between FOB and farm-gate crop prices, this application attaches a transportation cost module to the basic SIMPLE-G model. In the basic structure, the gridded farm-gate crop price equals the national crop price (FOB price), so the transportation cost is not incorporated. In the transportation cost module, we replace the relationship between farm-gate and FOB crop prices with the following equation:

Table 15.1 Data sources for SIMPLE-G-Brazil transportation application

Variable	Data source
Spatial level	
Cropland area	*MapBiomas* (2020)
Crop yield	Portmann et al. (2010), adjusted with crop yield at the microregion level from the Brazilian Institute of Geography and Statistics (IBGE) (Prado Siqueira 2022)
Irrigation status	Calculated based on municipality-level irrigation ratio from the agricultural census (IBGE 2019)
Transportation network	Ministry of Infrastructure of Brazil (2022)
Crop exporting port location	Victoria et al. (2021)
Carbon stock factor	Novaes et al. (2017), adjusted with tillage status from Fuentes-Llanillo et al. (2021)
Vehicle emission factor	Sims et al. (2014)
Aggregated level	
Cropland area	FAOSTAT (FAO 2021)
Crop output	FAOSTAT (FAO 2021)
Crop price	FAOSTAT (FAO 2021)
Population	World Bank open data (World Bank 2020)
Per Capita GDP	World Bank open data (World Bank 2020)
Crop demand	GTAP v.10 (Aguiar et al. 2019)
Input cost share	Hertel and Baldos (2016)

$$P^{\mathrm{FOB}} = P_i^{\mathrm{FG}} + P_i^{\mathrm{TC}}, \tag{15.1}$$

where P^{FOB} is the FOB price at the national level, P_i^{TC} is the transportation cost (i.e., the least cost to transport crops from a farm to any export port) of grid cell i (the estimation of P_i^{TC} is described in Sect. 4), and P_i^{FG} is the farm-gate crop price for farmers in grid cell i. This representation is similar to modeling the margin of transportation in the general equilibrium literature.

As discussed in Chap. 5, SIMPLE-G consists of linearized equations with percentage-change form variables, so Eq. 15.1 is also linearized as follows[2]:

$$p_i^{\mathrm{FG}} = (1 + \theta_i)p^{\mathrm{FOB}} - \theta_i p_i^{\mathrm{TC}}, \tag{15.2}$$

where p^{FOB}, p_i^{TC}, and p_i^{FG} refer to the percentage changes in FOB price, gridded transportation cost, and farm-gate crop price, respectively, and θ_i represents the share of transportation cost relative to the total FOB price in the baseline dataset. Equation

---

[2] Recall that we use uppercase letters to represent variables in level form and lowercase letters to represent variables in linearized form.

15.2 provides an updated relationship between farm-gate crop price, FOB crop price, and transportation cost. With the transportation cost module, the farm-gate price increases with the FOB crop price (positive p^{FOB}) and with a reduction in transportation costs (negative p_i^{TC}), while the magnitude of those impacts now depends not only on changes in p^{FOB} or p_i^{TC} but also on the share of transportation cost in crop price decomposition. This representation is essential for research questions when the change in transportation costs is spatially heterogeneous, and an explicit consideration of transportation costs therefore provides a more precise analysis.

3.3   Cropland Supply Module for SSA and Conservation Policy

In addition to the module of transportation cost, we also update the cropland supply module in SIMPLE-G-Brazil for two purposes: applying the restriction of cropland expansion from conservation policies and allowing researchers to conduct a SSA with respect to the parameters in cropland supply elasticity estimation.

In the basic SIMPLE-G structure, the gridded cropland supply is depicted with the following equation:

$$q_i^{\text{Land}} = \eta_i^{\text{Land}} p_i^{\text{Land}} + s_i^{\text{Land}}, \tag{15.3}$$

where q_i^{Land} and p_i^{Land} are the percentage changes in the quantity and rent of cropland in grid cell i, respectively; η_i^{Land} is the supply elasticity of cropland in grid cell i; and s_i^{Land} is an exogenous slack variable. When the cropland supply module is not introduced, η_i^{Land} is estimated with the following equation from Haqiqi et al. (2023) and read into the model as a constant:

$$\eta_i^{\text{Land}} = \alpha \sqrt{\text{Precip}_i} + \beta \text{Landshr}_i \sqrt{\text{Precip}_i} + \lambda, \tag{15.4}$$

where Precip_i is the average annual precipitation (mm/year), Landshr_i is the share of cropland in the total area of grid cell i at baseline, and α, β, and λ are parameters estimated from regression (λ is fixed at 0 in estimation). This simple formula gives a narrow range from 0 to slightly above 1 for the land supply elasticity.

However, this basic land supply function faces two challenges. First, land supply in Brazil is controlled not only by cropland rent, p_i^{Land}, precipitation, Precip_i, and the current cropland occupancy, Landshr_i, but also by the policy constraint of the natural vegetation protection law (Metzger et al. 2019). If a grid cell's current cropland occupancy has reached or exceeded the maximum of cropland share from the policy constraint, further expansion is prohibited. Such policy impacts should be captured in the cropland supply system. Second, the estimation of α and β from regression provides not only their value but also the uncertainty of estimation as the confidence intervals, and we wish to understand the extent to which our simulation results from

SIMPLE-G-Brazil are sensitive to the estimation of cropland supply elasticity parameters α and β.

To overcome these two challenges, we embed the cropland supply module in SIMPLE-G-Brazil, which allows the cropland supply elasticity, η_i^{Land}, to be calculated as

$$\eta_i^{\text{Land}} = \frac{\max((\text{Maxlandshr}_i - \text{Landshr}_i), 0)}{\max(|\text{Maxlandshr}_i - \text{Landshr}_i|, \text{tiny})}$$
$$\times \left(\alpha \sqrt{\text{Precip}_i} + \beta \text{Landshr}_i \sqrt{\text{Precip}_i} \right) + \lambda, \qquad (15.5)$$

where Maxlandshr_i is the maximum of cropland share allowed by the conservation policy and tiny refers to a sufficiently small positive value (this application uses tiny $= 0.000001$). In Eq. 15.5, if the current cropland share has reached or exceeded the maximum allowed share (i.e., $\text{Landshr}_i \geq \text{Maxlandshr}_i$), we have max $((\text{Maxlandshr}_i - \text{Landshr}_i), 0) = 0$, but $\max(|\text{Maxlandshr}_i - \text{Landshr}_i|, \text{tiny}) = \text{tiny}$, so $\frac{\max((\text{Maxlandshr}_i - \text{Landshr}_i), 0)}{\max(|\text{Maxlandshr}_i - \text{Landshr}_i|, \text{tiny})} = 0$ and $\eta_i^{\text{Land}} = 0$. Otherwise, when $\text{Landshr}_i < \text{Maxlandshr}_i$, then $\frac{\max((\text{Maxlandshr}_i - \text{Landshr}_i), 0)}{\max(|\text{Maxlandshr}_i - \text{Landshr}_i|, \text{tiny})} = \frac{(\text{Maxlandshr}_i - \text{Landshr}_i)}{|\text{Maxlandshr}_i - \text{Landshr}_i|} = 1$ and Eq. 15.5 converts to Eq. 15.4, so the cropland supply elasticity will be calculated using the basic approach. Thus, Eq. 15.5 provides an automatic control for external policy restrictions.

Furthermore, calculating the cropland supply elasticity within the model allows us to conduct a SSA with respect to the uncertainties in parameters α and β with the help of the RunGEM software. RunGEM is a component of the GEMPACK software package that facilitates the running of SIMPLE-G models and provides tools ("Tools – SSA wrt parameters") for conducting sensitivity analyses given the mean and range of parameters. In this study, we apply triangular distributions of α and β with their estimated values as means and their 95% confidence intervals as ranges to estimate the 95% confidence intervals of the simulation results, which are reported as the error bar in Sect. 5.[3] For detailed instructions on using RunGEM to conduct an SSA, please refer to the application's "Readme Use RunGEM for SSA" folder.

4   Experimental Design

In this study, we designed three scenarios to be estimated with SIMPLE-G-Brazil. In all three scenarios, we apply the shock of a global crop price increase of 10%, while the three scenarios differ from each other with respect to the domestic transportation costs.

---

[3] For detailed information on SSA and other SSA approaches, please refer to Wang et al. (2024).

Business-as-usual (BAU) In this scenario, we consider a counterfactual scenario in which the PNL2035 policy has not been implemented and the transportation infrastructure and costs remain at baseline (2017) levels. In other words, this scenario contains the pure effects of a global crop price change.

Infrastructure Expansion at a Low Level Here, we define a low-level infrastructure expansion scenario (thereby referred to as the "low scenario") as the case in which all expansions that were under construction by the time when PNL2035 was launched are completed, but no new projects are implemented. The associated transportation cost reduction will be incorporated into the experiment together with the global crop price change.

Infrastructure Expansion at a High Level In contrast with the low scenario, we define a high-level infrastructure expansion scenario (thereby referred to as the "high scenario") as the case in which all the infrastructure expansion proposed in PNL2035, regardless of construction status, will be completed. Under the high scenario, further transportation cost reduction is expected on the basis of the low scenario, and the interactive effects between domestic transportation cost reduction and global crop price change are also greater.

As PNL2035 focuses on the expansion of the railway network, Fig. 15.4 illustrates the railway network at baseline and at its expected expansion under the low and high scenarios, respectively (Fig. 15.4a). The transportation cost and its reduction due to infrastructure expansion under the low and high scenarios (Fig. 15.4b–d), calculated based on Wang et al. (2024), are incorporated into the transportation cost module in order to simulate the interactive effects on farm-gate crop price and corresponding responses of crop supply, the extensive margin (cropland expansion), and the intensive margin (yield growth or multicropping). Finally, to assess the environmental impacts of transportation and agricultural production, we further simulate the impacts on carbon balance, measured as the change in CO_2 in million metric tons (Mt). We consider two drivers of carbon balance change: direct emissions of greenhouse gases from road and railway vehicles and indirect emissions from the loss of carbon stock when land with natural vegetation is converted to cropland. All simulation results are represented as the percentage changes or level changes with respect to the 2017 baseline.

5   Results and Discussion

This section briefly describes the findings of these three scenarios, both extracting expected aggregate insights and illuminating the fertile ground for future research revealed by this gridded analysis. While the aggregated results provide robust validation of theoretical frameworks, the spatial heterogeneity unveiled by the gridded results demands further exploration.

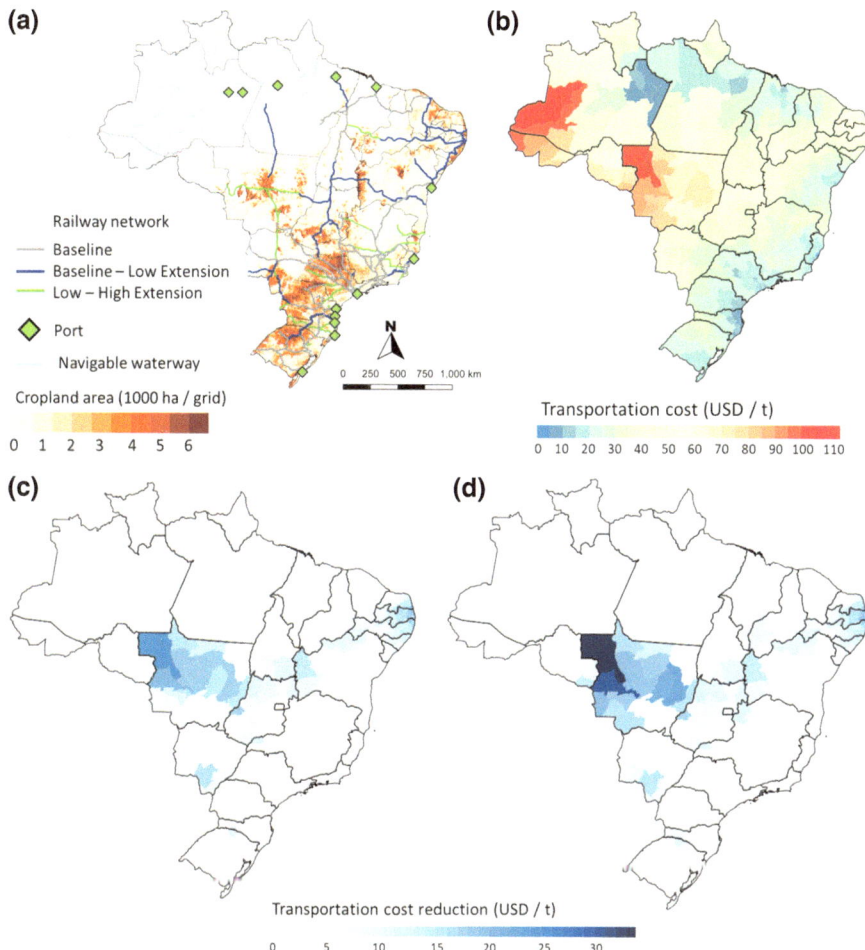

Fig. 15.4 Railway extension plan from PNL2035 and its contribution to reducing transportation costs. (**a**) The railway network in 2017 (*gray lines*) and the expansions planned in the PNL2035 low (*blue lines*) and high (*both blue and green lines*) scenarios. (**b**) Calculated transportation cost from microregion to exporting ports at the 2017 baseline. (**c**) Reduction in transportation costs under the PNL2035 low scenario. (**d**) Reduction in transportation costs under the PNL2035 high scenario. (Adapted and modified from Wang et al. (2024))

5.1 Global-to-Local Interactive Effects on Farm-Gate Crop Price

When the global crop price shock and the domestic transportation cost reduction shocks are applied in SIMPLE-G Brazil, their direct impact is an increase in the farm-gate crop price. At the national level, the crop price in Brazil increases due to the increase in the global crop price but decreases with the reduction in

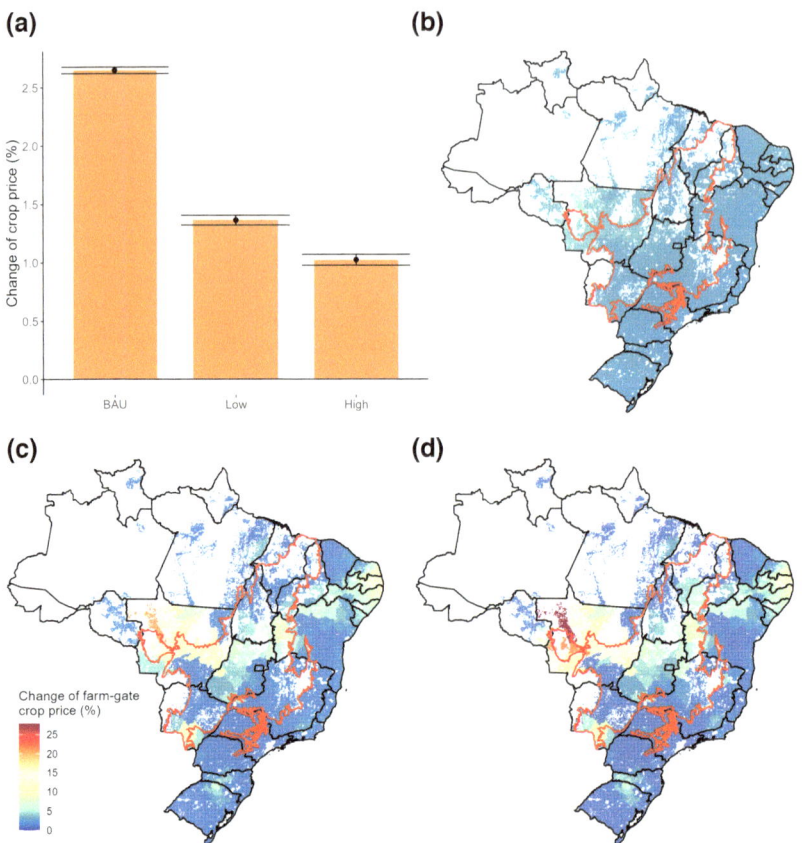

Fig. 15.5 (**a**) Percentage change in national crop price levels under the business-as-usual (BAU), low, and high transportation expansion scenarios. Gridded farm-gate crop price received under (**b**) the BAU scenario, (**c**) the low scenario, and (**d**) the high scenario. Error bars show the 95% confidence intervals from land supply elasticity estimation. The *red boundary* refers to the Cerrado biome and the *black boundaries* indicate state boundaries

transportation costs and the subsequent increase in supply (Fig. 15.5a). The reduction in transportation costs increases the profitability of farmers, encouraging them to expand crop production at the margin. Consequently, national crop production rises, resulting in higher equilibrium crop production but lower equilibrium crop prices, consistent with our graphical analysis (Fig. 15.2).

At the grid cell level, the interactive impacts of these two shocks cause spatially heterogeneous changes in local farm-gate crop prices, as shown in Fig. 15.5b–d. When only the global crop price shock is implemented under the BAU scenario, all grid cells in Brazil experience a slight increase in the farm-gate crop price received, similar to the change in the national crop price. However, when the reduction in transportation costs is incorporated, states in the Cerrado biome benefit from a much higher increase in farm-gate crop prices (10–25%), while the increase in farm-gate crop prices in the south/southeast becomes even smaller compared with the BAU scenario. That is, whereas an increase in the global crop price brings an almost uniform stimulus to all of Brazil, the reduction in domestic transportation costs mainly improves the profitability of farming in the Cerrado region but does little to benefit farming in the south/southeast region. This change in relative spatial profitability causes a shift in the pattern of national crop production and land use.

5.2 Spillover Effects on Crop Production and Land Use

Figure 15.6 illustrates the simulated percentage change in state-level crop output and its decomposition by the extensive margin (change in cropland area), intensive margin (change in crop yield due to intensive nonland input use or multicropping), and their interactions. Figure 15.6 highlights the results from the top eight crop-producing states in Brazil. These eight states accounted for 81% of the national cropland area and 92% of national crop production in 2017, so they capture the overall pattern of agricultural impacts. The first group of states is located in the south/southeast region: São Paulo (SP), Paraná (PR), Rio Grande do Sul (RS), and Minas Gerais (MG)[4]; the second group is located in the core region of the inland Cerrado biome: Mato Grosso (MT), Goiás (GO), Mato Grosso do Sul (MS), and Bahia (BA). Within each group, states are ranked by their baseline crop production in descending order. Results from the remaining states (aggregated as "Rest of Brazil (ROB)") and the entire country are also reported.

Figure 15.6 reveals state-level responses and spillover effects on crop production and land use. Under the BAU scenario, all states show a moderate increase in crop production, driven by both extensive and intensive margins (Fig. 15.6a). However, results under the low scenario show that as the Cerrado biome gains a relative advantage in farming due to the increase in farm-gate crop prices, farmers in Cerrado further expand crop production both by cultivating more cropland and, more importantly, by attracting more mobile inputs (e.g., fertilizer, labor, and capital) (Fig. 15.6b). On the other hand, although the south/southeast region also experiences an increase in farm-gate crop prices, the magnitude of the price increase is much lower than that in the Cerrado region, making the south/southeast region relatively disadvantaged in farming and less competitive in attracting mobile inputs. States in the south/southeast region exhibit reductions in crop output across both the intensive and extensive margins, which indicates that the expansion of transportation infrastructure will not only improve the connectivity of the Cerrado biome and benefit its agricultural production, but also cause spillover effects in the south/southeast region. Furthermore, the increase in crop production in Brazil is greater in the low scenario than in the BAU scenario, indicating that the gains from Cerrado overwhelm the

---

[4]While SP and MG belong to the southeast region of Brazil, they also partially overlap with the Cerrado biome.

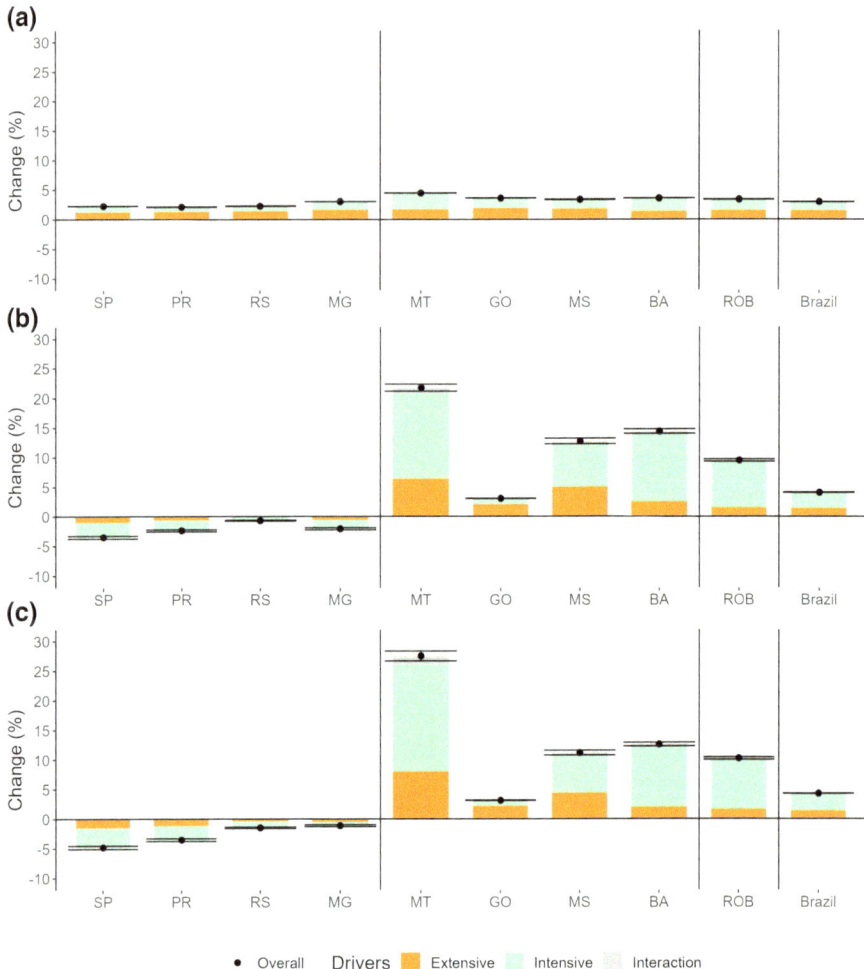

Fig. 15.6 Simulated changes in crop production and the decomposition by extensive margins, intensive margins, and their interactions at the state and national level under (**a**) the BAU scenario, (**b**) the low scenario, and (**c**) the high scenario. Error bars show 95% confidence intervals from the land supply elasticity estimation, as obtained by using the SSA feature of RunGEM

losses in the south/southeast region, also consistent with the graphical analysis (Fig. 15.2). Finally, the simulation results under the high scenario are similar to those under the low scenario but with greater magnitudes, which is due to the fact that transportation infrastructure under the high scenario includes all constructions under low scenario, with further expansion.

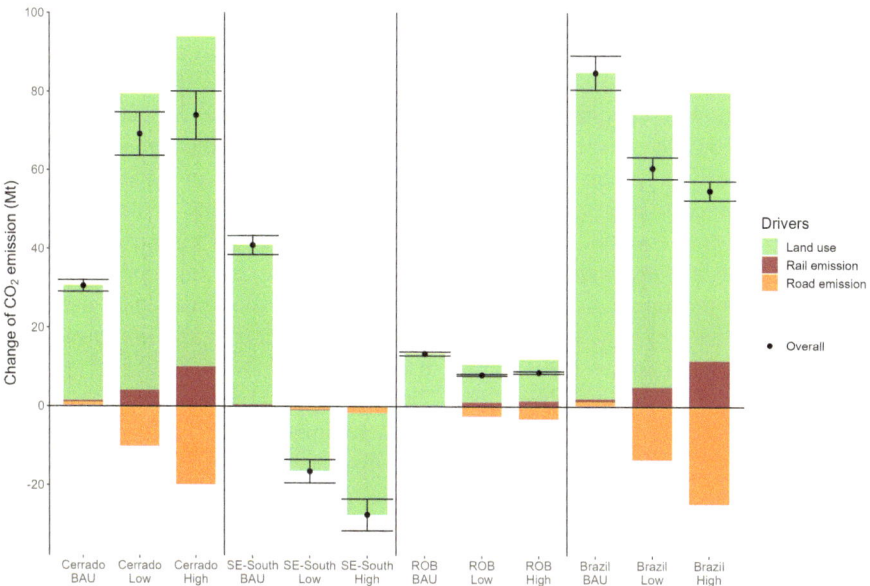

Fig. 15.7 Simulated changes in greenhouse gas emissions and the decomposition by drivers. Error bars show 95% confidence intervals from the land supply elasticity estimation. We further aggregate the first group of states in Fig. 15.6 into the "south/southeast"(SE-South) subnational region and aggregate the second group of states into the "Cerrado" subnational region

5.3 Environmental Impacts on Carbon Balance

In addition to the impacts on crop price, crop production, and land use, the global crop price and transportation infrastructure have considerable direct and indirect impacts on carbon emissions in Brazil. Figure 15.7 reports the net changes in greenhouse gas emissions at the national and subnational levels (Cerrado, south/southeast, ROB); their decomposition by direct emission from vehicles (for road and rail transportation); and indirect emissions from the loss of carbon stock due to cropland expansion. Under the BAU scenario, since all states in Brazil experience crop production growth and cropland expansion (Fig. 15.6a), the global crop price shocks cause a net increase in CO_2 emissions, mainly from the carbon stock loss from cropland replacing natural vegetation. When the expansion of transportation infrastructure is also incorporated into the simulation under the low and high scenarios, a considerable amount of the national crop production pattern shifts from the south/southeast region to the Cerrado biome (Fig. 15.6b and c), which causes carbon emissions from land use change in the Cerrado to become more than double of the amount under BAU scenario, while the land use-related carbon emissions in the south/southeast region are overturned and become negative. Moreover, as PNL2035 markedly extends the railway network but barely improves the roadway network, more crops are transported over more cost-efficient and carbon-efficient railways instead of roadways. As a result, under the low and high scenarios, CO_2 emissions from railway transportation increase in the Cerrado biome, but this effect is fully countered by the much higher reduction of CO_2 emissions from roadways. In total, national CO_2 emissions decrease from 84.9 Mt under the BAU scenario to 60.6 Mt under the low scenario and 54.9 Mt under the high scenario. The implication of this carbon balance analysis is that an increase in global crop prices would boost crop production in Brazil, but it also threatens deforestation and increases carbon emissions. The expansion of transportation infrastructure would exacerbate local carbon emissions in the Cerrado biome, but this impact could be offset by the reduction of cropland demand in the south/southeast region due to the spillover effect. Thus, the expanded infrastructure facilitates Brazil's crop supply to the global level in a more carbon-efficient way.

This research has significant implications for policymakers and stakeholders striving to balance economic vitality and environmental responsibility. By understanding the intricate dance between global markets, domestic infrastructure, and carbon emissions, we can design policies that unlock the economic potential of Brazilian agriculture while safeguarding the country's precious natural heritage.

5.4 Limitations and Future Directions

Several avenues for future research emerge from the limitations encountered in this chapter's analysis. To fully unpack the complexities explored in this chapter, it is crucial to acknowledge and address the following limitations. First, the expansion of transportation infrastructure would reduce transportation costs not only for crops but also for mobile inputs such as fertilizers, machinery, labor, and other capital inputs. However, this analysis does not include the role of infrastructure expansion in reducing farm input costs. In Wang et al. (2024), we extend the analysis with additional scenarios on the impact of transportation infrastructure on factor mobility and find that when the impacts of infrastructure on factor mobility are considered, our conclusions do not change qualitatively, but the magnitude of the results is strengthened quantitatively with better factor mobility. Additionally, farm-gate crop prices can be influenced by factors other than global crop prices and the expansion of infrastructure networks, including the market power of transporters, road status, and congestion. In this chapter, we consider only the global price and transportation infrastructure shocks to illustrate the capacity of SIMPLE-G for studying the interaction between global and local level shocks. We encourage interested readers to build on this framework and further extend this analysis or generalize it to similar research questions.

Acknowledgments and Competing Interests This research was supported by NSF projects "CNH2-L: Uncovering Metacoupled Socio-Environmental Systems (DEB-1924111)" and "GLASSNET (OISE-2020635)" as well as the project "Projeto Rural Sustentável – Cerrado" (IDB Technical Cooperation #BR-T1409) from the Inter-American Development Bank and

FAPESP's Project "Integração cana-de-açúcar/pecuária: modelagem e otimização" (FAPESP, 2017/11523-5).

The findings and conclusions presented in this chapter are those of the authors and should not be construed to represent any official determination or policy of the US Department of Agriculture (USDA), or the National Science Foundation (NSF). Furthermore, we declare that there is no conflict of interest related to this work.

References

Aguiar, Angel, Maksym Chepeliev, Erwin L. Corong, Robert McDougall, and Dominique van der Mensbrugghe. 2019. The GTAP data base: Version 10. *Journal of Global Economic Analysis* 4: 1–27. https://doi.org/10.21642/JGEA.040101AF.

Bicudo Da Silva, Ramon Felipe, Mateus Batistella, Emilio Moran, Otávio Lemos De Melo Celidonio, and James D.A. Millington. 2020. The soybean trap: Challenges and risks for Brazilian producers. *Frontiers in Sustainable Food Systems* 4: 12. https://doi.org/10.3389/fsufs.2020.00012.

FAO. 2021. *FAOSTAT: Food and agriculture data*. Rome: Food and Agriculture Organization of the United Nations. http://faostat.fao.org/. Accessed 4 Feb 2024.

Fliehr, Olivia, Yelto Zimmer, and Linda H. Smith. 2019. Impacts of transportation and logistics on Brazilian soybean prices and exports. *Transportation Journal* 58: 65–77. https://doi.org/10.5325/transportationj.58.1.0065.

Fuentes-Llanillo, Rafael, Tiago Santos Telles, Dimas Soares Junior, Thadeu Rodrigues De Melo, Theodor Friedrich, and Amir Kassam. 2021. Expansion of no-tillage practice in conservation agriculture in Brazil. *Soil and Tillage Research* 208: 104877. https://doi.org/10.1016/j.still.2020.104877.

Haqiqi, Iman, Danielle S. Grogan, Marziyeh Bahalou Horeh, Jing Liu, Uris L.C. Baldos, Richard Lammers, and Thomas W. Hertel. 2023. Local, regional, and global adaptations to a compound pandemic-weather stress event. *Environmental Research Letters* 18: 035005. https://doi.org/10.1088/1748-9326/acbbe3.

Hertel, Thomas W., and Uris Lantz C. Baldos. 2016. Attaining food and environmental security in an era of globalization. *Global Environmental Change* 41: 195–205. https://doi.org/10.1016/j.gloenvcha.2016.10.006.

Horridge, J.M., Michael Jerie, Dean Mustakinov, and Florian Schiffmann. 2018. *GEMPACK manual*. Victoria University, Centre of Policy Studies/IMPACT Centre.

Instituto Brasileiro de Geografia e Estatística (IBGE). 2010. Population density map 2010. https://geoftp.ibge.gov.br/cartas_e_mapas/mapas_do_brasil/sociedade_e_economia/mapas_murais/densidade_populacional_2010.pdf. Accessed 4 Feb 2024.

———. 2019. Censo agropecuário 2017. https://sidra.ibge.gov.br/pesquisa/censo-agropecuario/censo-agropecuario-2017/resultados-definitivos. Accessed 4 Feb 2024.

MapBiomas. 2020. Project MapBiomas—Collection v4.1 of Brazilian land cover & use map series. https://brasil.mapbiomas.org/. Accessed 4 Feb 2024.

Meade, Birgit, Estefania Puricelli, William D. McBride, Constanza Valdes, Linwood Hoffman, Linda Foreman, and Erik Dohlman. 2016. *Corn and soybean production costs and export competitiveness in Argentina, Brazil, and the United States*, Economic Information Bulletin EIB-154. Washington, DC: US Department of Agriculture, Economic Research Service. https://doi.org/10.22004/ag.econ.262143.

Mendes Dos Reis, João Gilberto, Pedro Sanches Amorim, José António Sarsfield Pereira Cabral, and Rodrigo Carlo Toloi. 2020. The impact of logistics performance on Argentina, Brazil, and the US soybean exports from 2012 to 2018: A gravity model approach. *Agriculture* 10: 338. https://doi.org/10.3390/agriculture10080338.

Metzger, Jean Paul, Mercedes M.C. Bustamante, Joice Ferreira, Geraldo Wilson Fernandes, Felipe Librán-Embid, Valério D. Pillar, Paula R. Prist, Ricardo Ribeiro Rodrigues, Ima Célia G. Vieira, and Gerhard E. Overbeck. 2019. Why Brazil needs its legal reserves. *Perspectives in Ecology and Conservation* 17: 91–103. https://doi.org/10.1016/j.pecon.2019.07.002.

Ministry of Infrastructure of Brazil. 2022. NLP 2035: Brazil's national logistics plan to 2035. https://ontl.epl.gov.br/wp-content/uploads/2022/05/PNL-English-version.pdf.

Novaes, Renan M.L., Ricardo A.A. Pazianotto, Miguel Brandão, Bruno J.R. Alves, André May, and Marília I.S. Folegatti-Matsuura. 2017. Estimating 20-year land-use change and derived CO_2 emissions associated with crops, pasture and forestry in Brazil and each of its 27 states. *Global Change Biology* 23: 3716–3728. https://doi.org/10.1111/gcb.13708.

Portmann, Felix T., Stefan Siebert, and Petra Döll. 2010. MIRCA2000—Global monthly irrigated and rainfed crop areas around the year 2000: A new high-resolution data set for agricultural and hydrological modeling. *Global Biogeochemical Cycles* 24: 2008GB003435. https://doi.org/10.1029/2008GB003435.

Prado Siqueira, Renato. 2022. sidrar: An interface to IBGE's SIDRA API (version 0.2.9).

Sims, Ralph, Roberto Schaeffer, Felix Creutzig, Xochitl Cruz-Núñez, Márcio D'Agosto, Dimitriu Delia, M.J. Meza, et al. 2014. Transport. In *Climate change 2014: Mitigation of climate change. Contribution of working group III to the fifth assessment report of the intergovernmental panel on climate change*, ed. Intergovernmental Panel on Climate Change, 599–670. Cambridge, UK: Cambridge University Press. https://doi.org/10.1017/CBO9781107415416.014.

Souza, Carlos M., Julia Z. Shimbo, Marcos R. Rosa, Leandro L. Parente, Ane A. Alencar, Bernardo F.T. Rudorff, Heinrich Hasenack, et al. 2020. Reconstructing three decades of land use and land cover changes in Brazilian biomes with Landsat archive and earth engine. *Remote Sensing* 12: 2735. https://doi.org/10.3390/rs12172735.

Tiller, Kara Carroll, and Jean-Claude Thill. 2017. Spatial patterns of landside trade impedance in containerized South American exports. *Journal of Transport Geography* 58: 272–285. https://doi.org/10.1016/j.jtrangeo.2017.01.001.

Valdes, Constanza, ed. 2022. Brazil's momentum as a global agricultural supplier faces headwinds. *Amber Waves*. https://doi.org/10.22004/ag.econ.338866.

Victoria, Daniel de Castro, Ramon Felipe Bicudo da Silva, James D.A. Millington, Valeri Katerinchuk, and Mateus Batistella. 2021. Transport cost to port though Brazilian federal roads network: Dataset for years 2000, 2005, 2010 and 2017. *Data in Brief* 36: 107070. https://doi.org/10.1016/j.dib.2021.107070.

Wang, Zhan, Geraldo B. Martha Jr, Jing Liu, Cicero Z. Lima, and Thomas W. Hertel. 2024. Planned expansion of transportation infrastructure in Brazil has implications for the pattern of agricultural production and carbon emissions. *Science of The Total Environment* 928:172434. https://doi.org/10.1016/j.scitotenv.2024.172434

World Bank. 2020. World Bank Open Data. https://data.worldbank.org/. Accessed 4 Feb 2024.

Chapter 16
Global Groundwater Sustainability and Virtual Water Trade

Iman Haqiqi, Chris J. Perry, and Thomas W. Hertel

1 Introduction

The overexploitation of open-access resources is a significant issue in many locations across the world, and it is caused mainly by the fact that the cost of agricultural production does not include the negative environmental externalities of resource use. One way to reduce these externalities is by imposing regulatory restrictions on the use of natural resources in agriculture. However, these regulations can reduce the competitiveness of the local agricultural sector. This may, in turn, induce new sustainability pressures on distant land and water resources. Therefore, policymakers need to carefully consider the consequences of conservation policies before undertaking them at broad scale.

Sustainability policies can have a significant impact on how agricultural production is distributed across different regions. One way of achieving sustainability is by encouraging the cultivation of water-intensive crops in areas with abundant sustainable water resources, thereby increasing agricultural activity in those regions that can best support it. However, the effect of sustainability policies on the spatial

This chapter is a slightly revised version of a paper originally published as Haqiqi, Iman, Chris J. Perry, and Thomas W. Hertel. 2022. When the virtual water runs out: Local and global responses to addressing unsustainable groundwater consumption. *Water International* 47: 1060–1084. https://doi.org/10.1080/02508060.2023.2131272.

Data availability statement: The files needed to replicate this application are available at https://gtap.agecon.purdue.edu/simple-g/

I. Haqiqi (✉) · T. W. Hertel
Center for Global Trade Analysis, Department of Agricultural Economics, Purdue University, West Lafayette, IN, USA
e-mail: ihaqiqi@purdue.edu

C. J. Perry
Independent Researcher, London, UK

distribution of agricultural production also depends on the possibility of input substitution. For instance, sustainability policies that incentivize farmers to use less water can lead to agricultural production that makes more intensive use of capital and other agricultural inputs by encouraging the adoption of new agricultural technologies. For instance, cultivating new drought-resistant crop varieties can be economically viable, even if doing so results in higher production costs, as it helps minimize the reduction in production.

In contrast to Chap. 12, which focuses on the Western United States, here we focus on groundwater withdrawals at the global level. Local groundwater restrictions can alter patterns of global production, thereby altering trade patterns and inducing changes in virtual water trade. SIMPLE-G provides an ideal framework for studying these relationships. The next section introduces the concept of virtual water, which was originally introduced by J. A. "Tony" Allan (2011).

1.1 Virtual Water

Allan's seminal insight, captured in the phrase "virtual water," was that trade in agricultural commodities was effectively the transfer of huge quantities of water from producing countries to consuming countries. For example, producing 1 kg of wheat or rice involves the transpiration of about 1000 kg of water, so the location of this production can have a significant impact on groundwater abstraction. Thus, the virtual water trade refers to the implicit flow of embodied water within traded goods and services. Allan's main interest was in the consequences of these transfers: the extent to which political stability is preserved in countries that rely on imported "virtual water" to ensure food security. This linkage has two dimensions: First, the avoidance of "water wars" as the import of virtual water makes it unnecessary to "capture" the underlying natural resource itself, and second, enhanced food security as a direct consequence of this trade. Few rulers of fragile democracies—and nondemocracies—could survive long when the shops have no bread (Dizard 2022).

1.2 Contribution

In this chapter, we explore another perspective on virtual water: Many food-exporting countries also "export" large quantities of groundwater that are "mined" from their aquifers. Stated bluntly, these nations are exporting their environment, free of charge, to food-importing countries. Applying this insight to policy evalua-tions by raising concerns about food security as a result of water sustainability policies will have consequences; through a set of models and calculations within a multiscale framework, we explore the impacts of a "best case" scenario in which groundwater consumption is constrained to a "sustainable" equilibrium by prudent management. What changes are likely in the local and global patterns of commodity production, and what changes in prices can be expected as a consequence of such "good governance"? SIMPLE-G offers the opportunity for a holistic approach to incorporating these spillover effects of market-mediated land and water conservation policies. This approach can provide a more accurate reflection of the impact of groundwater policies on production and land use and promote the sustainable use of resources within a comprehensive framework.

1.3   Sustainable Groundwater Use: Some Definitions

What does sustainability mean for groundwater? Groundwater systems are often highly complex, comprising multiple layers that are sometimes directly linked, sometimes partially linked, and sometimes completely independent; these layers are affected by the nature of the medium(s) through which, and in which, groundwater is stored, surface topology, land cover, lakes, rivers, and artificial interventions such as irrigation systems. Here, the objective is not to address these physical variables in detail—although in any specific case, they are fundamentally important.

The commonly adopted definition of sustainable groundwater use is that abstraction should not exceed recharge. This apparently logical formulation is at best misleading (Bredehoeft 2002); useful analysis of the status of a groundwater system—and the consequences of overabstraction—requires a rather more careful specification of the scenario being evaluated.

A groundwater system is in *equilibrium* when inflows are equal to outflows. Inflows include natural recharge from rainfall, additional infiltration of water imported via surface irrigation systems, return flows from local groundwater abstractions, and lateral inflow from surrounding hills or aquifer systems. Outflows include evaporation from wet surfaces and capillary rise, transpiration by plants, lateral flows toward streams or surrounding aquifers, and pumping. When these two sets of flows are equal, the water table remains essentially constant when observed over a period of years, with seasonal and annual variations due to, for example, actual rainfall patterns, and associated pumping rates. Yet even from this simple formulation, we immediately see that any increase in net abstraction from an aquifer (e.g., by introducing a new tube well) must have implications for one or more of the outflows. The water table must reach a new equilibrium at a lower level at which one or more outflows are reduced—lateral outflows to streams, water consumption by natural vegetation, flows to surrounding aquifers, or evaporation from waterlogged soil. When this has happened, inflows and outflows again equate, and stability is reestablished.

Already, the definition of "sustainable" becomes more complicated than is implied by the simple rule that abstraction should be less than recharge. (Indeed, if abstraction is less than recharge, the water table will rise over time.) In the example just described, *equilibrium* is restored after an increase in abstraction; again, the observed status of the groundwater system will remain constant over a period of years. Yet the impacts of the reduced outflows to vegetation and local streams may cause damage to downstream wetlands or aspects of the natural landscape that were previously supported. Environmentalists might well argue that these are not "environmentally sustainable" outcomes, as generally interpreted; rather, they are a new, less desirable, equilibrium. Downstream users (e.g., irrigators, fishermen, and ferry operators) will have similarly negative views of the new "sustainable" scenario.

As the process of increased groundwater demand for irrigation and other sectors unfolds, we pass through a continuum of equilibria, each of which meets the stability test where inflows are equal to outflows (see, e.g., Chinnasamy and Agoramoorthy 2016). Eventually, all the natural outflows fall to zero, and incremental abstraction is supported by a continuous reduction in aquifer storage—a scenario that is literally "unsustainable" because eventually the aquifer will be depleted, salinized, or otherwise rendered unfit for use. During this stage, there are no points of equilibrium. The water table continues to fall until the final equilibrium eventually arrives: The aquifer is depleted to the point at which use is restricted to whatever recharge still, sporadically and unpredictably, reaches the saturated zone. The "mining" element, in this scenario, necessarily reduces to zero.

Why is this more complex perspective on "sustainable" groundwater use important? It is commonplace to report that because measured abstraction is less than recharge there is "net availability" of groundwater for development (see, e.g., Chatterjee and Ray 2016). This position is apparently based on the assumption that the "available" resource is currently disappearing somewhere; such a conclusion is rarely (never?) associated with a reportedly *rising* water table. Alternatively, it might be assumed that the outflows that are impacted by further abstraction have no value. In this phase of development, the properly evaluated "sustainable" yield is or should be a *political* decision based on trade-offs among alternative water allocation regimes; as such, it is an issue of governance to which information (e.g., measurements, observations, and modeling) contributes to political decisions that determine allocations. Here, we include in the spectrum of possible political interventions the decision to do nothing, thereby allowing the tragedy of the commons (Hardin) to unfold. To our knowledge, this policy has never been made explicit, but this scenario is not rare. Before 2014, California's aquifers were being used excessively and without regulation. However, following the introduction of the Groundwater Sustainability Act, the Groundwater Sustainability Agencies were required to take up local control. They gradually implemented regulations, bringing 98% of Californians under their umbrella. Despite challenges, this has been a significant step toward securing California's groundwater resources.

In fact, all water resource development (e.g., diversion from rivers, construction of dams, water harvesting, deforestation, reforestation, and conversion to new irrigation technologies) involves some degree of reallocation of water among users and uses. Food security and rural employment increase when irrigation water is diverted from estuaries and wetlands. While such developments have vastly increased food production and hence have benefited society generally, they are not without costs; it is the role of governments, through political processes, to evaluate the underlying trade-offs and intervene appropriately.

In sum, groundwater development has four stages: (1) the natural state of precipitation, vegetation, and runoff; (2) progressive human interventions that expand use (e.g., agriculture and domestic water supply) at the expense of other outflows while maintaining equilibrium between inflows and outflows; (3) groundwater mining, where outflows exceed inflows and the aquifer is progressively depleted; and (4) effective depletion of the aquifer, often associated with (a) subsidence of the land surface, which in many areas is already ongoing and substantial (Galloway and Burbey 2011); (b) soil compaction, such that infiltration is restricted and storage of soil moisture in the profile is reduced; and (c) extended time for any infiltration to reach the saturated zone, due to a combination of soil compaction and the ever-increasing depth to the water table. Unfortunately, many aquifers around the world are in the third stage of progressive depletion; in the absence of interventions by the relevant authorities, they will automatically progress to the fourth stage, effective depletion, and irreparable damage.

In this chapter, we assess the impacts of restricting groundwater abstraction to sustainable levels—halting the "mining" of groundwater and returning to stage 2, maintaining equilibrium between inflows and outflows. This intervention results in the redistribution of production and trade and associated local and global price changes. Achieving this goal would constitute improved governance, requiring interventions by the relevant authorities to reduce groundwater abstraction in many areas by defining allocation policies, introducing laws that reflect those policies, and providing institutional arrangements to enforce those laws.

These interventions will be politically challenging, and some countries will fail. In the following analysis, we attempt to evaluate the local and global economic costs associated with this move to "good governance." Such governance would allow for continued, controlled use of aquifers as temporary buffer storage and preservation of resources that can be assigned to priority uses during periods of severe drought. The alternative of allowing aquifer depletion and loss of buffer function will undoubtedly be far more costly—an issue to be addressed in a further study.

1.4   The State of the World's Groundwater

Various analysts have assembled data and models to assess groundwater status globally and for major regions (Siebert and Döll 2008, 2010; Wada et al. 2012, 2014; Famiglietti 2014). While estimates vary depending on the methodology and the time period analyzed, there is consistency among all analysts that current irrigation in major agricultural areas is dependent on groundwater abstraction rates that deplete the underlying aquifers.

In addition to these concerns, the rate of increase is remarkable. According to Wada et al. (2014, p. 14), "the current degree of non-sustainable use may compromise the future livelihoods of millions of people and their living standards": Global aquifer depletion almost tripled between 1960 and 2010 and is projected to almost double again by 2099 (although whether these quantities of water actually exist must be questioned). Among individual countries, the increased rate of depletion between 1960 and 2010 was most pronounced in India (almost tripling) and the United States, China, and Pakistan (roughly doubling). In Saudi Arabia, fortunately, an outlier, consumption increased sixfold—mostly due to the now-abandoned policy of pursuing self-sufficiency in wheat, but abstraction is still driven by substantial production of irrigated fodder. According to Famiglietti (2014, p. 948), "vanishing groundwater will translate into major declines in agricultural productivity and energy production, with the potential for skyrocketing food prices and profound economic and political ramifications." The next sections address the likely impacts—locally and globally— of the reductions in water supply that would stabilize aquifers.

2   SIMPLE-G Version Employed in This Chapter

To capture the local, national, and global impacts of and responses to sustainability groundwater restrictions, we employ a globally gridded version of SIMPLE-G that contains around 1.3 million grid cells at a resolution of 5 arcmin, comprising squares with a width of 9.26 km at the equator. As in the SIMPLE-G-US model used in Chap. 12, each grid cell represents a distinct unit of agricultural production on which competition for land and water resources plays out within and between rainfed and irrigated crop production. It also includes differentiated markets at subregional and grid-cell levels. In the face of groundwater sustainability restrictions, the model solves for the gridded equilibrium level of irrigated and nonirrigated land as well as for the extraction of groundwater and surface water for crop production. It also determines the new equilibrium for local, national, and global crop prices and for domestic use and international trade in crops.

The production at each grid cell follows a nested Constant Elasticity of Substitution functional form as detailed in Fig. 16.1. This figure illustrates how various agricultural inputs are combined, reflecting the imperfect substitutability between them. Specifically, groundwater, surface water, and irrigation equipment are first combined into a composite "water input." This composite water input is then combined with land to form "augmented land." Finally, all remaining inputs are grouped into a single category, with a distinct representation of labor demand.

2.1   Product Differentiation

The SIMPLE-G model does not model individual farmers' behavior but rather depicts the outcome at the aggregate level in each grid cell for all crops aggregated into a single composite commodity. The crop output of each grid cell is differentiated from that of other grid cells, reflecting differences in crop composition. This product differentiation applies at both the national and regional levels. The SIMPLE-G model used here has 17 consumption regions (derived from SIMPLE, see Part III)

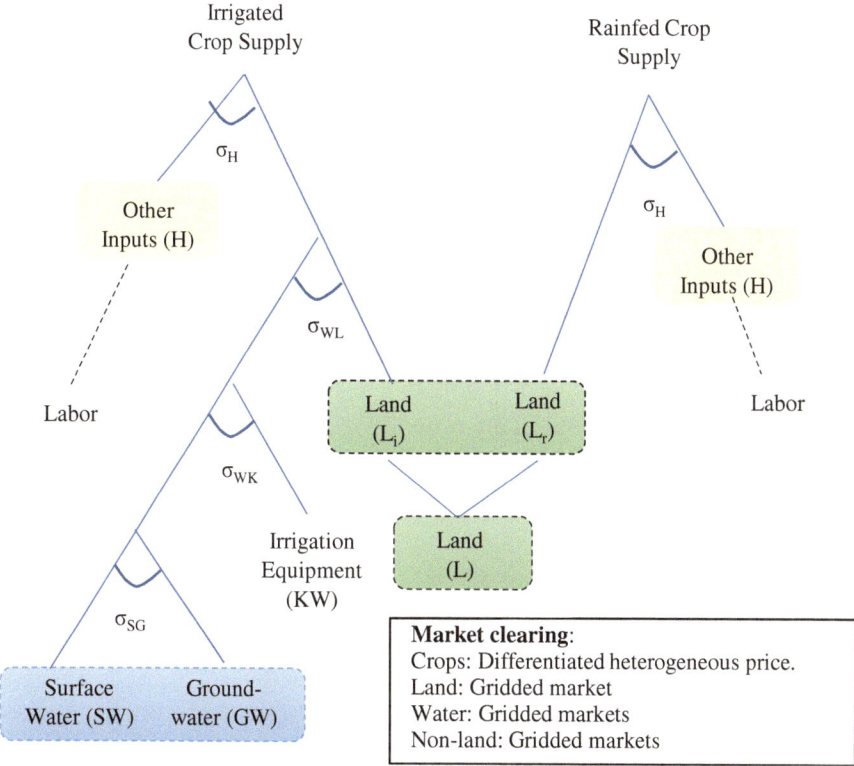

Fig. 16.1 Overall gridded production structure of the SIMPLE-G model The model determines the equilibrium local prices, land use, water use, and agricultural demand and supply for other inputs, including labor

and 136 subregions, of which 120 are individual countries. In any given grid cell, yield, water use, and nitrogen fertilizer use in crop production are weighted averages over all crops. However, we distinguish between rainfed and irrigated crop production within each grid cell.

2.2   Virtual Water Trade

This version of SIMPLE-G introduces virtual water trade to represent the water embedded in the international trade of agricultural commodities. There is a rich body of literature around the concept, applications, and modeling of virtual water trade (Allan 1997, 2003, 2011; Dalin et al. 2017; Rosa et al. 2019). Here, changes in virtual water trade are distinguished by source, and these changes follow the corresponding changes in local production and regional exports (Haqiqi et al. 2022).

2.3   Labor Market Outcomes

In light of the findings in Chap. 13, we add labor market outcomes to this version of SIMPLE-G. It is crucial to consider the effects of sustainability policies on the labor market, especially for farm workers. Sustainability policies can have a range of impacts on the labor market. Certain sustainability policies, such as those that target labor-intensive agricultural activities, can result in job loss. Farmworkers are particularly susceptible to this because the agricultural sector already relies heavily on mechanization and new technologies. On the other hand, some sustainability policies can create new jobs. However, farm workers may not have the skills or resources to make the sectoral or geographic transition to these new jobs. Here, we add a simple model of labor employment in crop production without explicitly modeling movement across economic sectors. In this version, the labor market follows the composite of human system inputs. The mobility and rigidity of the farm labor force are spatially heterogeneous and informed by travel time datasets.

3   Experiment Design

To focus our analysis on the impact of groundwater sustainability restrictions, we assume no changes in regional population, income, or biofuel demands. However, the demand responses to changes in food prices are captured in the interaction of demand and supply in the regional and global markets. Additionally, we assume no change in local climate conditions; this assumption allows us to assess the impact of groundwater restrictions taken on their own.

3.1   Groundwater Restriction Scenario

To permit a unified analysis of groundwater sustainability policies, we pair information from a hydrological model with the SIMPLE-G-Global model (Fig. 16.2). Several studies have successfully linked these two models (e.g., Woo et al. 2022), shedding light on the impacts of surface water scarcity (Liu et al. 2017) and the yield impacts of compound hydroclimatic extremes (Haqiqi et al. 2023). Here, we follow Grogan et al. (2017a, b) and employ the outputs of the water balance model (WBM), a validated and widely used macroscale hydrological model (Wisser et al. 2010; Grogan 2016; Grogan et al. 2022). Consumptive water requirements are calculated by crop and growth stage based on soil moisture, temperature, and irrigation status. The sources of irrigation in the WBM can include reservoirs, rivers, shallow groundwater, and nonrenewable groundwater if available. The model also considers nonbeneficial consumption and return flows through irrigation runoff, baseflow between groundwater and surface water, and percolation of irrigation return flows

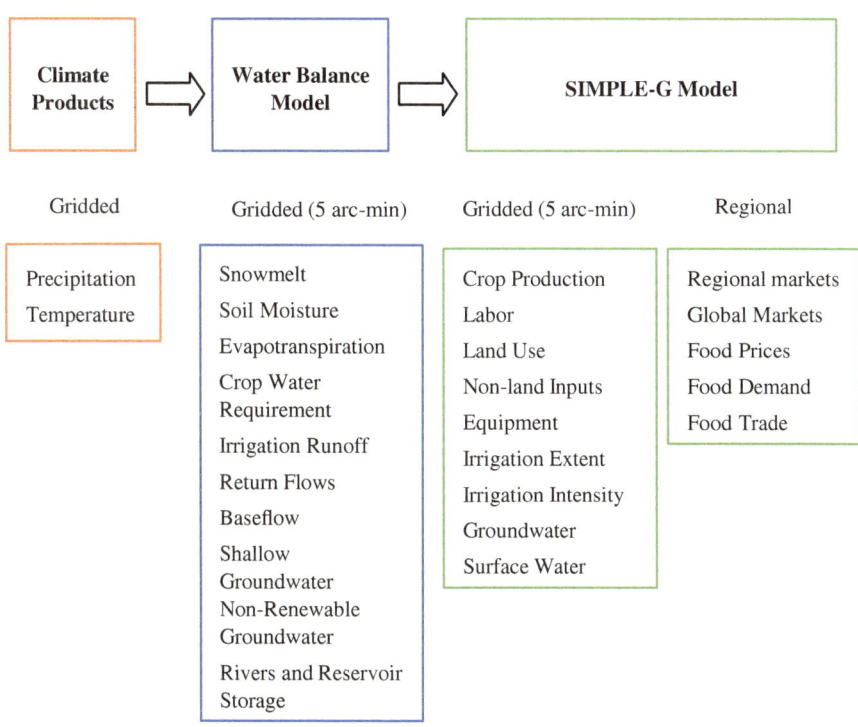

Fig. 16.2 Employing hydroclimatic information to inform the economic model about irrigation water availability and nonrenewable groundwater irrigation The water balance model (WBM) is a global gridded framework for modeling water mass balance for each grid cell. It simulates the vertical water exchange between the atmosphere and land as well as the horizontal flow of water through river networks, baseflow, and runoff. Daily precipitation and temperature data from climate products and information regarding agriculture and water demand at each location are used as inputs for the WBM. The simulated changes in surface water and groundwater volumes are fed into the SIMPLE-G model, which models decisions about land and water use. The regional outcomes are determined via the interaction of gridded crop outputs and regional and global market responses

to shallow groundwater. The multiscale nature of this model, which builds from the grid cell to global hydrology, makes it appropriate for this study.

We follow Grogan et al. (2016, 2017a, b) to determine the sustainable level of groundwater abstraction, determining the maximum allowable level of groundwater use such that there is zero nonrenewable groundwater use, thereby ensuring a stable equilibrium level of groundwater. For each grid cell, we calculate the required reduction in abstractions to achieve this objective. This level is calculated by running the WBM over the period from 1980 to 2009, while not allowing abstraction from nonrenewable resources. The sustainable level of abstraction varies each year depending on weather conditions. To avoid excessive complexity and in order to focus on long-run conditions, we take the 30-year long-term average as the basis for determining the sustainable level of abstraction. We utilize historical global observational climate products based on ERA-interim (European Centre for Medium-

Range Weather Forecasts-reanalysis, Dee et al. 2011), MERRA (Modern-Era Retrospective Analysis for Research and Applications, Rienecker et al. 2011), NCEP (National Centers for Environmental Prediction, Saha et al. 2014), and UDEL (University of Delaware, Willmott and Matsuura 2001). The 30-year average sustainable level of abstraction is used to inform the economic model about long-run groundwater availability. There are other possible definitions of sustainability, and we will discuss the implications of this choice below.

3.2   Interoperability and Consistency

Interoperability refers to the ability of different models to communicate and exchange data with each other. In multisystem analyses of agriculture that involve hydrology and climate, interoperability is important because it enables researchers to combine different datasets and model insights, resulting in more comprehensive and realistic simulations of the complex interactions among agriculture, climate, and other systems. It is crucial to ensure temporal and spatial consistency of the exchanged information, and this is often challenging when it comes to combining economic and biophysical models.

One challenge we face is that SIMPLE-G-Global is constructed based on economic and agricultural information circa 2017. In other words, crop production, cropped area, yields, and prices reflect more recent market conditions than are reflected in historical weather data. In addition, the size of SIMPLE-G grid cells is 5 arcmin, while the historic WBM runs are for global 30 arcmin grid cells. Having determined the long-run historic sustainable level of groundwater withdrawal by grid cell, we take the current conditions of water resources based on recent WBM runs from 2012 to 2018 based on GLEAM v3 (Global Land Evaporation Amsterdam Model, Martens et al. 2017), which uses 5 arcmin grid cells. This allows for consistency with the 2017-based economic model. We assume that all the smaller (5 arcmin) grid cells follow their underlying larger grid cells in terms of nonrenewable groundwater rates. For this study, we reconstruct the SIMPLE-G global model based on the WBM land use and land cover dataset.

Table 16.1 summarizes the current situation for a set of major groundwater users in the world. Overall, 27% of total irrigation water consumption is linked to nonrenewable resources. However, this is substantially higher in some regions—62% in Iran and 39% in India. Compared with total global crop water demand, the nonrenewable groundwater contribution is around 6% (although this figure is 43% for Iran and 23% for India). The outputs of the WBM in terms of groundwater and surface water use in crop production are in line with those of other studies (Siebert and Döll 2010; Chapagain and Hoekstra 2011; Gleeson et al. 2012; Bierkens and Wada 2019; Mekonnen and Hoekstra 2020).

Figure 16.3 shows the distribution of the cropped areas that rely on irrigation groundwater around the year 2017. "Green" water is that provided to crops by rainfall; "blue" water comprises irrigation water from surface and groundwater.

Table 16.1 Total crop water demand and contribution of nonrenewable resources for major groundwater users circa 2017

	Total crop water demand	Green water contribution		Blue water contribution		Nonrenewable groundwater	
	(km³ year⁻¹)	(km³ year⁻¹)	%	(km³ year⁻¹)	%	(km³ year⁻¹)	%
India	493	203	41	289	59	112	23
China	438	223	51	215	49	64	15
Pakistan	101	2	2	98	98	21	21
US	428	374	87	55	13	13	3
Iran	6	2	31	4	69	3	43
Mexico	48	36	76	12	24	3	5
World	3,848	2,934	76	915	24	244	6

Source: Outputs of the global water balance model based on GLEAM v3. Aggregated from 5 arcmin grid cells. Authors' calculations are based on Grogan (2016) and Grogan et al. (2017a, b, 2022). The percentage values show the share of total crop water consumption

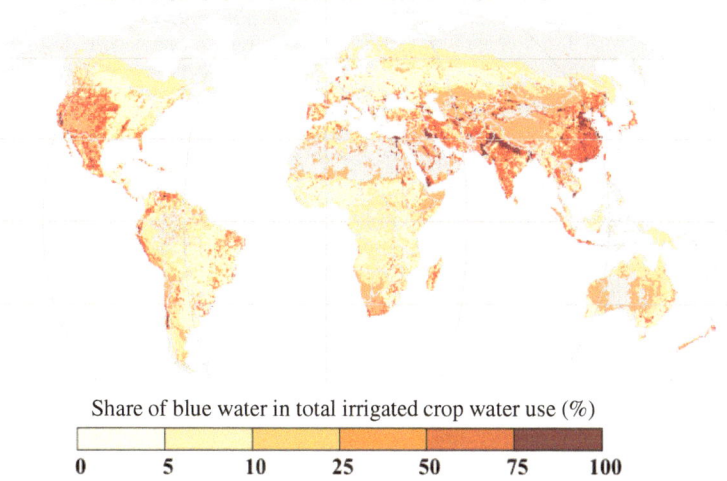

Share of blue water in total irrigated crop water use (%)

0   5   10   25   50   75   100

Fig. 16.3 Spatial distribution of irrigation showing the share of blue water in total gridded crop water consumption around the year 2017 *Red* indicates that little green water is available to crops. Therefore, if there is crop production in these locations, the area must be irrigated

Figure 16.4 shows groundwater hotspots across the globe. These are most prominent in the United States, China, India, Pakistan, and the Middle East, where crop water demand in some grid cells is obtained almost entirely from nonrenewable resources. As seen here, the problem of nonrenewable groundwater abstraction tends to be quite concentrated.

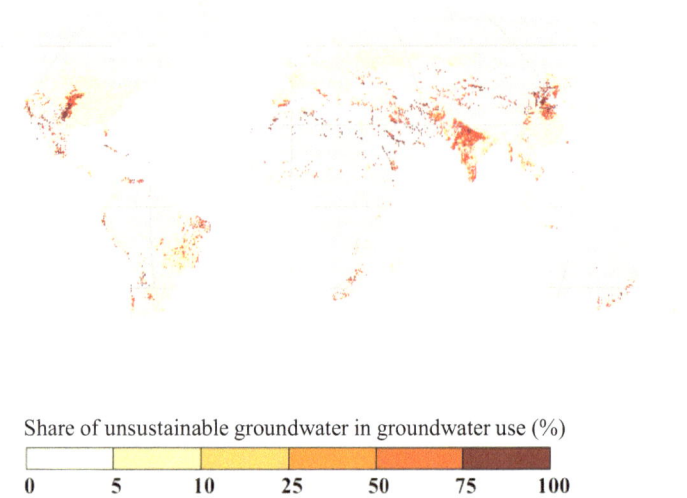

Share of unsustainable groundwater in groundwater use (%)

| 0 | 5 | 10 | 25 | 50 | 75 | 100 |

Fig. 16.4 Spatial distribution of unsustainable groundwater illustrating the share of unsustainable groundwater in total gridded irrigation water consumption around the year 2017 *Red* indicates that crop production relies mainly on unsustainable groundwater. *Gray* indicates areas not irrigated

3.3 Implementing Groundwater Restrictions in the SIMPLE-G Model

As discussed previously, the groundwater sustainability limits are calculated by the WBM considering the hydrological relationships and connections within and across grid cells. To allow for a deeper understanding of the distinct forces at work when groundwater sustainability restrictions are imposed, we first run the SIMPLE-G model assuming no change in surface water use, irrigation intensity, or irrigation extent. This restricted scenario generates the largest possible impacts on food prices and food consumption.

While informative, this simplified scenario does not capture adaptation to the sustainability policy, nor does it capture spillover effects. Therefore, we consider additional scenarios in which the system responds to these direct impacts. The first line of response involves a change in the composition of irrigation water resources. Depending on availability and relative costs, the model measures the change in surface water withdrawals as cost-minimizing producers attempt to reduce the impacts of groundwater restrictions on production. Then, the model considers changes in irrigation intensity in terms of the water supplied and used by crops. At the level of a grid cell, this can be due to a change in the composition of crops within the grid cell or it can reflect a shift to a different deficit irrigation strategy that is not explicitly modeled. All these changes can affect the rate of groundwater recharge and thus alter the initial sustainability limit. These changes also have implications for downstream surface water availability. To capture the downstream effects, we include another sustainability scenario with iterations between the SIMPLE-G economic model and the WBM hydrological model. In this scenario, decisions regarding surface water, irrigation intensity, and irrigation extent are transferred to the WBM, which then provides implications for surface water availability downstream and a revised required change in groundwater restrictions. We will show why these return flows are important for determining the spatial pattern of production at the local level.

4 Results

Restricting groundwater consumption alters global agricultural production. If there are no other sources of water, which is the case for many hotspots of unsustainable groundwater abstraction, the immediate direct impact is a fall in production. According to our calculations, the consequence of this "first round" of responses to the groundwater sustainability restriction is a 12.3% reduction in global crop production (around two billion tons of corn-equivalent crops). Despite comprising only 6% of global water consumption, the proportionate reduction in production is twice as large due to the relatively high productivity of groundwater. However, as the food system responds to this groundwater constraint, the final impact on global crop production is greatly moderated and limited to only −0.2% corresponding to a change of −28.4 million tons in corn-equivalent crop production.

Figure 16.5 decomposes the change in global crop production due to groundwater sustainability restrictions. We label them as follows: "no adaptation" (the direct effects), "substitution" of surface water, "rainfed conversion," "relocation," and "global trade." The initial margin of adjustment involves substituting surface water for groundwater, where feasible, and utilizing other farm inputs more intensively (bar b). This adaptation results in a 3.5% moderation in the global crop output decline. The next farm-level response considered is the conversion of irrigated land to dryland crop production, where rainfall is adequate. Despite rainfed production having a lower yield (or simply not being possible in some locations), in the aggregate, these results suggest that rainfed substitution can offset one-third of the damage at the global level (a 3.18% increase in Fig. 16.5, bar c). The consequences of conversion from irrigated to rainfed crop production depend on the relative yields, changes in local land rents, and the demand for alternative land uses. The next margin of response in Fig. 16.5 (bar d) is due to the national relocation of production. Rising prices in countries where groundwater is restricted motivate the expansion of production in other suitable locations, especially in regions with strong, price-inelastic, domestic demand (e.g., South Africa, India, and China). The extent of expansion in other grid cells depends on biophysical and economic conditions in each grid cell, but at a global level, these adjustments offset about one-sixth of the direct impact (2.1% increase in Fig. 16.5, bar d). Similarly, international trade in agricultural production offsets the initial fall in production through exports and imports and the ensuing trade in "virtual water." (Bar e offsets 3.3% of the global

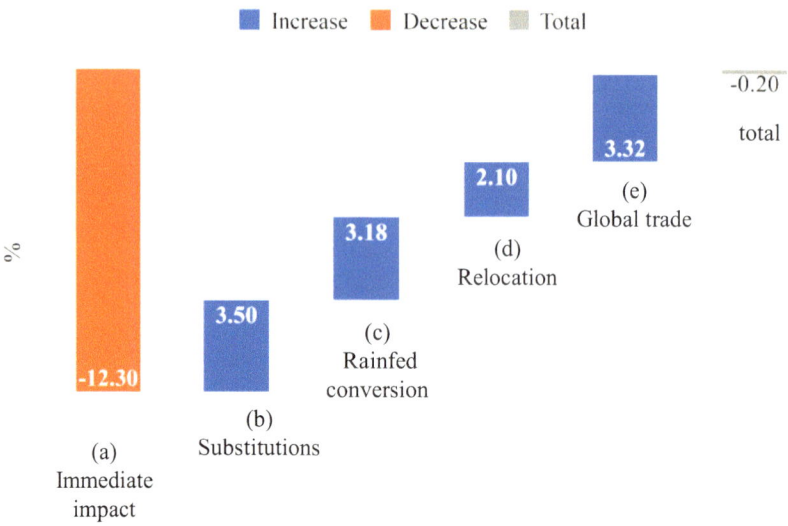

Fig. 16.5 Decomposition of global crop response to global groundwater sustainability restrictionsThe leftmost (*red*) bar shows the immediate/direct impact prior to adaptation. The subsequent bars incorporate successive market-mediated adaptations: (**b**) substitution of surface for groundwater as well as other economic responses by farmers, (**c**) conversion of irrigated to rainfed area, (**d**) expansion of domestic production in response to higher prices, and (**e**) price-induced expansion in other regions

production decline). Depending on and product similarity and trade flexibility, the contribution of relocation can increase to 4.5% while trade contribution may decline to 1.9%. See the application files for different closures. Columns 2 and 4 in Table 16.4 provide more detail on the change in virtual water trade due to the groundwater sustainability policy.

At the global level, implementing the sustainability policy implies a 27% reduction in groundwater consumption (244 km^3), corresponding to a 6% fall in total water consumption in agriculture from all sources (rainfall, surface water, and groundwater). Due to multiscale responses, the long-run impact on crop prices is quite modest (0.4%). However, the price increases are much larger in some local markets. The analysis projects a long-run increase in global surface water withdrawal of 3.42%. On the other hand, despite the fact that global irrigated production has declined by 1.03% (−68 million tons), rainfed production has increased by 0.42% (+39 million tons). This nearly offsets half of the global reduction in irrigated output. This is also reflected in the cropped area: changes of around −0.43% in the irrigated area (−1.4 million ha) and + 0.21% in the rainfed area (+2.6 million ha), leading to an increase in the global cropped area of 0.08% or + 1.2 million ha (Table 16.3, columns 2–8).

The long-run production impacts are small at the country level, as similar adaptations can occur within a country. The findings suggest that in China, India, the United States, Pakistan, and Iran, the total annual production will decline by −12.5, −6.0, −3.0, −2.6, and −2.1 million tons of corn-equivalent crops, respectively. Additionally, the overall long-run employment impacts are positive for many countries. According to the model output, China, India, Indonesia, Bangladesh, and Turkey will experience 296.6, 236.9, 66.7, 57.5, and 34.6 thousand new agricultural jobs, respectively, due to spillovers and expansions in croplands. The largest negative employment effects are in Pakistan and Vietnam, where crop employment falls by −36.0 and −24.8 thousand jobs, respectively.

4.1 Spatial Distribution of the Impacts

This section illustrates the final changes in the equilibrium levels of input demand and crop supply in response to the global enforcement of sustainable groundwater use. These are equilibrium outcomes of the model and reflect interactions between local and global agricultural markets. The spatial heterogeneity of the impacts is due to both heterogeneity in the magnitude of the reduction in groundwater and biophysical differences across locations. The findings suggest a significant change in the pattern of irrigated production, water consumption, irrigated cropped area, and employment in irrigated agriculture as a consequence of the sustainable groundwater policy.

4.1.1 Impact on Cropped Area

Figure 16.6 shows the spatial distribution of the impact on cropped areas. Red and yellow indicate contraction, and green indicates expansion as a percentage change. The grid cells with a high dependency on unsustainable groundwater would face a sharp reduction in irrigated cropped area, coupled with an increase in rainfed cropland. The net change for such areas—which include California's Central Valley, the US Ogallala Aquifer, the Mississippi River Basin, parts of India and Pakistan, parts of China, North Africa, and the Middle East—is a decline in cropped area. The analysis projects a slight increase in both irrigated and nonirrigated areas for the rest of the world.

4.1.2 Impact on Surface Water

Figure 16.7 shows the spatial distribution of the impact of the groundwater conservation policy on surface water irrigation as a percentage change. Surface water irrigation increases in almost all irrigated areas, with the greatest increases occurring in the hotspots of nonrenewable groundwater use. As this is a percentage change, the magnitude of the change in surface water irrigation depends on availability and current uses. The increase in surface water irrigation reflects the substitution of

(a)

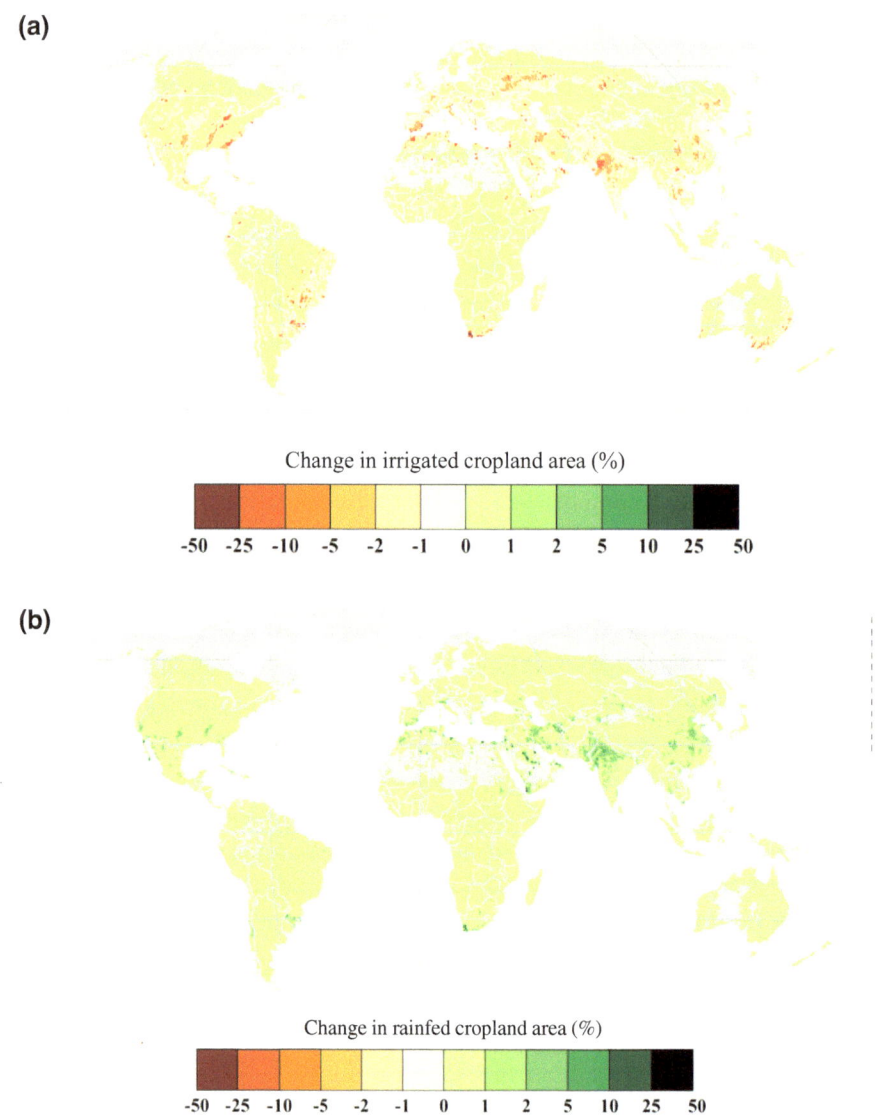

Fig. 16.6 Percentage change in (a) irrigated and (b) rainfed cropped areas in response to global groundwater sustainability restrictions produced by the water balance model

groundwater or a change in the location and scale of production. These changes can be decomposed into a "substitution effect" and a "scale effect." If the substitution effect is dominant, the percentage change in surface water use in the targeted areas is positive and more than compensates for the decline in irrigated areas. If the scale

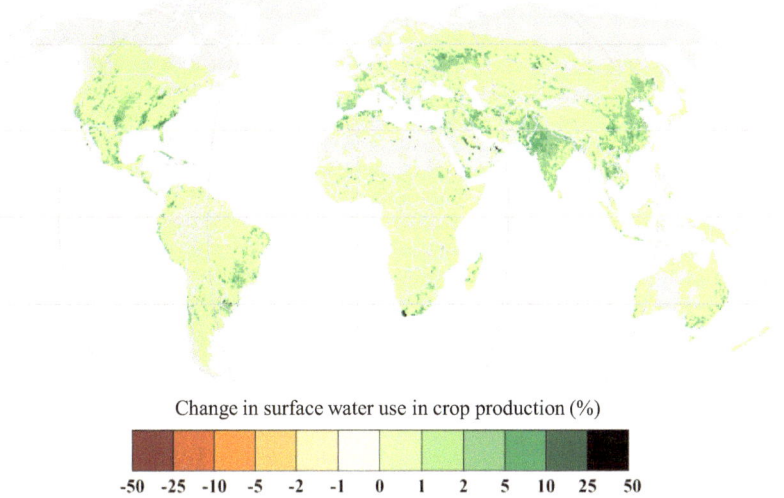

Change in surface water use in crop production (%)

Fig. 16.7 Global changes in surface water irrigation in response to global groundwater sustainability restrictions The map shows the percentage change in surface water use in irrigation. The impact on surface water can be decomposed into a "substitution effect" and a "scale effect." If the substitution effect is dominant, the change in surface water use in targeted areas is positive and compensates for the decline in irrigated areas. If the scale effect is dominant (less irrigated area), the change in surface water use in the targeted area is negative. Dark green indicates an increase in surface water use with a dominant substitution effect, and light green indicates the locations with an increase in surface water irrigation due to a positive scale effect, absent the substitution effect

effect is dominant (i.e., less irrigated area overall), then the change in surface water use in the targeted area is negative. (Haqiqi et al. (2022) explore the extreme case in which the expansion of surface water irrigation is restricted in targeted areas.)

The maps illustrate the conversion to rainfed production in targeted locations: A clear increase in rainfed areas is projected for most of the grid cells, with large reductions in irrigated areas. Both maps show a moderate increase in irrigated and rainfed areas (*light green*). The expansion in global cropland area and conversion to rainfed production are important channels of adjustment and economic responses to groundwater sustainability policy.

4.1.3   Impact on Production

Figure 16.8 illustrates the impact of enforcing sustainable groundwater use on equilibrium irrigated and rainfed crop production. These changes are the consequences of the changes in water use, land conversion, employment, local rents, crop prices, farm revenues, and other local and global changes. In general, irrigated production declines in hotspots with unsustainable groundwater use, while rainfed production increases. The increase in surface water use in some dry locations may

(a)

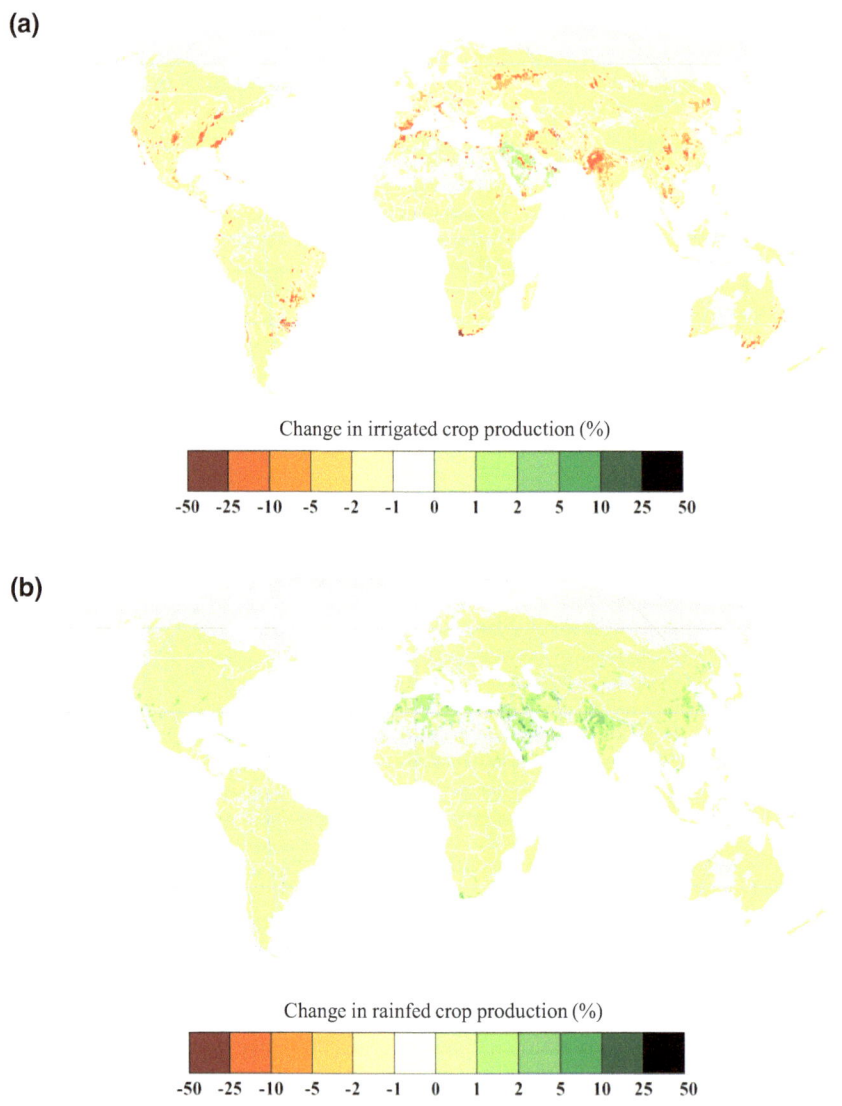

Fig. 16.8 Percentage change in (**a**) irrigated and (**b**) rainfed crop production in response to global groundwater sustainability restrictions *Red* and *yellow* areas indicate a reduction in production and *green* areas indicate an increase in production. While irrigated production in the unsustainable groundwater hotspots is substantially reduced, these hotspots also show the biggest percentage increase in rainfed production. This conversion to rainfed agriculture compensates for some of the immediate impacts on irrigated production. In addition, *light green* on the irrigated and rainfed maps indicates an expansion of crop production in nontargeted locations. While the patterns are similar to changes in the cropped area, the percentage change in production is smaller because rainfed crops have lower yields

cause an increase in irrigated production next to the targeted areas or may not be enough to recover irrigation production (e.g., in the Middle East).

4.2 Regional Outcomes

Restricting groundwater irrigation has diverse implications for regional crop production, land use, and food prices. Table 16.2a reports the long-run results at the regional scale, and Table 16.2b reports the absolute changes. In terms of percentage changes, the largest price increases arise in the Middle East, North Africa, and South Africa. In general, rainfed production increases, while irrigated production declines. In countries facing less severe restrictions, conversion to rainfed and surface water-fed cropping (driven by higher local prices) can fully offset the decline in irrigated production.

In absolute terms, the biggest declines in crop production are predicted in China, India, the Middle East, and North Africa, where crop production will decrease by − 12.5, −8.8, −3.6, and −2.5 million tons, respectively.

4.3 Implications for Virtual Water Trade

Groundwater restrictions are expected to alter agricultural trade patterns (Table 16.3). The biggest percentage change in agricultural trade flows is expected for North Africa, South Africa, and the Middle East, where production is sharply reduced and imports rise. The model suggests that trade will play an important role in meeting the demand for food in these regions. (Table A6 in Haqiqi et al. (2022) reports country-level results.) The largest percentage declines in production are estimated to occur in Saudi Arabia, Jordan, South Africa, Oman, and Iran, where production drops by 19.0%, 10.5%, 7.6%, 4.8%, and 2.8%, respectively (see Haqiqi et al. 2022, Table A6). In these countries, there is little possibility of using surface water as a substitute for groundwater (indeed, rivers in these areas are often overabstracted), and there is little room for expanding rainfed agriculture at competitive production costs. On the contrary, production is expected to increase in Turkey by 1.7%, as access to more land and water is less challenging. In larger countries such as India, China, and the United States, the relocation occurs predominantly within the country (Fig. 16.6).

Table 16.4 shows the impacts of groundwater sustainability restrictions on virtual trade in blue water. For all the regions listed in this table, exports of virtual blue water decrease, with the biggest declines occurring in the United States and South Asia. Note that this is largely due to the expansion of rainfed agriculture, both through conversion from irrigation and from cropped area expansion. Expanding rainfed crop production likely leads to a rise in virtual trade in green water, which we have not quantified in this study. In summary, the blue water (groundwater and

Table 16.2 Long-term regional impacts of restricting groundwater irrigation to renewable resources

Panel A. Percentage change in crop price, production, and area, by region

Region	Crop price (% change)	Crop production (% change)			Cropped area (% change)		
		Total	Irrigated	Rainfed	Total	Irrigated	Rainfed
East Europe	0.23	0.12	−0.71	0.21	0.07	−0.29	0.09
North Africa	1.10	−0.72	−1.21	0.91	0.19	−0.41	0.43
Sub-Saharan Africa	0.13	0.06	−0.46	0.11	0.06	−0.14	0.06
South America	0.29	0.10	−0.50	0.28	0.11	−0.18	0.14
Brazil	0.28	0.12	−1.37	0.33	0.12	−0.77	0.20
Australia and New Zealand	0.34	0.09	−0.90	0.35	0.11	−0.37	0.15
Europe	0.34	0	−1.46	0.35	0.09	−0.62	0.18
South Asia	0.80	−0.40	−1.20	0.74	0.04	−0.54	0.55
Central America	0.38	−0.15	−0.94	0.35	0.09	−0.36	0.24
South Africa	0.63	−1.53	−3.95	0.58	0.11	−1.61	0.32
Southeast Asia	0.24	−0.01	−0.71	0.21	0.06	−0.35	0.16
Canada	0.32	0.36	−0.24	0.37	0.14	−0.13	0.15
United States	0.43	−0.26	−1.45	0.43	0.09	−0.69	0.25
China	0.64	−0.27	−0.91	0.65	0.07	−0.32	0.46
Middle East	1.00	−0.58	−1.19	0.76	0.14	−0.30	0.43
Japan and Korea	0.10	0.01	−0.11	0.11	0.02	−0.08	0.09
Central Asia	0.40	−0.14	−0.36	0.29	0.07	−0.07	0.11
World	0.4	−0.2	−1.1	0.4	0.1	−0.5	0.2

Panel B. Actual change in crop price, production, and area, by region

Region	Change in crop production (1000 MT)			Change in cropped area (1000 ha)		
	Total	Irrigated	Rainfed	Total	Irrigated	Rainfed
East Europe	889	−549	1,390	146	−38	183
North Africa	−2,497	−3,241	738	53	−34	87
Sub-Saharan Africa	783	−568	1,331	130	−13	143
South America	591	−747	1,285	78	−16	95
Brazil	1,108	−1,628	2,709	76	−40	117
Australia and New Zealand	129	−286	400	34	−9	43
Europe	39	−3,303	3,328	96	−80	176
South Asia	−8,820	−15,781	6,576	87	−541	628
Central America	−620	−1,484	869	38	−37	74
South Africa	−1,266	−1,381	276	16	−25	40
Southeast Asia	−80	−2,506	2,381	69	−91	161
Canada	635	−20	617	55	−1	56
United States	−2,983	−6,100	3,189	137	−190	327
China	−12,522	−24,971	11,874	98	−217	315
Middle East	−3,581	−4,992	1,469	81	−67	148
Japan and Korea	11	−87	97	1	−3	4
Central Asia	−193	−326	128	13	−3	16
World	−28,374	−67,968	38,655	1,209	−1,404	2,613

Table 16.3 Trade impacts of restricting groundwater irrigation (percentage change in supply to the domestic market, exports, and imports)

Region	% Change in supply		% Change in demand
	Domestic	Exports	Imports
East Europe	0.02	0.54	−0.45
North Africa	−0.56	−2.83	1.96
Sub-Saharan Africa	0.01	0.78	−0.75
South America	−0.01	0.34	−0.28
Brazil	0.01	0.38	−0.29
Australia and New Zealand	−0.01	0.18	−0.10
Europe	−0.05	0.10	−0.11
South Asia	−0.35	−1.66	1.09
Central America	−0.14	−0.18	−0.02
South Africa	−0.99	−2.29	0.76
Southeast Asia	−0.05	0.40	−0.45
Canada	0.09	0.52	−0.16
United States	−0.16	−0.43	0.12
China	−0.27	−1.07	0.72
Middle East	−0.42	−2.43	1.78
Japan and Korea	−0.01	0.83	−0.82
Central Asia	−0.14	−0.21	−0.02

Table 16.4 Impact of groundwater sustainability policy on virtual blue water exports by source

	%		10^6 m^3	
	Surface water	Groundwater	Surface water	Groundwater
East Europe	1.41	−11.90	59	−152
North Africa	4.87	−48.39	103	−249
Sub-Saharan Africa	1.42	−8.96	25	−74
South America	2.12	−14.26	144	−431
Brazil	2.84	−17.61	152	−551
Australia and New Zealand	1.99	−18.02	52	−170
Europe	2.29	−23.14	238	−928
South Asia	5.26	−26.74	779	−2,984
Central America	3.23	−21.42	239	−647
South Africa	6.30	−28.41	88	−141
Southeast Asia	3.47	−23.54	246	−845
Canada	1.25	−10.30	10	−25
United States	3.62	−27.65	885	−2,935
China	3.54	−21.20	65	−228
Middle East	4.39	−31.38	138	−299
Japan and Korea	0.76	−5.40	2	−6
Central Asia	1.48	−14.71	48	−171
World	3.35	−23.48	3271	−10,835

surface water) virtual trade declines by 10.8 billion m^3 in the wake of this sustainability policy.

4.4   Implications for Employment

The change in employment is calculated assuming a gridded labor market with no explicit modeling of labor mobility across grid cells. Labor is part of the nonland composite input assuming a fixed proportions production process (recall Fig. 16.1). Thus, the demand for labor input follows the changes in demand for the nonland composite input due to changes in intensification. Figure 16.9 illustrates the spatial pattern of change in employment. The pattern is different from that of changes in cropped area and production due to the varying contributions of nonland factors of production by country. The initial numbers for gridded employment were obtained from subregional rates of labor per hectare of cropland. Haqiqi et al. (2022) provide more information on the average number of workers per hectare and per ton of corn-equivalent output in rainfed and irrigated agriculture.

Table 16.5 provides the model results for the aggregated impacts on employment in crop production activities around the world. The findings suggest that employment in irrigated agriculture will decline by about 1.5 million jobs globally. However, employment is expected to increase by 2.5 million jobs in areas with more extensive rainfed agriculture, creating a net gain of one million new jobs globally. Overall, these results suggest an increase of more than half a million jobs in China and South Asia, where domestic cropping area increases with the move to more rainfed production. This is due to the expansion of more extensive rainfed crop production, which requires more labor to work the increased land area. Finally, note that the average worker per hectare of land will be different for the marginal land endogenously added or removed from crop production due to substitutions between the nonland and land–water composites. As the land–water composite becomes scarcer and more costly, an intensification is expected, increasing the average number of workers per hectare of land.

4.5   Sensitivity to Adaptation Margins

Assessing the sensitivity of these findings to key parameter values reveals an interesting finding. The strength of any given individual margin of adaptation is dependent on the other margins. For example, when surface water substitution is limited, there is a stronger price response, and we observe larger adaptation responses from trade and relocation. Alternatively, when substitution in trade is less possible or conversion to rainfed is not feasible, the global irrigated cropped area tends to respond more strongly and there are bigger responses in surface water use. As a consequence, while the component parts of Fig. 16.5 are quite sensitive to

(a)

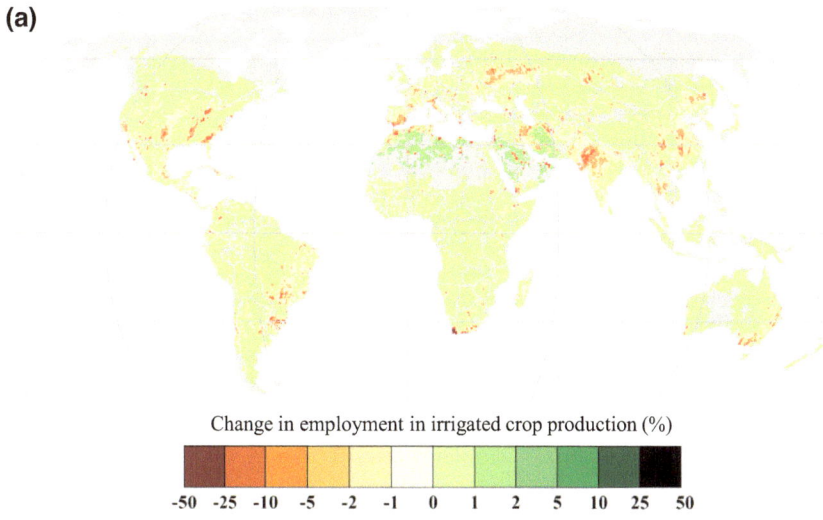

Change in employment in irrigated crop production (%)

-50   -25   -10   -5   -2   -1   0   1   2   5   10   25   50

(b)

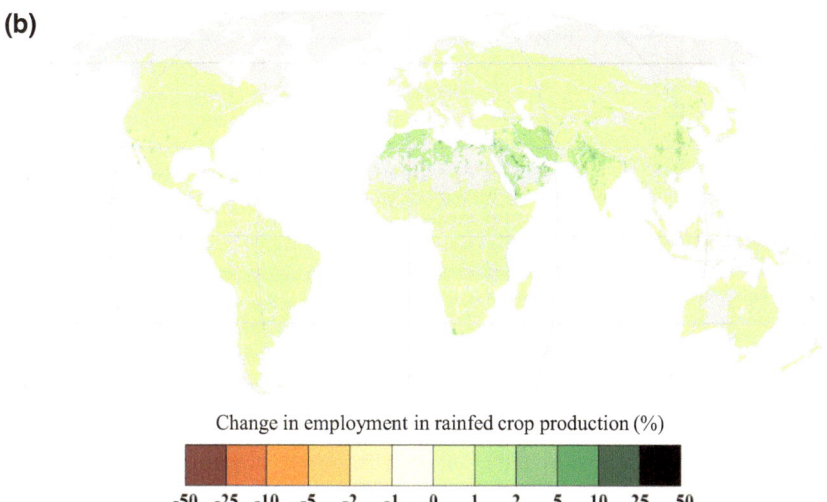

Change in employment in rainfed crop production (%)

-50   -25   -10   -5   -2   -1   0   1   2   5   10   25   50

Fig. 16.9 Global changes in agricultural employment in (**a**) irrigated and (**b**) rainfed crop production (%) in response to global groundwater sustainability restrictions *Red* and *yellow* areas indicate a reduction in employment and green indicates increases in employment. While the patterns of change are similar to those of changes in cropped area and production, the reductions in employment are generally smaller, while the increases in employment are larger than the percentage change in cropped area

Table 16.5 Impact of groundwater sustainability policy on employment in crop production by practice and region

	% Change			Change in number of jobs		
	Total	Irrigated	Rainfed	Total	Irrigated	Rainfed
East Europe	0.16	−0.40	0.25	6,107	−1,979	8,086
North Africa	0.28	−0.26	1.03	28,500	−15,631	44,131
Sub-Saharan Africa	0.12	−0.04	0.12	154,544	−1,790	156,336
South America	0.18	−0.15	0.34	7,979	−2,246	10,226
Brazil	0.23	−0.96	0.34	32,697	−11,238	43,935
Australia and New Zealand	0.32	−0.24	0.37	582	−38	620
Europe	0.22	−0.45	0.36	13,417	−4,827	18,244
South Asia	0.11	−0.69	0.83	289,680	−858,248	1,147,920
Central America	0.20	−0.44	0.39	14,867	−7,281	22,148
South Africa	0.31	−1.73	0.55	2,885	−1,654	4,538
Southeast Asia	0.09	−0.40	0.24	83,296	−83,868	167,200
Canada	0.37	0.01	0.38	690	1	690
United States	0.23	−0.83	0.45	4,233	−2,626	6,863
China	0.14	−0.41	0.69	296,608	−425,672	722,272
Middle East	0.52	−0.16	0.88	46,601	−4,980	51,581
Japan and Korea	0.06	−0.01	0.11	662	−52	714
Central Asia	0.17	−0.09	0.36	10,355	−2,271	12,626
World	0.13	−0.52	0.50	993,701	−1,424,400	2,418,129

model parameters, the final impact on global production of eliminating individual margins of adaptation is less than 1%, and the increase in price for more rigid assumptions is also less than 1%. Haqiqi et al. (2022) provide details about these sensitivity analyses and the interactions among the various adaptation margins.

5   Discussion and Conclusion

Stress on groundwater is expected to increase due to income and population growth, and global warming is expected to increase agronomic water requirements. The rising number of drying wells and increasing instances of land subsidence around the world have led to increased attention being given to sustainable groundwater use (Befus et al. 2017; Jasechko and Perrone 2020, 2021; Klasic et al. 2022). However, little is known about the local and global implications of restricting groundwater use to sustainable levels. This study sheds light on the implications of such a scenario on land use and agricultural production.

Recent studies evaluating the impact of restricting groundwater irrigation on agriculture are either geographically limited (United States) or limited in their treatment of global impacts (Baldos et al. 2020; Graham et al. 2021). This study demonstrates that the global impacts of groundwater restrictions on global food availability are largely ameliorated in the long run due to local, regional, and global adaptations. We have also added to this literature by expanding the analysis to include virtual water trade and employment impacts of groundwater governance. The decline in unsustainable groundwater irrigation is accompanied by a decline in virtual trade in blue water, while employment in agriculture increases in the wake of expanding rainfed production. However, even in the long run, local impacts can be very significant for locations with a high dependency on unsustainable groundwater. In some locations, production and employment decline by up to 90%. In locations without alternative sources of employment and food supply, this would have significant local impacts, leading to more than one million jobs lost and likely more than one million families affected. These local impacts will have adverse impacts on the broader economy, a point that our partial equilibrium analysis ignores by holding income constant. Addressing these economy-wide impacts of sustainability policies requires a general equilibrium model, which would typically be implemented at the national level (Calzadilla et al. 2010; Golub and Hertel 2012; Liu et al. 2014).

Our findings suggest that restricting groundwater irrigation could have a significant impact on local irrigated production in hotspots of nonrenewable groundwater irrigation. This directly causes a substantial reduction in agricultural production and employment for these communities. However, the impact of global groundwater sustainability policies is likely to be overestimated if the dynamic responses to restricted groundwater consumption are ignored. We show that these responses at the local and global levels will lessen the negative impact on food prices and production at the regional and global levels. Changes in relative prices at different scales will motivate changes in decisions and market outcomes, including compositional effects at the local level and changes in surface water irrigation, the location of crop production, irrigation extent and intensity, and the international trade in food commodities. Of course, these adjustments are costly and can cause further environmental issues. Overall, our findings suggest that the land-use and deforestation implications of a global groundwater sustainability policy are small and that the long-run impact on food production is less than 1% at the global level. The long-run change in global cropland is also limited to +0.1%, corresponding to +1.2 million hectares of cropland. Expansion as a result of deforestation and cropping of marginal lands could lead to environmental degradation. In addition, as rainfed zones tend to be richer in carbon and biodiversity (Taheripour et al. 2013), other environmental implications of this policy should be studied more carefully, with ecological concerns in mind. This is a critical finding when comparing the likely benefits and costs of such a policy.

Our decomposition framework highlights the possible adaptation mechanisms and demonstrates why ignoring market-mediated responses may lead to the overestimation of the costs of adopting sustainability policies. In addition, the findings are important for understanding the implications of sustainable groundwater use for the virtual trade of blue water. Because of the likely changes in farmers' decisions and economic responses, it is necessary to consider wider market responses when evaluating the impacts of conservation policies. Global environmental and agricultural models that neglect the economic modeling of international trade may overestimate the losses by not accounting for changes in exports and imports based on economic decisions.

Considering the socioeconomic responses in the studies involving impact analyses of sustainability policy will face another key challenge. While sustainability issues are typically very localized and thus require high-resolution land- and water-use models, analyzing the social responses and economic decisions at a fine scale is difficult due to lack of information. Unfortunately, to the best of our knowledge, few global models are capable of high-resolution economic modeling. For example, the GLOBIOM model from the International Institute for Applied Systems Analysis (Valin et al. 2013) is a recursive-dynamic optimization model. However, due to its large size and complexity, the model is not resolved at the level of individual grid cells; rather, it is applied to representative groupings of grid cells, making analysis of local responses to sustainability restrictions more challenging. Similarly, the Global Change Analysis Model (GCAM) provides a framework for modeling groundwater resources and land use, deploying a range of supply- and demand-driven adaptive responses (Turner et al. 2019a, b), but it is implemented at the river-basin level. However, high-resolution analysis is important due to the heterogeneity of land and water use among grid cells within aquifers (e.g., the sustainability restrictions and responses would be different between the Northern Plains and southern regions of the Ogallala Aquifer in the United States). Here, SIMPLE-G introduces one of the first frameworks capable of modeling land use and water use decisions at the grid-cell level while also considering grid-cell-specific responses within local, regional, and global markets. This was possible thanks to close collaboration with hydrologists in the Water Systems Analysis Group of the University of New Hampshire. Further research is required to improve the ability of global models to integrate economic and environmental sciences considering cross-system feedback.

The reliance on virtual water trade carries the risk that many countries will not intervene to restore groundwater stability but rather will avoid the political and administrative challenges of regulation, safe in the knowledge (or more likely, hope) that virtual water will continue to compensate for reduced local production. This route will eventually generate far more severe local (and global) consequences. Intervening to restore stability, which we have endeavored to model in this study, reduces the *average* water available from groundwater but crucially preserves the flexibility and emergency buffer resources that renewable groundwater provides. If mining continues, that function will disappear because aquifers will effectively run dry, salinizing or becoming so deep that recharge no longer reaches the saturated zone in a useful timeframe or the cost of abstraction becomes prohibitive. The reduction in water availability implied in that scenario—a loss of both volume and flexibility—will have far more profound consequences locally and globally. The implication of this is that while introducing governance "now" is relatively cheap, failure to do so will be very expensive.

With increasing risks of megadroughts in the future, groundwater resources play an important role by acting as a buffer in times of extreme drought when there is no surface water available. They provide a temporary source for essential needs (including drinking water). By incurring modest production losses today, nations can reserve their groundwater resources for a future time when they will be more valuable. This intertemporal substitution is an important topic for future exploration.

In closing, this study has provided insights into the long-run costs and benefits of groundwater sustainability policy. However, we did not model the transition costs and long-run benefits of sustainable groundwater use. During the transition period—and while farmers adjust their decisions based on market outcomes—there can be a high reduction in local production, employment, and farm income. We find that aggregate employment will increase in the long run, ensuring that communities that use water sustainably will avoid the risk of running out of groundwater when it is most needed. However, there are still uncertainties about the actual ramifications of ambitious global sustainability policies. How big are the unintended consequences that might occur from local water conservation, potentially spurring deforestation and excess fertilizer use in distant corners of the world? Can rigidities within the labor market reduce the effectiveness of even the most well-intentioned policy efforts? These questions call for future research to shed light on the different benefits and costs of sustainability policies and on the synergies and trade-offs between sustainability goals within planetary boundaries.

Acknowledgments and Competing Interests   This work was supported by the National Science Foundation CBET-1855937, CBET-1805808, OISE-2020635, and DEB-1924111 and the US Department of Agriculture, National Institute of Food and Agriculture, grant #2019–67,023-29,679.

The findings and conclusions presented in this chapter are those of the authors and should not be construed to represent any official determination or policy of the US Department of Agriculture (USDA), the US government, the Department of Energy (DOE), or the National Science Foundation (NSF). Furthermore, we declare that there is no conflict of interest related to this work.

References

Allan, Tony. 1997. "Virtual water": A long term solution for water short Middle Eastern economies? In *Festival of science*. Leeds: British Association for the Advancement of Science.

Allan, John Anthony. 2003. Virtual water – The water, food, and trade nexus. Useful concept or misleading metaphor? *Water International* 28: 106–113. https://doi.org/10.1080/02508060.2003.9724812.

Allan, Tony. 2011. *Virtual water: Tackling the threat to our planet's most precious resource*. London: Bloomsbury Publishing.

Baldos, C. Uris Lantz, Iman Haqiqi, Thomas W. Hertel, Mark Horridge, and J. Liu. 2020. SIMPLE-G: A multiscale framework for integration of economic and biophysical determinants of sustainability. *Environmental Modelling & Software* 133: 104805. https://doi.org/10.1016/j.envsoft.2020.104805.

Befus, Kevin M., Scott Jasechko, Elco Luijendijk, Tom Gleeson, and M. Bayani Cardenas. 2017. The rapid yet uneven turnover of Earth's groundwater. *Geophysical Research Letters* 44: 5511–5520. https://doi.org/10.1002/2017GL073322.

Bierkens, Marc F.P., and Yoshihide Wada. 2019. Non-renewable groundwater use and groundwater depletion: A review. *Environmental Research Letters* 14: 063002. https://doi.org/10.1088/1748-9326/ab1a5f.

Bredehoeft, John D. 2002. The water budget myth revisited: Why hydrogeologists model. *Groundwater* 40: 340–345. https://doi.org/10.1111/j.1745-6584.2002.tb02511.x.

Calzadilla, Alvaro, Katrin Rehdanz, and Richard S.J. Tol. 2010. The economic impact of more sustainable water use in agriculture: A computable general equilibrium analysis. *Journal of Hydrology* 384: 292–305. https://doi.org/10.1016/j.jhydrol.2009.12.012.

Chapagain, A.K., and A.Y. Hoekstra. 2011. The blue, green and grey water footprint of rice from production and consumption perspectives. *Ecological Economics* 70: 749–758. https://doi.org/10.1016/j.ecolecon.2010.11.012.

Chatterjee, Rana, and Ranjan Kumar Ray. 2016. A proposed new approach for groundwater resources assessment in India. *Journal of the Geological Society of India* 88: 357–365. https://doi.org/10.1007/s12594-016-0498-2.

Chinnasamy, Pennan, and Govindasamy Agoramoorthy. 2016. India's groundwater storage trends influenced by tube well intensification. *Ground Water* 54: 727–732. https://doi.org/10.1111/gwat.12409.

Dalin, Carole, Yoshihide Wada, Thomas Kastner, and Michael J. Puma. 2017. Groundwater depletion embedded in international food trade. *Nature* 543: 700–704. https://doi.org/10.1038/nature21403.

Dee, D.P., S.M. Uppala, A.J. Simmons, P. Berrisford, P. Poli, S. Kobayashi, U. Andrae, et al. 2011. The ERA-Interim reanalysis: Configuration and performance of the data assimilation system. *Quarterly Journal of the Royal Meteorological Society* 137: 553–597. https://doi.org/10.1002/qj.828.

Dizard, John. 2022. Ukraine war disrupts global market for grains. *Financial Times*, February 26.

Famiglietti, J.S. 2014. The global groundwater crisis. *Nature Climate Change* 4: 945–948. https://doi.org/10.1038/nclimate2425.

Galloway, Devin L., and Thomas J. Burbey. 2011. Review: Regional land subsidence accompanying groundwater extraction. *Hydrogeology Journal* 19: 1459–1486. https://doi.org/10.1007/s10040-011-0775-5.

Gleeson, Tom, Yoshihide Wada, Marc F.P. Bierkens, and Ludovicus P.H. Van Beek. 2012. Water balance of global aquifers revealed by groundwater footprint. *Nature* 488: 197–200. https://doi.org/10.1038/nature11295.

Golub, Alla A., and Thomas W. Hertel. 2012. Modeling land-use change impacts of biofuels in the GTAP-BIO framework. *Climate Change Economics* 3: 1250015. https://doi.org/10.1142/S2010007812500157.

Graham, Neal T., Gokul Iyer, Mohamad I. Hejazi, Son H. Kim, Pralit Patel, and Matthew Binsted. 2021. Agricultural impacts of sustainable water use in the United States. *Scientific Reports* 11: 17917. https://doi.org/10.1038/s41598-021-96243-5.

Grogan, Danielle. 2016. *Global and regional assessments of unsustainable groundwater use in irrigated agriculture*. Dissertation, Durham, NH: University of New Hampshire. https://www.proquest.com/dissertations-theses/global-regional-assessmentsunsustainable/docview/1802824016/se-2.

Grogan, Danielle S., Dominik Wisser, Alex Prusevich, Richard B. Lammers, and Steve Frolking. 2017a. The use and re-use of unsustainable groundwater for irrigation: A global budget. *Environmental Research Letters* 12: 034017. https://doi.org/10.1088/1748-9326/aa5fb2.

———. 2017b. The use and re-use of unsustainable groundwater for irrigation: A global budget. *Environmental Research Letters* 12: 034017. https://doi.org/10.1088/1748-9326/aa5fb2.

Grogan, Danielle S., Shan Zuidema, and Stanley Glidden. 2022. *Water balance model (WBM) open source release version 1.0.0 ancillary data (version 1.0.0)*. University of New Hampshire Earth Systems Research Center. https://doi.org/10.34051/d/2022.2.

Haqiqi, Iman, Chris J. Perry, and Thomas W. Hertel. 2022. When the virtual water runs out: Local and global responses to addressing unsustainable groundwater consumption. *Water International* 47: 1060–1084. https://doi.org/10.1080/02508060.2023.2131272.

Haqiqi, Iman, Danielle S. Grogan, Marziyeh Bahalou Horeh, Jing Liu, Uris L.C. Baldos, Richard Lammers, and Thomas W. Hertel. 2023. Local, regional, and global adaptations to a compound pandemic-weather stress event. *Environmental Research Letters* 18: 035005. https://doi.org/10.1088/1748-9326/acbbe3.

Jasechko, Scott, and Debra Perrone. 2020. California's Central Valley groundwater wells run dry during recent DROUGHT. *Earth's Futures* 8: e2019EF001339. https://doi.org/10.1029/2019EF001339.

———. 2021. Global groundwater wells at risk of running dry. *Science* 372: 418–421. https://doi.org/10.1126/science.abc2755.

Klasic, Meghan, Amanda Fencl, Julia A. Ekstrom, and Amanda Ford. 2022. Adapting to extreme events: Small drinking water system manager perspectives on the 2012–2016 California Drought. *Climatic Change* 170: 26. https://doi.org/10.1007/s10584-021-03305-8.

Liu, Jing, Thomas W. Hertel, Farzad Taheripour, Tingju Zhu, and Claudia Ringler. 2014. International trade buffers the impact of future irrigation shortfalls. *Global Environmental Change* 29: 22–31. https://doi.org/10.1016/j.gloenvcha.2014.07.010.

Liu, Jing, Thomas W. Hertel, Richard B. Lammers, Alexander Prusevich, Uris Lantz C. Baldos, Danielle S. Grogan, and Steve Frolking. 2017. Achieving sustainable irrigation water withdrawals: Global impacts on food security and land use. *Environmental Research Letters* 12: 104009. https://doi.org/10.1088/1748-9326/aa88db.

Martens, Brecht, Diego G. Miralles, Hans Lievens, Robin van der Schalie, Richard A.M. de Jeu, Diego Fernández-Prieto, Hylke E. Beck, Wouter A. Dorigo, and Niko E.C. Verhoest. 2017. GLEAM v3: Satellite-based land evaporation and root-zone soil moisture. *Geoscientific Model Development* 10: 1903–1925. https://doi.org/10.5194/gmd-10-1903-2017.

Mekonnen, Mesfin M., and Arjen Y. Hoekstra. 2020. Sustainability of the blue water footprint of crops. *Advances in Water Resources* 143: 103679. https://doi.org/10.1016/j.advwatres.2020.103679.

Rienecker, Michele M., Max J. Suarez, Ronald Gelaro, Ricardo Todling, Julio Bacmeister, Emily Liu, Michael G. Bosilovich, et al. 2011. MERRA: NASA's modern-era retrospective analysis for research and applications. *Journal of Climate* 24: 3624–3648. https://doi.org/10.1175/JCLI-D-11-00015.1.

Rosa, Lorenzo, Davide Danilo Chiarelli, Tu Chengyi, Maria Cristina Rulli, and Paolo D'Odorico. 2019. Global unsustainable virtual water flows in agricultural trade. *Environmental Research Letters* 14: 114001. https://doi.org/10.1088/1748-9326/ab4bfc.

Saha, Suranjana, Shrinivas Moorthi, Wu Xingren, Jiande Wang, Sudhir Nadiga, Patrick Tripp, David Behringer, et al. 2014. The NCEP climate forecast system version 2. *Journal of Climate* 27: 2185–2208. https://doi.org/10.1175/JCLI-D-12-00823.1.

Siebert, Stefan, and Petra Döll. 2008. *The Global Crop Water Model (GCWM): Documentation and first results for irrigation crops*, Frankfurt hydrology paper 7. Frankfurt: Goethe University.

———. 2010. Quantifying blue and green virtual water contents in global crop production as well as potential production losses without irrigation. *Journal of Hydrology* 384: 198–217. https://doi.org/10.1016/j.jhydrol.2009.07.031.

Taheripour, Farzad, Thomas W. Hertel, and Jing Liu. 2013. The role of irrigation in determining the global land use impacts of biofuels. *Energy, Sustainability and Society* 3: 4. https://doi.org/10.1186/2192-0567-3-4.

Turner, Sean W.D., Mohamad Hejazi, Katherine Calvin, Page Kyle, and Sonny Kim. 2019a. A pathway of global food supply adaptation in a world with increasingly constrained groundwater. *Science of the Total Environment* 673: 165–176. https://doi.org/10.1016/j.scitotenv.2019.04.070.

Turner, Sean W.D., Mohamad Hejazi, Catherine Yonkofski, Son H. Kim, and Page Kyle. 2019b. Influence of groundwater extraction costs and resource depletion limits on simulated global nonrenewable water withdrawals over the twenty-first century. *Earth's Future* 7: 123–135. https://doi.org/10.1029/2018EF001105.

Valin, Hugo, Petr Havlík, Niklas Forsell, Stefan Frank, Aline Mosnier, Daan Peters, Carlo Hamelinck, Matthias Spöttle, and Maarten van den Berg. 2013. Description of the GLOBIOM (IIASA) model and comparison with the MIRAGE-BioF (IFPRI) model. *Crops* 8: 10.

Wada, Yoshihide, L.P.H. Van Beek, and Marc F.P. Bierkens. 2012. Nonsustainable groundwater sustaining irrigation: A global assessment. *Water Resources Research* 48: 2011WR010562. https://doi.org/10.1029/2011WR010562.

Wada, Y., D. Wisser, and M.F.P. Bierkens. 2014. Global modeling of withdrawal, allocation and consumptive use of surface water and groundwater resources. *Earth System Dynamics* 5: 15–40. https://doi.org/10.5194/esd-5-15-2014.

Willmott, Cort J., and Kenji Matsuura. 2001. *Terrestrial air temperature and precipitation: Monthly and annual time series (1950–1999) (version 1.02).* Newark: University of Delaware Center for Climatic Research.

Wisser, D., B.M. Fekete, C.J. Vorosmarty, and A.H. Schumann. 2010. Reconstructing 20th century global hydrography: A contribution to the Global Terrestrial Network- Hydrology (GTN-H). *Hydrolology and Earth System Science* 14: 1–24. https://doi.org/10.5194/hess-14-1-2010.

Woo, Jungha, Lan Zhao, Danielle S. Grogan, Iman Haqiqi, Richard Lammers, and Carol X. Song. 2022. C3F: Collaborative container-based model coupling framework. In *PEARC '22: Practice and experience in advanced research computing.* Boston: Association for Computing Machinery. https://doi.org/10.1145/3491418.3530298.

Chapter 17
Interplay Between the Pandemic and Environmental Stressors

Iman Haqiqi, Danielle S. Grogan, Marziyeh Bahalou, Jing Liu, Uris Lantz C. Baldos, Richard Lammers, and Thomas W. Hertel

1 Introduction

In this chapter, we aim to provide an introduction to multisystem dynamics. Our primary focus is to demonstrate how SIMPLE-G can be integrated with Earth system models to generate more accurate and holistic evaluations of the impacts of multiple shocks on agriculture. We look at the case of a compound pandemic–weather extreme event. Through this example, we showcase how Earth system models can offer multiple inputs to SIMPLE-G, including changes in crop yields and water availability. We also demonstrate how SIMPLE-G can calculate the amount of greenhouse gas (GHG) emissions released by the food system, which is crucial for assessing climate change. Moreover, we delve into how SIMPLE-G can connect the dynamics of hydroclimatic systems to food security outcomes. By examining the relationship between hydroclimatic conditions and crop growth and incorporating

---

This chapter is a slightly revised version of a paper originally published as Haqiqi, Iman, Danielle S. Grogan, D. S., Marziyeh Bahalou Horeh, Jing Liu, Uris L. C. Baldos, Richard Lammers, and Thomas W. Hertel. 2023. Local, regional, and global adaptations to a compound pandemic-weather stress event. *Environmental Research Letters* 18(3): 035005. https://doi.org/10.1088/1748-9326/acbbe3.

Data availability statement: The files needed to replicate this application are available at https://gtap.agecon.purdue.edu/simple-g/

I. Haqiqi (✉) · J. Liu · U. L. C. Baldos · T. W. Hertel
Center for Global Trade Analysis, Department of Agricultural Economics, Purdue University, West Lafayette, IN, USA
e-mail: ihaqiqi@purdue.edu

D. S. Grogan · R. Lammers
Earth Systems Research Center, University of New Hampshire, Durham, NH, USA

M. Bahalou
Department of Industrial Engineering, Purdue University, West Lafayette, IN, USA

market connections, we can better understand how changes in weather patterns can impact food security and identify potential adaptation strategies.

Short-term extreme events can threaten global food security through negative impacts on food production, food purchasing power, and agricultural economic activity. While the impacts of weather-based extremes such as droughts and floods have been well studied (Konapala et al. 2020; Pokhrel et al. 2021), societal disasters like pandemics are only more recently being evaluated (Diffenbaugh et al. 2020; Mishra et al. 2021; McDermott and Swinnen 2022; Vos et al. 2022). The potential for additional global pandemics, along with the increasing frequency of weather extremes (Pokhrel et al. 2021), also requires us to consider the impacts of compound pandemic–weather events and potential adaptation options.

The COVID-19 pandemic posed remarkable challenges to food security, impacting local, regional, and global food security through several mechanisms. Under- and unemployment-induced income losses led to reduced food purchasing power for many households, decreases in agricultural production were caused by restricted population movements, and supply chain disruptions resulted in losses to global economic production (Diffenbaugh et al. 2020; World Bank 2020; Haqiqi and Horeh 2021).

Widespread weather-based impacts on crop yields can also reduce food production and agricultural economic activity (Boyer et al. 2013; Kim et al. 2019). Droughts have a particularly negative impact on food security in developing countries (Morton 2007; Cooper et al. 2019) and have been linked not only to increased malnourishment in those countries but also to conflicts and social instability (Hsiang et al. 2013; Gleick 2014; Cooper et al. 2019).

From a food security perspective, the world was fortunate that weather outcomes during the COVID-stressed year of 2020 were largely favorable to food production. But what if the pandemic had coincided with global-scale water stress events? There have been studies of individual shocks to human and environmental systems on food security (Fujimori et al. 2018; Mora et al. 2018; Chateau et al. 2020; De Lima et al. 2021; Haqiqi et al. 2021), but very few studies have examined the interaction and compound impacts of multiple, simultaneous shocks to food security and environmental outcomes (Chateau et al. 2020; Mishra et al. 2021; Smith et al. 2021).

Here, we evaluate the impacts of compound stresses resulting from a pandemic such as COVID-19 coinciding with widespread drought and heat waves. We further quantify the role of interdependent local, regional, and global adaptation options in response to these compound stresses. This is accomplished through the lens of SIMPLE-G, linked to hydrologic and crop yield models, which are in turn driven by climate model outputs (Fig. 17.1). We build on the global gridded model from Chap. 16 and extend it to incorporate additional systems. Using this framework, we look at both undernourishment outcomes and GHG emissions for the compound disaster scenario and adaptation options.

The multimodel structure includes climate data (gray), hydrologic modeling (blue), crop yield modeling (green), and economic modeling (yellow and peach). Daily weather data allow the hydrologic model to simulate changes in agricultural water requirements and availability; the crop yield model uses these same weather

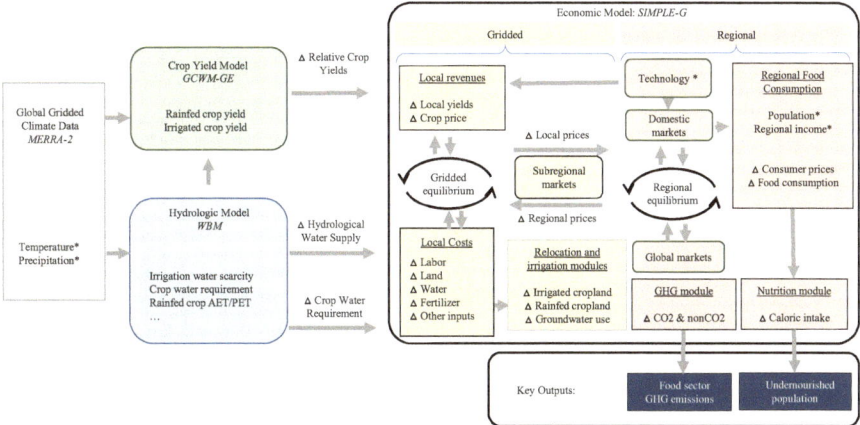

Fig. 17.1 Method overview

data, along with outputs from the hydrologic model, to simulate weather-induced changes in crop yields. The economic model uses pandemic-based shocks to income, along with the water supply and crop yield changes from the other two models, as inputs to a coupled fine-scale gridded (yellow) and regional (peach) equilibrium supply–demand system. The full modeling system produces the key metrics evaluated here: changes in food sector greenhouse gas (GHG) emissions and undernutrition as a result of the combined pandemic and weather shocks. Asterisks (*) indicate exogenous inputs to the system.

2 SIMPLE-G Model Used in This Chapter

In this chapter, we introduce a new version of SIMPLE-G that is more appropriate for short-run analyses. We employ two sets of SIMPLE-G-Global models (Haqiqi et al. 2022) that are designed to evaluate the impact of multiple shocks. The first model is that used in Chap. 16, which incorporates local production and resource use decisions that respond to market prices. This means that the model considers the market dynamics of land, water, and other inputs for both irrigated and rainfed production. The second model, called SIMPLE-G-Global-Climate, is a new version of SIMPLE-G that provides a short-run evaluation of the impact of multiple shocks. In this model, a critical assumption is that local land and water use decisions have already been made in the short run, and local production is affected by weather conditions through yield and water availability. This model considers the environmental factors that influence agricultural production, such as weather extremes, to provide a more accurate assessment of the short-run impact of shocks.

We evaluate key outputs from this multimodel system: changes in GHG emissions from the food sector and changes in undernourished populations in developing countries. Undernourished population is defined as the number of people who do not consume the minimum caloric requirement, as defined by the Food and Agriculture Organization of the United Nations (FAO 2021). GHG emissions are computed as the sum of all emissions from different stages of production in the food system and include CO_2, N_2O, and CH_4. We further evaluate multiscale endogenous adaptation responses to reduce negative food security impacts, as detailed in Haqiqi et al. (2023).

The SIMPLE-G framework is particularly well suited to this multisystem analysis of compound extremes since the economic decisions play out at a fine scale but are also spatially connected through regional and global markets. This distinguishes SIMPLE-G from most economic equilibrium models, which operate at the regional level. Because SIMPLE-G offers a combined gridded and regional economic model framework, it is well-positioned to make the best use of the fine-scale gridded results coming from the hydrologic and crop yield models. These gridded results are important to include in economic modeling because impacts on water supply and crop yields can vary greatly within a single region. Averaging these grid-cell-level biophysical model results prior to economic modeling would fail to capture changes in the spatial patterns of agricultural production that rely on local natural resources and are impacted by fine-scale weather events.

The SIMPLE-G-Global model used in this study determines crop production, land use, and irrigation water demand for each of the 1.3 million grid cells (recall Chap. 16). These grid cells are connected through agricultural markets and their relative production costs and land and water availability determine the scale of production at each location. We also use the nutrition module within SIMPLE which was developed by Baldos and Hertel (2014) based on Neiken (2003). In addition, we use a GHG emission module to estimate changes in emissions from the food system (CO_2, N_2O, and CH_4) based on the GTAP Power Data Base (Aguiar et al. 2019; Chepeliev 2020).

3 Experiment Design

To evaluate the impacts of the compound disaster posed by a co-occurring global pandemic and a global low-yield agricultural year, we need to connect the biophysical agricultural system with the global economic system. To do this, we use a coupled model framework that integrates gridded physical models from hydrology and agronomy with equilibrium models from economics (Fig. 17.1). These models are (1) a process-based gridded hydrologic model that simulates changes in agricultural water requirements and availability based on daily weather inputs, (2) a gridded crop yield model emulator that simulates changes in irrigated and rainfed crop yields based on both daily weather and hydrologic model output, and (3) a gridded-regional economic model that integrates both the pandemic-related shocks to income and the agricultural shocks from weather as simulated by the other two models.

3.1 Hydrologic Model

The hydrologic model used here is the water balance model (WBM), a global gridded hydrological model that simulates the land surface component of the water cycle based on daily weather inputs (Grogan et al. 2022). The WBM has been used and validated globally (Wisser et al. 2010; Grogan et al. 2017), including in multimodel systems coupled with economic models (Zaveri et al. 2016; Liu et al. 2017; Haqiqi et al. 2021). The WBM simulates both the vertical exchange of water between groundwater, surface soil moisture, and the atmosphere as well as the horizontal transport of water through runoff and stream networks. The WBM represents many anthropogenic interactions with the hydrological cycle, including irrigation (Wisser et al. 2010) and agricultural land use (Grogan et al. 2017). Here, the WBM is used to simulate changes in water supply and changes in crop water requirements.

3.2 Crop Yield Model

This chapter examines the interactions between multiple, punctuated extreme shocks (e.g., pandemics, short-term droughts, and heat stress). Note that we are not looking at long-term climate changes and regime shifts that can permanently move human and environmental systems to a new state. To simulate yearly changes in crop yields due to weather impacts, we develop an emulator from the crop yield component of the Global Crop Water Model (Siebert and Döll 2010), which we call the Global Crop Water Model General Emulator (GCWM-GE). The yields are estimated based on beneficial degree days between 10 °C and 30 °C and harmful degree days above 30 °C as well as evaporative stress for five climate zones. While this tool does not capture yield damage due to frost, extreme winds, flooding, pests, or disease, it has the great advantage of eliminating the need for a crop-by-crop calculation of yield while modeling the relative pattern of crop yields in a statistical framework. Overall, GCWM-GE is capable of explaining around 50% of yield variation circa 2000 (Haqiqi et al. 2023).

3.3 Scenario Design

To simulate a compound pandemic and weather crisis, we construct three shocks that are simultaneous inputs to the economic modeling system. These shocks comprise changes to key values within the economic framework that cause the system to fall out of equilibrium, triggering multiple adjustments in the global economy. Table 17.1 lists all shock values at the regional level, and Fig. 17.2 shows the global, weather-based crop yield shocks at the grid-cell level.

Table 17.1 Pandemic and weather shock scenario design

Region	Pandemic shocks	Weather shocks		Other shocks
	Gross domestic product (GDP) shock (%)	Water resources shock[a] (%)	Crop yield shock[b] (%)	Additional shock to crop yields[c] (%)
East Europe	−5.35	−0.36	−5.02	4.89
North Africa	−6.30	−3.52	−1.22	13.87
Sub-Saharan Africa	−5.58	−1.69	−8.94	10.78
South America	−8.78	0.46	5.47	−0.45
Brazil	−8.78	−2.23	6.17	−1.76
Australia and New Zealand	−6.51	−5.28	4.80	−5.93
Europe	−7.76	−3.00	−11.10	8.59
South Asia	−8.94	1.94	5.15	−8.42
Central America	−8.78	−0.73	−0.68	1.96
South Africa	−5.58	−14.39	−12.57	6.54
Southeast Asia	−5.70	−7.86	−1.32	5.67
Canada	−5.00	−1.06	−4.78	9.74
United States	−5.00	1.11	0.50	−1.41
China	1.17	−2.84	5.22	1.32
Middle East	−6.30	5.66	1.26	7.76
Japan and Korea	−4.73	−2.55	−1.14	−4.95
Central Asia	−5.35	18.94	1.69	−0.16

[a] These shocks are applied at the grid-cell level. Regional aggregate values are reported in this table for the purpose of describing the shock scenario. See Fig. 17.2 for the grid-cell shock to rainfed crop yields

[b] Crop yield shocks are estimated by the crop model at the grid-cell level.

[c] Other yield shock is the portion of the change in reported regional average yield that is not captured by the crop model. This shock is applied at the country level

3.3.1 Pandemic Shocks

The economic shocks from a pandemic are represented here as changes in regional income due to lockdowns and changes in productivity due to supply chain disruptions. We use the estimated impacts of the COVID-19 pandemic in 2020 on income and production from the World Bank (2020) *Global Economic Prospects* report as inputs to the regional economic model. We believe that this is indicative of the global economic impact that a future pandemic might have.

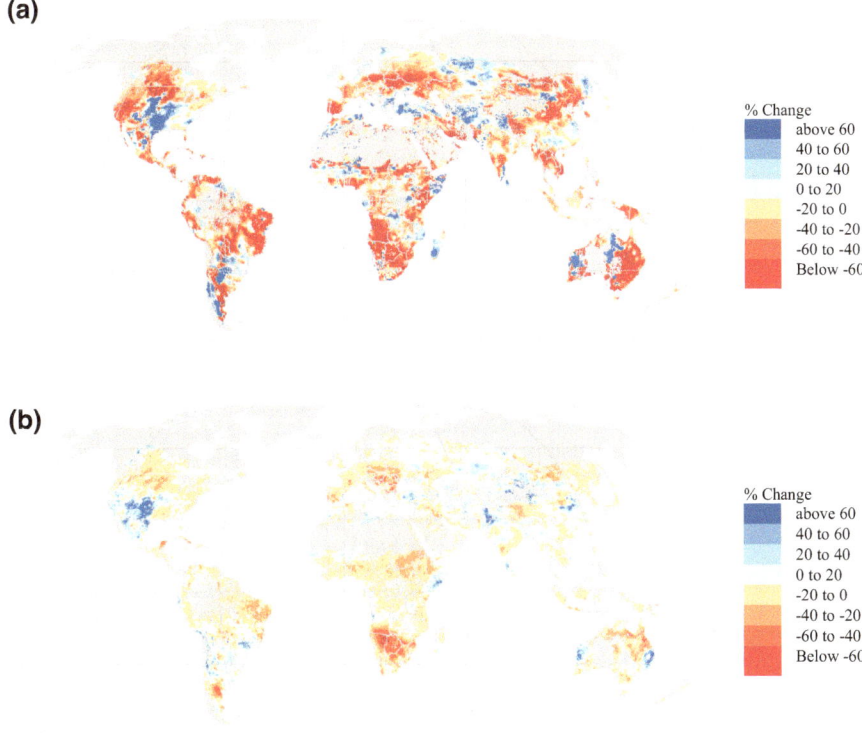

Fig. 17.2 Grid cell shock to (**a**) rainfed crop yields and (**b**) surface water supply due to weather impacts The gridded yield shocks are calculated using the global yield emulator model, while the shock on surface water is based on the outputs of the water balance model

3.3.2   Weather Shocks

While there is interannual variability in global agricultural production, 2015 was a particularly poor year for crop production compared with the recent past (2000s and 2010s), largely due to concurrent droughts in key breadbasket regions that year, according to the World Bank (2020) crop production index. To simulate this weather shock, both the hydrologic model and the crop yield model were used to simulate outcomes over 2012–2018 based on daily input weather data. The shocks to water resources and crop yields are computed as the difference between the model output for 2015 and the average output over the 7-year simulation period. These shocks are applied at the grid-cell level and are assumed to represent the kind of adverse weather events that could exacerbate the impacts of a pandemic.

3.3.3 Additional Crop Yield Shocks

Weather extremes explain only a portion of the variation in crop yield anomalies (Ray et al. 2015; Vogel et al. 2019). In addition to heat and water stress, there are many other factors affecting food production that are not captured in the GCWM-GE or the WBM, including human system components (e.g., political changes, economic policies, labor strikes, conflicts, migration, technology, growth, and recession) and environmental factors (e.g., frost, flooding, pests, and disease). We use annual reported agricultural production and yields from the World Bank by country to produce a residual change in crop yields. These observed outcomes from 2015, which lie outside the predictions of the crop yield model, are treated as an additional yield shock.

3.3.4 Adaptations to Shocks

The SIMPLE-G gridded economic model encompasses the equilibrium adaptation responses considered in this study at the local (irrigation), regional (production pattern), and global levels. Specifically, these adaptation options include surface water and groundwater substitution for irrigation on existing irrigated cropland, changes in the amount of water applied to land, rainfed-to-irrigated (or irrigated-to-rainfed) land conversions, expansion or contraction of cropland, relocation of cropland, input substitutions in the livestock and processed food sectors, consumer responses to price changes, dietary changes, and international trade. Water, land, and fertilizer inputs are treated as exogenous and unchanging in the no-adaptation scenario.

3.4 Decomposition Methods

We use the GEMPACK software utility to decompose the results in two ways. First, we utilize GEMPACK's "subtotal" feature (Harrison and Pearson 1996; Harrison et al. 2000) to identify the relative importance of the pandemic, weather, and other shocks on each of our outcome metrics. Second, we decompose the relative influence of local water, regional land, and global trade adaptation options on undernutrition metric outcomes by successively turning these options off in the model simulations.

4 Results

We assess the impact of compound events for adaptation and no-adaptation scenarios. Comparing the two scenarios highlights the value of global adaptation channels (e.g., changes in international trade) and local adaptation measures (e.g., irrigation).

4.1 Change in GHG Emissions in the Absence of Adaptation

Globally, the compound pandemic and weather shocks lower GHG emissions from the food sector by 11.5%, with the largest changes occurring in South Africa and Europe (Fig. 17.3a). These reductions are due to relatively larger negative shocks, lower agricultural production levels, and reduced food transportation. This finding is consistent with observations made during the first year of the COVID-19 pandemic (Diffenbaugh et al. 2020). However, GHG emissions from the food sector increase slightly in the Middle East, China, and Central Asia. These increases are caused by higher crop production and transportation occurring in regions with the capacity to produce more food under the combined shock scenario. In all three of the regions showing increases in GHGs, total shocks to crop yields (weather and additional shocks) are positive, allowing them to become key suppliers in the short run.

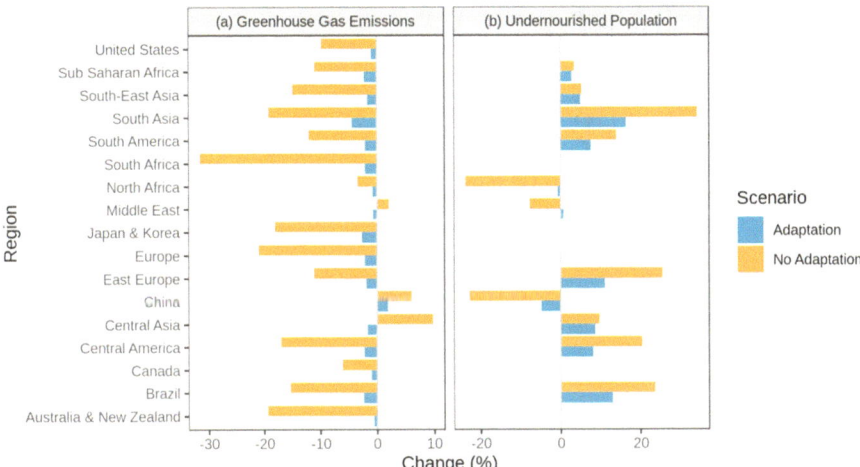

Fig. 17.3 Percentage change in (**a**) greenhouse gas emissions from the agricultural sector and (**b**) undernourished populations in developing nations due to the compound shocks, with and without adaptations The magnitude of the impacts on both emissions and food security varies regionally, although most regions experience a decline in emissions and an increase in the undernourished population. Comparing the adaptation (*blue*) to the no-adaptation (*yellow*) scenarios shows that adaptations reduce the impact of the compound disaster in all cases. In Table 2 and Fig. S3, Haqiqi et al. (2023) report results for the undernourished population in units of millions of people. Note that wealthier regions have no estimate of the change in undernourished population as the FAO methodology is only appropriate for low-income regions; these omitted regions are the United States, South Africa, Japan, Korea, Europe, Canada, Australia, and New Zealand

4.2 Change in Food Security in the Absence of Adaptation

Without adaptations, global food security declines significantly. The number of undernourished people in developing countries increases by 12.51%, from 426 million to 479 million people (Table 17.2). Notably, there are changes in both directions due to regional variations; we arrive at the net change of 53 million people through an increase of 66 million people in negatively affected regions and a decrease of 13 million people in positively affected regions (Table 17.2). The regional variation is considerable, with changes of up to $\pm20\%$. Decomposition of the shocks' relative impacts on nutritional outcomes (Fig. 17.4a) shows that all three shocks (pandemic, weather, and additional) contribute to a decline in the number of undernourished people in China. For all eight regions with increases in undernourished populations, the pandemic shock (red bar) is the largest contributor to the food security crisis (Fig. 17.4a).

Table 17.2 Undernourished population by region in the initial condition (no disaster) and change in undernourished population due to the compound disaster with and without adaptation

	Initial undernourished population (million people)	Compound disaster impact, no adaptation (million people)	Compound disaster impact, with adaptation (million people)	Change due to adaptation (million people)
Sub-Saharan Africa	119.58	4.16	3.40	−0.77
Southeast Asia	39.88	2.12	2.02	−0.10
South Asia	148.85	51.19	24.49	−26.70
South America	9.43	1.32	0.72	−0.60
North Africa	2.34	−0.56	−0.02	0.54
Middle East	15.97	−1.25	0.14	1.38
East Europe	6.50	1.66	0.72	−0.94
China	50.16	−11.45	−2.44	9.00
Central Asia	9.29	0.90	0.80	−0.10
Central America	14.84	3.04	1.20	−1.84
Brazil	9.18	2.17	1.20	−0.97
Total	426.01	53.30	32.21	−21.09

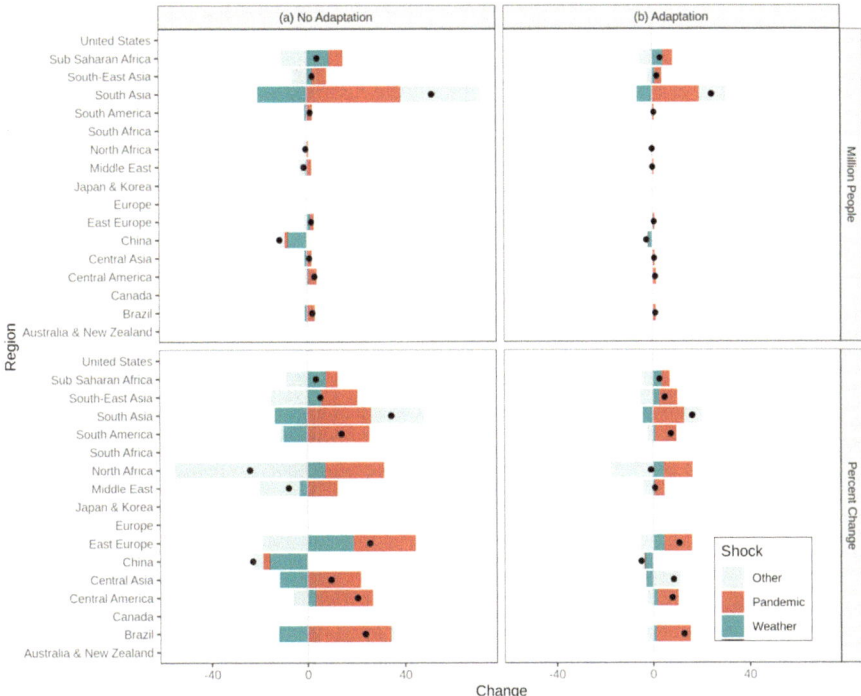

Fig. 17.4 Decomposition of the drivers of change in undernourished populations under the (**a**) no-adaptation scenario and (**b**) adaptation scenario The top row shows undernourished population changes in units of millions of people, and the bottom panel shows results as percentage changes. For each region, the impact of component shocks on the final outcome is shown; individual impacts can be positive or negative. The total resulting outcome is shown with a black dot. Note that developed regions have no estimates of change in undernourished population; these regions are the United States, South Africa, Japan, Korea, Europe, Canada, Australia, and New Zealand

4.3   Changes in GHG Emissions Under Adaptation

When adaptations are included, GHG emissions from the agricultural sector decline by 1.8% globally in the compound disaster scenario. This is a smaller decline than that under the no-adaptation scenario, as the adaptations help reduce the shocks' negative impacts on agricultural production through higher volumes of international trade and transportation. There are two exceptions: Central Asia and the Middle East, where declines in GHG emissions from the no-adaptation scenario switch to increases in GHG emissions when adaptation options are allowed to play out (Fig. 17.3).

4.4   Changes in Food Security Under Adaptation

In the scenario of a combined shock including adaptations, there is a 7.6% increase in the undernourished population in developing nations. While an increase of ~32 million people is an adverse development, it is an improvement over the increase of 53.3 million people that occurred in the no-adaptation scenario (Table 17.2). Adaptations prevent the compound disaster from causing undernourishment for approximately 21 million people in developing nations, with most of the change occurring in South Asia (Figs. 17.3b and 17.4). Adaptations have significant impacts on reducing undernourishment and mitigating the adverse consequences of compound disasters within and across regions.

The reduced impact and spatial distribution of undernourished populations in this scenario, compared with those in the no-adaptation scenario, arise from factors at the global, regional, and local levels. With global international trade, farmers with relatively more favorable weather conditions can benefit from higher prices in local and global markets, earning higher profits while simultaneously producing more food for the global market (thus pushing down prices) and reducing the malnourishment caused by high food prices. Local irrigated farms will also benefit from higher yields for irrigated crops compared with rainfed crops, especially under heat stress. Farmers with access to groundwater may benefit more when responding to surface water scarcity and higher local prices. Additionally, consumers in regions that experience large reductions in domestic production can buffer against high increases in local prices by importing food from overseas.

In the three regions where food security improves under the no-adaptation scenario, the outcome is less favorable under adaptation: In the Middle East, where undernourishment fell by -8% under no adaptations but increased by $+1\%$ with adaptations, North Africa (-24% to -1%), and China (-23% to -5%). While all three adaptations contribute to this reversal, the largest driver is global trade (Fig. 17.5). As an example of one of the few regions with positive shocks to crop yields, the Middle East (mostly Türkiye) is able to produce more food in both the no-adaptation and adaptation scenarios; even without any adaptations, the positive crop yield shock causes higher food production. Under the no-adaptation scenario, the produced food could only be supplied domestically (within the Middle East). With adaptations, this extra food was exported from Türkiye in the Middle East region to take advantage of the higher global crop prices. (Global crop prices increase by 8.84% in the adaptation scenario.) This price rise leads farmers in the Middle East to benefit financially from exports but causes regional consumers to suffer from the price increase, boosting undernourishment.

Positive values indicate that the adaptations increase the undernourished population relative to the no-adaptation scenario; negative values show that adaptations lead to improvements in food security by reducing the number of undernourished people. Each adaptation can cause positive or negative changes; the net change in each region is shown with a black dot. Note that developed regions have no estimate

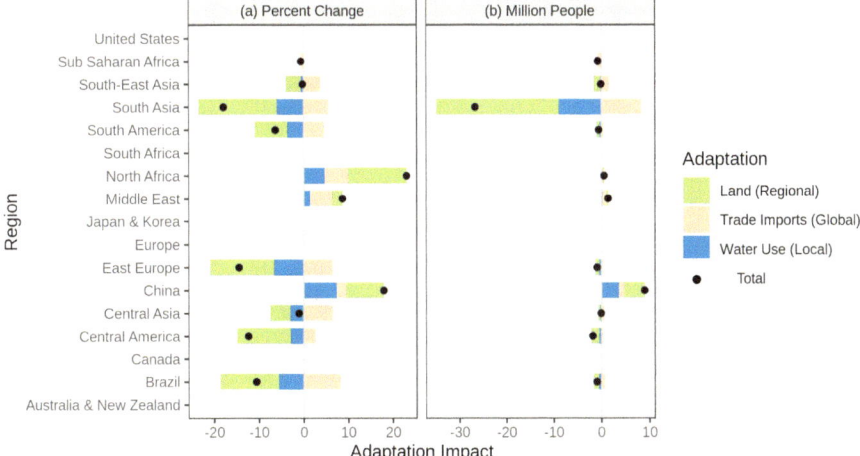

Fig. 17.5 Decomposition of adaptation contributions to changing undernourished populations compared with the no-adaptation scenario in (**a**) percentage change and (**b**) millions of people

of change in undernourished population; these regions are the United States, South Africa, Japan, Korea, Europe, Canada, Australia, and New Zealand.

Overall, global trade adaptations have the largest impact on changes in the undernourished population (Fig. 17.5), accounting for 57% of the improvement in food security when compared with the no-adaptation scenario. In total, global trade reduced undernourished populations by approximately 12 million people compared with numbers under the no-adaptation scenario, and regional variation shows that this net change is due to an increase of around six million and a decrease of around 18 million in the number of undernourished people (Fig. 17.5). The next largest adaptation impact is from land adaptation at the regional level, which accounts for 36% of the total adaptation impact; local water adaptation accounts for the remaining 7%.

5   Discussion

The COVID-19 pandemic and climate change are two major threats facing humanity. Extreme weather events such as droughts, floods, heat waves, and storms can have devastating effects on food security, water availability, infrastructure, and ecosystems. When these events happen alongside a pandemic, they create compound weather–pandemic stresses that can overpower the ability of individuals, communities, and nations to cope with them. This section discusses the likely impacts of this compound stress and evaluates its implications for global and local food security.

5.1   Model Limitations

This study uses a linked modeling framework; each model has its own limitations, uncertainties, and biases. A general discussion of these limitations can be found in the model-specific literature of the hydrologic model (Wisser et al. 2010; Zuidema et al. 2020; Grogan et al. 2022), the global crop water model (Siebert and Döll 2010; Haqiqi et al. 2023), and the SIMPLE-G model (see Part III) and related applications (Baldos et al. 2023; Haqiqi 2023; Liu et al. 2023; Ray et al. 2023).

The most important model limitation to discuss here is the assumptions about the distribution of undernourished population. It is likely that this model underestimates the food security outcomes from the combined pandemic and weather events. We believe this to be the case because of our assumptions that the income shock is homogeneous across the entire population and that within-country income distributions do not change. In reality, individuals living in poverty tend to suffer greater impacts under both pandemic-type income shocks and all types of food price increases (Hoogeveen and Lopez-Acevedo 2021; World Bank 2022). This suggests that a compound pandemic–weather shock would disproportionately impact people in the lower range of the income distribution and that this income distribution would shift so that a larger share of the population would occupy the lower range.

The modeling framework used here provides novel insights into some key impacts of the combined shock implemented. However, no model can capture all social, economic, and environmental impacts, the combined shock evaluated here would inevitably have impacts on parts of the system not captured by our modeling framework. These impacts include—but are certainly not limited to—health metrics other than undernourishment, migration spurred by changing socioeconomic conditions, economic impacts on the nonagricultural sectors of the global economy, and changing ecosystem services and their economic impacts as patterns of crop production change.

5.2   Adaptation Codependencies

We simulated three adaptations here: changes to local water use, changes to regional cropland use, and changes in global trade. These options span the local-to-global scales at which agricultural activity and the global food market adjust to shocks, and it may seem that such disparate scales operate separately; however, this modeling framework was built because each scale interacts with the others, aggregating local changes up to global level and global changes in turn impacting local action. While we decompose the relative impacts of each of these adaptations, it is important to note that implementing any of these impacts independently within the modeling framework would not result in the changes reported by the decomposition, because they interact with one another. Increased exports to the global market are driven in part by increased production; this extra production will not occur unless regional and/or local adaptations provide greater inputs of land and/or water. Conversely, local adaptations aimed at increasing production will only occur if global food prices make such extra investments in land and water profitable.

5.3 Compound Shock Interactions

The compound shock scenario evaluated here was designed to stress multiple interacting components of the human–earth system. While we decompose the relative importance of each of the two hazards (Fig. 17.4), this decomposition does not tell us the impact of each hazard alone, nor does it show whether the interaction of the two hazards reduces or increases the impacts. To test whether the interaction effect is synergistic or additive, we have calculated the magnitude of this effect (Haqiqi et al. 2023). Isolating individual stressors and regions reveals that the interaction effects can be synergistic (e.g., when looking at prices) or antagonistic (e.g., when looking at undernourishment), as shown in Fig. 17.6. However, this is not a general rule: Rather, the direction and magnitude of the interaction effect depend on location-specific elasticity parameters. Food price elasticities are important in responses to supply-side shocks (weather-driven) and demand shifts (pandemic-driven). For more details, refer to Haqiqi et al. (2023).

Figure 17.6 illustrates the interaction effect if we isolate only one region, SSA (Sub-Saharan Africa), and two shocks (pandemic or weather). The scenarios show a reduction in gross domestic product (GDP) (5% or 10%) and/or crop yields (10%) only in SSA. (**a**) A indicates the magnitude of the impact of the weather-only shock and B indicates the interaction effect. (**b**) C indicates the magnitude of the impact of the weather only shock and D indicates the interaction effect.

5.4 Environmental Impacts

There has been discussion in the literature about the potential for the pandemic to reduce negative anthropogenic impacts on the environment, especially GHG emissions (Diffenbaugh et al. 2020; Khan et al. 2021; Smith et al. 2021; Kumar et al. 2022). Our findings show that GHG emission reductions from the agricultural sector are modest in the face of combined pandemic and weather shocks. Under the no-adaptation scenario, we find an 11.5% reduction in GHG emissions from the agriculture sector. As this sector accounts for approximately 30% of annual GHG emissions, this represents a reduction of 3–4% in total global GHG emissions. Forward-looking economic studies suggest that the short-term declines in GHG emissions, even when considering larger emissions reductions from the transportation sector, will quickly return to pre-COVID levels and potentially even exceed those levels as countries relax emissions policies in an attempt to reverse economic losses (Smith et al. 2021), while others point to the dramatic short-term changes in

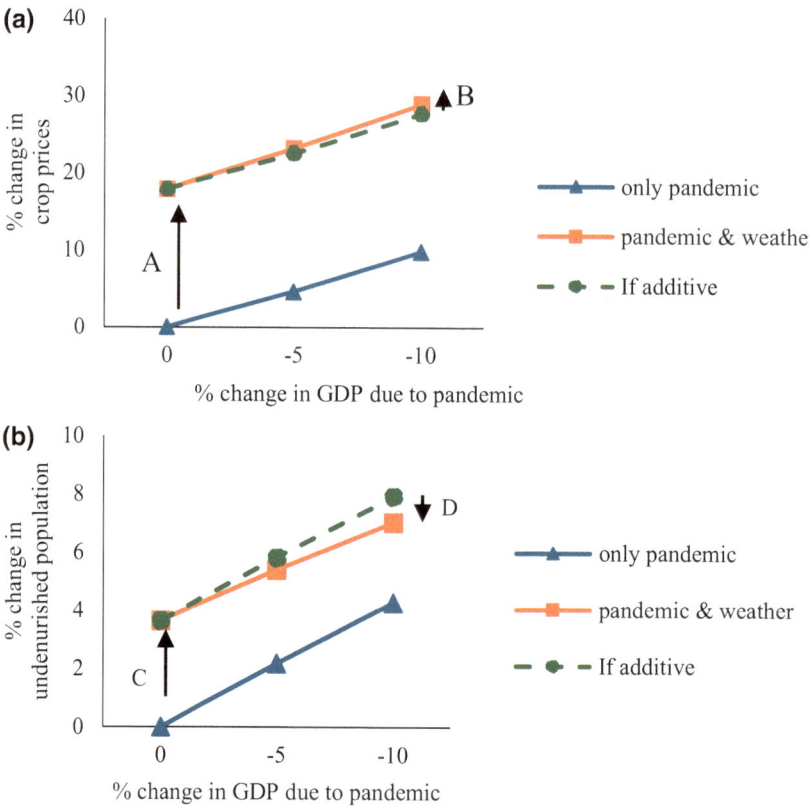

Fig. 17.6 The interaction effect of weather and pandemic on (**a**) crop prices and (**b**) undernourishment outcomes in Sub-Saharan Africa

GHG emissions and other pollutants as evidence that anthropogenic impacts can be reversed (Le et al. 2020; Khan et al. 2021). We add to this discussion the point that GHG emissions are not independent of adaptations. These adaptations are critical for reducing food security impacts on an annual time scale, and both the environment and food security should be evaluated in a cohesive framework. Echoing recent literature that considers both the economy and weather extremes together, more aggressive policy interventions will be required to maintain decreases in GHG emissions while recovering economically from the pandemic (Cheng et al. 2021; D'Orazio 2021; Smith et al. 2021), as pandemics and weather are short-term shocks while climate change occurs on decadal (approximately 30-year) time scales. Furthermore, while there has been a focus on how the pandemic has altered environmental indicators like GHG emissions, we argue that it is also important to consider the reverse: Protecting the environment and natural resources like land and water during nonstress years can make a difference in our ability to quickly adapt to short-term stresses. For example, one component of the water adaptation simulated here is the ability to switch from surface water to groundwater for irrigation. This is only possible if aquifer resources are protected against depletion during nonstressed years (recall Chap. 16) and if policies that enable annual-scale investment in infrastructure like groundwater wells and pumps are enacted.

6 Conclusions

The COVID-19 pandemic did not coincide with major weather shocks affecting food production. However, with increasing weather extremes, it is possible that such a global economic shock could coincide with severe environmental stresses. To investigate the potential consequences for food security and sustainability, we employ a global gridded biophysical-economic framework that permits us to explore the interactions between economic and nature-induced stresses on the global food system. The results illustrate the significance of adaptation options in the case of a pandemic combined with environmental stresses. Employing this modeling framework, we can identify hotspots of food insecurity vulnerable to compound shocks.

We find that, of the three adaptations implemented, global trade has the largest beneficial impact on food security. While irrigation is also important at the local level, international trade can buffer the impact of future rainfall and irrigation shortfalls (Liu et al. 2014). While global trade was the largest driver of food security improvements between the no-adaptation and adaptation scenarios, increased trade could not have occurred if there had been no adaptation in demand and supply capacity. Effective adaptation requires adaptations across scales, from local to global.

Even with adaptation, there will be a significant increase in the number of undernourished people in South Asia, Sub-Saharan Africa, and Central and South America in the presence of such a compound pandemic–weather event. These results are relevant to policy, as they highlight the significance of economic and financial relief to combat hunger and undernutrition caused by compound short-term stresses. Given the likelihood of environmental stressors and future pandemics, we can decrease the future costs of such compound events by mitigating climate change and ensuring that resources like land and water—along with the investments and the infrastructure (Lobell et al. 2013; Baldos et al. 2020a, b) required to make full use of those resources—are available. Finally, the adverse effects of pandemics and environmental shocks can be mitigated through global trade most effectively when paired with adaptive local and regional resource use.

We have explored the complex relationship between compound pandemic–environment extremes and their significant impact on food security and emissions. However, many intriguing questions need further investigation. For instance, what are the effects of changes in food loss and waste combined with a pandemic on a global scale? What about the human element? Will rising temperatures slow the productivity of agricultural labor? Finally, can sustainable groundwater governance act as a resilient buffer, supporting our food systems against future extreme shocks? These fascinating questions, which are still unanswered, are an invitation to the next chapter of future directions. It is clear that as we continue to explore the intricate linkage between environmental pressures and food security, we need both rigorous analyses and measurements of adaptation. Through continued research and innovative policies, we hope to create a more sustainable future where food security thrives and the environment is protected by sustainable practices.

Acknowledgments and Competing Interests This work was supported by the US Department of Energy, Office of Science, Biological and Environmental Research Program, Earth and Environmental Systems Modeling, MultiSector Dynamics under Cooperative Agreement DE-SC0022141. The authors also acknowledge support from the National Science Foundation INFEWS award #1855937: "Identifying Sustainability Solutions through Global-Local-Global Analysis of a Coupled Water-Agriculture-Bioenergy System."

The findings and conclusions presented in this chapter are those of the authors and should not be construed to represent any official determination or policy of the US Department of Agriculture (USDA), the US government, the Department of Energy (DOE), or the National Science Foundation (NSF). Furthermore, we declare that there is no conflict of interest related to this work.

References

Aguiar, Angel, Maksym Chepeliev, Erwin L. Corong, Robert McDougall, and Dominique van der Mensbrugghe. 2019. The GTAP data base: Version 10. *Journal of Global Economic Analysis* 4: 1–27. https://doi.org/10.21642/JGEA.040101AF.

Baldos, Uris Lantz C., and Thomas W. Hertel. 2014. Global food security in 2050: The role of agricultural productivity and climate change. *Australian Journal of Agricultural and Resource Economics* 58: 554–570. https://doi.org/10.1111/1467-8489.12048.

Baldos, Uris Lantz C., Keith Fuglie, and Thomas W. Hertel. 2020a. The research cost of adapting agriculture to climate change: A global analysis to 2050. *Agricultural Economics* 51: 207–220. https://doi.org/10.1111/agec.12550.

Baldos, Uris Lantz C., Iman Haqiqi, Thomas W. Hertel, Mark Horridge, and Jing Liu. 2020b. SIMPLE-G: A multiscale framework for integration of economic and biophysical determinants of sustainability. *Environmental Modelling & Software* 133: 104805. https://doi.org/10.1016/j.envsoft.2020.104805.

Baldos, Uris Lantz C., Maksym Chepeliev, Brian Cultice, Matthew Huber, Sisi Meng, Alex C. Ruane, Shellye Suttles, and Dominique Van Der Mensbrugghe. 2023. Global-to-local-to-global interactions and climate change. *Environmental Research Letters* 18: 053002. https://doi.org/10.1088/1748-9326/acc95c.

Boyer, J.S., P. Byrne, K.G. Cassman, M. Cooper, D. Delmer, T. Greene, F. Gruis, et al. 2013. The U.S. drought of 2012 in perspective: A call to action. *Global Food Security* 2: 139–143. https://doi.org/10.1016/j.gfs.2013.08.002.

Chateau, Jean, Erwin Corong, Elisa Lanzi, Caitlyn Carrico, Jean Fouré, and David Laborde. 2020. Characterizing supply-side drivers of structural change in the construction of economic baseline projections. *Journal of Global Economic Analysis* 5: 109–161. https://doi.org/10.21642/JGEA.050104AF.

Cheng, Yi, Haimeng Liu, Shaobin Wang, Xuegang Cui, and Qirui Li. 2021. Global action on SDGs: Policy review and outlook in a post-pandemic era. *Sustainability* 13: 6461. https://doi.org/10.3390/su13116461.

Chepeliev, Maksym. 2020. GTAP-power data base: Version 10. *Journal of Global Economic Analysis* 5: 110–137. https://doi.org/10.21642/JGEA.050203AF.

Cooper, Matthew W., Molly E. Brown, Stefan Hochrainer-Stigler, Georg Pflug, Ian McCallum, Steffen Fritz, Julie Silva, and Alexander Zvoleff. 2019. Mapping the effects of drought on child stunting. *Proceedings of the National Academy of Sciences* 116: 17219–17224. https://doi.org/10.1073/pnas.1905228116.

D'Orazio, Paola. 2021. Towards a post-pandemic policy framework to manage climate-related financial risks and resilience. *Climate Policy* 21: 1368–1382. https://doi.org/10.1080/14693062.2021.1975623.

De Lima, Cicero Z., Jonathan R. Buzan, Frances C. Moore, Uris Lantz C. Baldos, Matthew Huber, and Thomas W. Hertel. 2021. Heat stress on agricultural workers exacerbates crop impacts of climate change. *Environmental Research Letters* 16: 044020. https://doi.org/10.1088/1748-9326/abeb9f.

Diffenbaugh, Noah S., Christopher B. Field, Eric A. Appel, Ines L. Azevedo, Dennis D. Baldocchi, Marshall Burke, Jennifer A. Burney, et al. 2020. The COVID-19 lockdowns: A window into the Earth System. *Nature Reviews Earth & Environment* 1: 470–481. https://doi.org/10.1038/s43017-020-0079-1.

FAO. 2021. *FAOSTAT: Food and agriculture data*. Rome: Food and Agriculture Organization of the United Nations. Accessed 5 Feb 2024.

Fujimori, Shinichiro, Tomoko Hasegawa, Joeri Rogelj, Su Xuanming, Petr Havlik, Volker Krey, Kioshi Takahashi, and Keywan Riahi. 2018. Inclusive climate change mitigation and food security policy under 1.5 °C climate goal. *Environmental Research Letters* 13: 074033. https://doi.org/10.1088/1748-9326/aad0f7.

Gleick, Peter H. 2014. Water, drought, climate change, and conflict in Syria. *Weather, Climate, and Society* 6: 331–340. https://doi.org/10.1175/WCAS-D-13-00059.1.

Grogan, Danielle S., Dominik Wisser, Alex Prusevich, Richard B. Lammers, and Steve Frolking. 2017. The use and re-use of unsustainable groundwater for irrigation: A global budget. *Environmental Research Letters* 12: 034017. https://doi.org/10.1088/1748-9326/aa5fb2.

Grogan, Danielle S., Shan Zuidema, Alex Prusevich, Wilfred M. Wollheim, Stanley Glidden, and Richard B. Lammers. 2022. Water balance model (WBM) v.1.0.0: A scalable gridded global hydrologic model with water-tracking functionality. *Geoscientific Model Development* 15: 7287–7323. https://doi.org/10.5194/gmd-15-7287-2022.

Haqiqi, Iman. 2023. Quantifying the uncertainties in estimating the heterogeneous effects of carbon taxes on labor, land, water, and fertilizer use in US agriculture. In *Annual Meeting*. Washington, DC: Agricultural and Applied Economics Association.

Haqiqi, Iman, and Marziyeh Bahalou Horeh. 2021. Assessment of COVID-19 impacts on US counties using the immediate impact model of local agricultural production (IMLAP). *Agricultural Systems* 190: 103132. https://doi.org/10.1016/j.agsy.2021.103132.

Haqiqi, Iman, Danielle S. Grogan, Thomas W. Hertel, and Wolfram Schlenker. 2021. Quantifying the impacts of compound extremes on agriculture. *Hydrology and Earth System Sciences* 25: 551–564. https://doi.org/10.5194/hess-25-551-2021.

Haqiqi, Iman, Chris J. Perry, and Thomas W. Hertel. 2022. When the virtual water runs out: Local and global responses to addressing unsustainable groundwater consumption. *Water International* 47: 1060–1084. https://doi.org/10.1080/02508060.2023.2131272.

Haqiqi, Iman, Danielle S. Grogan, Marziyeh Bahalou Horeh, Jing Liu, Uris L.C. Baldos, Richard Lammers, and Thomas W. Hertel. 2023. Local, regional, and global adaptations to a compound pandemic-weather stress event. *Environmental Research Letters* 18: 035005. https://doi.org/10.1088/1748-9326/acbbe3.

Harrison, W. Jill, and Ken R. Pearson. 1996. Computing solutions for large general equilibrium models using GEMPACK. *Computational Economics* 9: 83–127. https://doi.org/10.1007/BF00123638.

Harrison, W. Jill, J. Mark Horridge, and K.R. Pearson. 2000. Decomposing simulation results with respect to exogenous shocks. *Computational Economics* 15: 227–249. https://doi.org/10.1023/A:1008739609685.

Hoogeveen, Johannes G., and Gladys Lopez-Acevedo, eds. 2021. *Distributional impacts of COVID-19 in the Middle East and North Africa region*. World Bank Publications.

Hsiang, Solomon M., Marshall Burke, and Edward Miguel. 2013. Quantifying the influence of climate on human conflict. *Science* 341: 1235367. https://doi.org/10.1126/science.1235367.

Khan, I., D. Shah, and S.S. Shah. 2021. COVID-19 pandemic and its positive impacts on environment: An updated review. *International journal of Environmental Science and Technology* 18: 521–530. https://doi.org/10.1007/s13762-020-03021-3.

Kim, Wonsik, Toshichika Iizumi, and Motoki Nishimori. 2019. Global patterns of crop production losses associated with droughts from 1983 to 2009. *Journal of Applied Meteorology and Climatology* 58: 1233–1244. https://doi.org/10.1175/JAMC-D-18-0174.1.

Konapala, Goutam, Ashok K. Mishra, Yoshihide Wada, and Michael E. Mann. 2020. Climate change will affect global water availability through compounding changes in seasonal precipitation and evaporation. *Nature Communications* 11: 3044. https://doi.org/10.1038/s41467-020-16757-w.

Kumar, Abhinandan, Pardeep Singh, Pankaj Raizada, and Chaudhery Mustansar Hussain. 2022. Impact of COVID-19 on greenhouse gases emissions: A critical review. *Science of the Total Environment* 806: 150349. https://doi.org/10.1016/j.scitotenv.2021.150349.

Le, Van Vang, Thanh Tung Huynh, Aykut Ölçer, Anh Tuan Hoang, Anh Tuan Le, Swarup Kumar Nayak, and Van Viet Pham. 2020. A remarkable review of the effect of lockdowns during COVID-19 pandemic on global PM emissions. *Energy Sources, Part A: Recovery, Utilization, and Environmental Effects*: 1–16. https://doi.org/10.1080/15567036.2020.1853854.

Liu, Jing, Thomas W. Hertel, Farzad Taheripour, Tingju Zhu, and Claudia Ringler. 2014. International trade buffers the impact of future irrigation shortfalls. *Global Environmental Change* 29: 22–31. https://doi.org/10.1016/j.gloenvcha.2014.07.010.

Liu, Jing, Thomas W. Hertel, Richard B. Lammers, Alexander Prusevich, Lantz C. Uris, Danielle S. Baldos, and Grogan, and Steve Frolking. 2017. Achieving sustainable irrigation water withdrawals: Global impacts on food security and land use. *Environmental Research Letters* 12: 104009. https://doi.org/10.1088/1748-9326/aa88db.

Liu, Jing, Laura Bowling, Christopher Kucharik, Sadia Jame, Lantz C. Uris, Larissa Jarvis Baldos, Navin Ramankutty, and Thomas W. Hertel. 2023. Tackling policy leakage and targeting hotspots could be key to addressing the "wicked" challenge of nutrient pollution from corn production in the U.S. *Environmental Research Letters* 18: 105002. https://doi.org/10.1088/1748-9326/acf727.

Lobell, David B., Uris Lantz C. Baldos, and Thomas W. Hertel. 2013. Climate adaptation as mitigation: The case of agricultural investments. *Environmental Research Letters* 8: 015012. https://doi.org/10.1088/1748-9326/8/1/015012.

McDermott, John, and Johan Swinnen. 2022. *COVID-19 and global food security: Two years later*. International Food Policy Research Institute.

Mishra, Ashok, Ellen Bruno, and David Zilberman. 2021. Compound natural and human disasters: Managing drought and COVID-19 to sustain global agriculture and food sectors. *Science of the Total Environment* 754: 142210. https://doi.org/10.1016/j.scitotenv.2020.142210.

Mora, Camilo, Daniele Spirandelli, Erik C. Franklin, John Lynham, Michael B. Kantar, Wendy Miles, Charlotte Z. Smith, et al. 2018. Broad threat to humanity from cumulative climate hazards intensified by greenhouse gas emissions. *Nature Climate Change* 8: 1062–1071. https://doi.org/10.1038/s41558-018-0315-6.

Morton, John F. 2007. The impact of climate change on smallholder and subsistence agriculture. *Proceedings of the National Academy of Sciences* 104: 19680–19685. https://doi.org/10.1073/pnas.0701855104.

Neiken, Loganaden. 2003. *FAO methodology for estimating the prevalence of undernourishment*. Rome: FAO.

Pokhrel, Yadu, Farshid Felfelani, Yusuke Satoh, Julien Boulange, Peter Burek, Anne Gädeke, Dieter Gerten, et al. 2021. Global terrestrial water storage and drought severity under climate change. *Nature Climate Change* 11: 226–233. https://doi.org/10.1038/s41558-020-00972-w.

Ray, Deepak K., James S. Gerber, Graham K. MacDonald, and Paul C. West. 2015. Climate variation explains a third of global crop yield variability. *Nature Communications* 6: 5989. https://doi.org/10.1038/ncomms6989.

Ray, Srabashi, Iman Haqiqi, Alexandra E. Hill, J. Edward Taylor, and Thomas W. Hertel. 2023. Labor markets: A critical link between global-local shocks and their impact on agriculture. *Environmental Research Letters* 18: 035007. https://doi.org/10.1088/1748-9326/acb1c9.

Siebert, Stefan, and Petra Döll. 2010. Quantifying blue and green virtual water contents in global crop production as well as potential production losses without irrigation. *Journal of Hydrology* 384: 198–217. https://doi.org/10.1016/j.jhydrol.2009.07.031.

Smith, L. Vanessa, Nori Tarui, and Takashi Yamagata. 2021. Assessing the impact of COVID-19 on global fossil fuel consumption and CO2 emissions. *Energy Economics* 97: 105170. https://doi.org/10.1016/j.eneco.2021.105170.

Vogel, Elisabeth, Markus G. Donat, Lisa V. Alexander, Malte Meinshausen, Deepak K. Ray, David Karoly, Nicolai Meinshausen, and Katja Frieler. 2019. The effects of climate extremes on global agricultural yields. *Environmental Research Letters* 14: 054010. https://doi.org/10.1088/1748-9326/ab154b.

Vos, Rob, John McDermott, and Johan Swinnen. 2022. COVID-19 and global poverty and food security. *Annual Review of Resource Economics* 14: 151–168. https://doi.org/10.1146/annurev-resource-111920-013613.

Wisser, D., B.M. Fekete, C.J. Vorosmarty, and A.H. Schumann. 2010. Reconstructing 20th century global hydrography: A contribution to the Global Terrestrial Network- Hydrology (GTN-H). *Hydrolology and Earth System Science* 14: 1–24. https://doi.org/10.5194/hess-14-1-2010.

World Bank. 2020. *Global economic prospects*. Washington, DC: World Bank.

———. 2022. *Poverty and shared prosperity 2022: Correcting course*. Washington, DC: World Bank.

Zaveri, Esha, Danielle S. Grogan, Karen Fisher-Vanden, Steve Frolking, Richard B. Lammers, Douglas H. Wrenn, Alexander Prusevich, and Robert E. Nicholas. 2016. Invisible water, visible impact: Groundwater use and Indian agriculture under climate change. *Environmental Research Letters* 11: 084005. https://doi.org/10.1088/1748-9326/11/8/084005.

Zuidema, Shan, Danielle Grogan, Alexander Prusevich, Richard Lammers, Sarah Gilmore, and Paula Williams. 2020. Interplay of changing irrigation technologies and water reuse: Example from the Upper Snake River Basin, Idaho, USA. *Hydrology and Earth System Sciences* 24: 5231–5249. https://doi.org/10.5194/hess-24-5231-2020.

Part V
Future Directions

Chapter 18
Future Directions: Policy Implications, Model Extensions, and Institutional Innovation

Iman Haqiqi, Thomas W. Hertel, Zhan Wang, Uris Lantz C. Baldos, Alfredo Cisneros-Pineda, and Jing Liu

This concluding chapter looks ahead to the future of the SIMPLE-G framework, exploring its potential for further development, policy applications, and wider adoption. Having applied the flexible theoretical framework to a wide range of applications, we hope that the reader can now appreciate how this powerful tool can contribute to our understanding of land and water sustainability challenges. By integrating economic and biophysical data and methods at a high spatial resolution, SIMPLE-G provides an ideal platform for analyzing complex interactions among human and natural systems. In addition, this framework can lead to important policy insights by quantifying unintended consequences and market-mediated spillover effects within a comprehensive global framework. This is a critical missing piece of many previous sustainability policy evaluations and impact assessments.

We hope that this introduction to the SIMPLE-G framework will be the beginning of your journey through research to a brighter and more sustainable future.

1 Policy Applications and Impact Assessment

Effective policies are at the heart of achieving sustainable development goals. We believe that the widespread adoption of SIMPLE-G will allow for novel policy applications and impact assessments related to food security and environmental sustainability. This includes policies and regulations to improve air and water quality, increase food security, reduce food loss and waste, promote research and development for sustainable agriculture, increase resilience, improve infrastructure,

I. Haqiqi (✉) · T. W. Hertel · Z. Wang · U. L. C. Baldos · A. Cisneros-Pineda · J. Liu
Center for Global Trade Analysis, Department of Agricultural Economics, Purdue University, West Lafayette, IN, USA
e-mail: ihaqiqi@purdue.edu

© The Author(s) 2025
I. Haqiqi, T. W. Hertel (eds.), *SIMPLE-G*,
https://doi.org/10.1007/978-3-031-68054-0_18

and encourage dietary shifts and related policy combinations. Here we look at several possible extensions and novel policy applications.

1.1 Inform Policy Decisions with Spatially Explicit Information

Most decision-making about the allocation of scarce land and water resources is performed at the local level. This approach is appropriate given the localized nature of the underlying physical and socioeconomic systems and the critical importance that local governance plays in guiding natural resource use. However, these decisions are rarely informed by broader "boundary conditions" that characterize the ways in which higher-level governance (e.g., state and national) and the global economy are likely to impinge on local resource demands. However, SIMPLE-G, with its integrative, multiscale, gridded representation of local, national, and global activity, can provide local decision-makers with a consistent set of boundary conditions to inform their decision-making. These conditions might include anticipated local cropland demand, groundwater withdrawals and depth, surface water availability, commodity prices, fertilizer costs, and labor market conditions. These boundary conditions are likely to vary depending on higher-level governance decisions and the global scenarios being considered, thereby providing useful ingredients for local planning exercises.

For example, SIMPLE-G can be used to identify policy options for promoting sustainable land and water management or reducing deforestation. This type of analysis can include rich spatial information about land types, cropland areas, yields, water requirements, and market connections. This framework can assist in the design of targeted interventions to improve water quality and fertilizer applications in areas at risk of water quality issues. It is also an appropriate framework for evaluating the effect of agricultural subsidies and taxes on environmental sustainability. By providing geospatial insights and understanding about complex interactions, SIMPLE-G can help identify actionable policies to transform ideals into reality.

1.2 Market-Mediated Spillover Effects

Another area in which SIMPLE-G can offer novel policy guidance relates to the spillover effects of local policies (Cisneros-Pineda et al. 2023). As shown in Chaps. 12 and 13, among others, localized policies are likely to generate market-mediated spillovers. That is, when output is curtailed—or enhanced—in one locality due to a policy intervention, prices increase (or decrease), and this market information is conveyed to other localities, where producers can be expected to respond accordingly. In general, these spillover effects will be stronger when policy interventions are more aggressive and the targeted region is more extensive.

We expect that these spillover effects will often moderate the gains from direct policy intervention. Thus, in the case of wetland restoration in the Midwestern United States, aimed at improving water quality, the ensuing price increase results in farmers elsewhere expanding production, partially offsetting direct gains (Chap. 14). When excessive groundwater pumping is curtailed to preserve an aquifer's long-run sustainability, the reduction in crop output results in expansion elsewhere, with potentially adverse environmental consequences (Chap. 12). SIMPLE-G can assist policymakers by anticipating these spillovers and allowing for the analysis of countervailing policies that have a broader spatial scope (e.g., the nitrogen fertilizer tax considered in Chap. 14).

Spillover effects are not restricted to market-driven effects. They may also be ecological in nature and relate to the surrounding spatial characteristics of the grid cell. Conservation actions in specific grid cells influence neighboring grid cells via species movement or ecosystem connections. For example, honeybees may take refuge in protected areas and then pollinate nearby cropland. A good example of neighboring spillover effects is the US Conservation Reserve Program, which creates conservation buffers around retired land that prevent runoff from polluting nearby water bodies. Grass filters and riparian buffers intercept contaminants before they can enter waterways. The disaggregation of SIMPLE-G into grid cells is useful not only for determining the market-mediated impacts of policies but also for assessing the social benefits deriving from the positive externalities of conservation actions.

1.3 Distributional Impacts

SIMPLE-G also enables users to evaluate the distributional impact of local conservation policies. By design, most conservation policies aim to take natural resources out of production, conserving them for nature and the enjoyment of future generations. However, doing so has a direct, and sometimes significant, impact on local economic activity. By disaggregating these impacts both spatially and by factors of production, SIMPLE-G can allow for robust analysis of the distributional impacts of conservation policies. In Chap. 13, Ray et al. show that a policy targeting excessive groundwater withdrawals can have significant impacts on local labor markets, depressing employment and wages. A particularly important point made by the authors is that the effectiveness of conservation policies is also likely to be influenced by the functioning of the labor market. If households and workers are reluctant to leave the region due to strong local ties to family and geography, then the attempted withdrawal of natural resources from market activity is likely to prove more challenging. Providing quantitative insights about equity implications of policies and shocks paves the way to design fair and just transformative strategies toward a sustainable future.

1.4 Assessing the Impact of Policies and Shocks on Food Security

SIMPLE-G is a powerful tool for assessing the impact of policies on food security. Baseline projections for the food sector draw out the implications of expected changes in population, income, and technology. By simulating changes in food production, prices, and trade flows arising from policy interventions (e.g., subsidies, tariffs, and land-use regulations), the model can identify vulnerable regions and populations susceptible to disruptions in the food system. These changes can stem from market-based policies (e.g., subsidies or taxes), land-use regulations (e.g., conservation programs or wetland restoration), climate effects (e.g., heat stress or water stress), or unprecedented events (e.g., pandemics or conflicts). This allows policymakers to evaluate the effectiveness of interventions designed to improve food security at the national and regional levels, ensuring targeted support for those most in need.

Recent research has also extended the SIMPLE framework to examine the double burden of caloric malnourishment, which refers to the fact that undernourishment and obesity coexist in many countries, localities, and even households. Lopez Barrera and Hertel (2023) incorporate both metrics using the SIMPLE framework and examine the implications of future scenarios and alternative sustainability policies for undernourishment, obesity, and related health burdens.

1.5 Assessing the Impact of the Food System on Environmental Sustainability

The SIMPLE-G framework is a valuable tool for analyzing the environmental consequences of agricultural policies and allows researchers to identify synergies and tradeoffs between sustainability goals by evaluating the impact of sustainability policies on food security. By quantifying changes in land use, water consumption, and greenhouse gas emissions triggered by food consumption decisions, the model can assess the impact of food system changes on land, water, biodiversity, and ecosystem services. These changes can be prompted by changes in all or part of the food supply chain (e.g., reduction in postharvest loss, consumer food waste, transportation, or storage). This type of analysis enables policymakers to identify policy options that promote sustainable agricultural practices, thereby minimizing environmental damage and safeguarding ecological integrity while ensuring food security. The contribution of SIMPLE-G to this literature can be through measuring the spillover effects and feedback between environmental and human systems, which is difficult to capture without a spatially resolved economic model.

1.6 Linkages from/to Other Models

The capabilities of SIMPLE-G can be further amplified through one-way linkages with other models, both as a recipient of inputs and as a provider of outputs, allowing researchers to leverage the strengths of each model to address complex sustainability challenges. For example, SIMPLE-G can benefit from the detailed hydrological insights provided by the water balance model (WBM), as described in Chap. 17. By incorporating WBM water availability estimates as an input, SIMPLE-G can produce more accurate simulations of agricultural production and land-use change responses to climate change and related policies. Similarly, integrating the outputs of Agro-IBIS, a process-based agroecological model, into SIMPLE-G can enhance its ability to capture the nuances of yield response to fertilizer application and leaching to aquatic systems (see Chap. 14). Agro-IBIS's detailed simulations of biophysical processes also contribute to a more realistic representation of substitution elasticities in SIMPLE-G. Alternatively, linking SIMPLE-G to the Environmental Impact and Sustainability Applied General Equilibrium (ENVISAGE) model unlocks new avenues for analysis. The ENVISAGE model can evaluate the economic impact of policy interventions, including carbon taxes. By including ENVISAGE data on the economic consequences of carbon taxes on fertilizer and energy prices, SIMPLE-G can incorporate the resulting changes in fertilizer prices and availability into its simulations, providing insights into the water quality co-benefits of climate policy (Zuidema et al. 2023).

One-way linkages to other models are an attractive way to provide a more complete picture of the impacts of policy interventions aimed at achieving the UN's Sustainable Development Goals related to food security, hunger eradication, and sustainable land management. By integrating geospatial economic models with other disciplines (e.g., environmental science, health, and energy), we gain a holistic understanding of critical interactions of these complex systems. These interdisciplinary connections allow us to address sustainability challenges more comprehensively.

1.7 Multiresolution and Multiscale

The resolution of production units in the SIMPLE-G framework can vary from grid cells (e.g., 250 meter, 5 arcmin, 15 arcmin, and 30 arcmin) to subregional, national, and aggregate regional production units. The versatility of this multiscale and multiresolution approach enables researchers to tailor the model to specific research questions and investigate scenarios with varying levels of detail. The critical rule for determining the optimal resolution is to capture the right amount of spatial heterogeneity required for the analysis. Excessive detail will result in challenging parameterization and potentially more difficult analysis. Insufficient detail for a given policy will limit the value of the ensuing analysis.

2   Model Extensions

SIMPLE-G can be extended in multiple ways. For example, authors can zoom in on specific regions by breaking them out from the global model. This unlocks local nuances and lets researchers tailor policies, pinpoint vulnerable populations, and assess impacts with greater accuracy. Exploring bilateral trade flows is another exciting frontier. Merging Global Trade Analysis Project (GTAP) data with SIMPLE-G could allow researchers to analyze specific trade partners and shocks, like the US-China soybean trade. Furthermore, livestock systems and diverse production factors (e.g., labor) have the potential for expansion. Finally, while SIMPLE-G aggregates crops, modeling multiple outputs within a grid cell is on the horizon. These extensions pave the way for even deeper dives into sustainability and food security challenges at local and global scales.

2.1   Extension to Other Regions

A natural way to extend the work presented in this book is to break out new regions. SIMPLE-G is designed with this possibility in mind and offers two avenues for doing so. One avenue would involve starting with a global version of SIMPLE-G, drawing on Chap. 17, for example, and further refining the global groundwater data, parameters, and policies for a particular country. However, in our experience, this is not the best approach. Using the full global gridded model in the context of a region-specific application is inefficient and, absent a global policy scenario, does not add value to the study. Rather, our preferred approach is to start with the aggregate regions in SIMPLE and break out the focus country. This is the approach taken in the applications focusing on the United States and Brazil in Part IV. In this way, researchers can preserve the global behavior in SIMPLE while focusing attention on the region of interest. The model is much faster to run and includes far less scope for computational problems.

Using a country-specific rather than a global dataset has several advantages when analyzing complex issues (e.g., food security and environmental sustainability) at regional scales. First, this approach increases accuracy and reliability: Country-specific datasets capture nuances and local variations that global datasets often miss. These nuances can impact food production and environmental pressures. In addition, country-specific data often undergo rigorous local validation and quality control processes, ensuring their accuracy and relevance to the specific context. This eliminates potential biases and inaccuracies inherent in global datasets aggregated from diverse sources and with less local validation. Second, the country-specific approach enables a deeper understanding of the specific dynamics within a particular region. This approach facilitates the identification of hotspots and populations most susceptible to food insecurity and environmental degradation. It allows for targeted policy interventions and strategies that address the region's unique needs and vulnerabilities effectively. By tracking policies and changes within a specific country or region, researchers can evaluate policies' effectiveness with greater accuracy. This facilitates evidence-based decision-making and informs future policy adjustments to ensure their effectiveness and impact.

The SIMPLE-G code is designed to facilitate this type of region-focused breakout. The only thing that changes is the mapping file that maps grid cells to regions. Open-source scripts are available for converting gridded data in raster format to GEMPACK-readable SIMPLE-G input files. This approach facilitates the construction of consistent gridded datasets and shocks for use in SIMPLE-G. We have also made advances in the development of country-specific databases in SIMPLE, which are publicly available (https://gtap.agecon.purdue.edu/simple-g/). Using this tool, users can change regional aggregation as well as assumptions about data and parameters to fit their own research work. In the future, we plan to link this regional dataset to the SIMPLE-G database workflow so that users can easily disaggregate regions into grid cells to facilitate their own global-local-global analysis.

2.2 Extension to Bilateral Trade Flows

In Part III of this book, we noted that the current SIMPLE-G framework includes a pooled global market for crop exports and imports (assuming that all crop exports enter the global market and that all crop imports come from the global market). This setting simplifies the international trade system and is the most suitable approach for modeling international trade in the long run, over which the bilateral geography of trade is less dominant. However, in the near term, empirical evidence suggests that current bilateral trade patterns are quite persistent. For example, the United States trades disproportionately with Canada, Mexico, China, and Europe. For this reason, it is important to offer users the option of modeling bilateral trade patterns explicitly (as is the case for the GTAP model mentioned previously). This can also have important policy implications. For example, during the US-China trade war, China imposed a retaliatory tariff on soybean imports from the United States in response to the US tariffs on Chinese imports. This sharply reduced US-China soybean trade but raised soybean imports from Brazil, China's other main soybean supplier. These bilateral relationships are not well captured when modeling trade via a global pool of commodities.

Given the need to analyze bilateral trade flows, an important future direction for extending SIMPLE-G is to combine its gridded supply system with a bilateral trade system. Given the ready availability of reconciled trade and tariff data from GTAP (Aguiar et al. 2019, 2022), we propose merging the bilateral trade in the GTAP crop database with the SIMPLE-G model. This extension will retain the advantages of SIMPLE-G for capturing the spatial heterogeneity of crop production and spillover effects at the local level and incorporate detailed GTAP data on regional-level trade in order to analyze more specific shocks and responses by trade partners.

2.3 Extension to Other Production Systems

The SIMPLE-G framework outlined in this book focuses on cropping systems as the gridded economic activity. However, it is entirely possible to add additional systems, provided that the necessary data can be obtained. Perhaps the most obvious are livestock systems, particularly those relying on pastureland. A large share of the world's agricultural lands comprises pasture, the demand for which is linked to ruminant livestock systems. Incorporating a gridded representation of ruminant livestock would allow users to explore issues involving the conversion of pastures to cropland and vice versa. Gridded data on ruminant and non-ruminant livestock are available (Gilbert et al. 2018). Incorporating these data would also allow users to explore environmental issues related to ruminant livestock production, including greenhouse gas emissions and water pollution, at a fine scale. To relate regional livestock production to pasture at the gridded level, one possible direction is to introduce the land allocation module from the GTAP-Agro-ecological Zone (GTAP-AEZ) model (Hertel et al. 2009, 2010), which presents the revenue-maximizing land-use decisions for crops, pasture, and forest. This modeling approach—together with gridded land-use datasets—would allow researchers to simulate the competition of land between crop and livestock production, which is necessary for researching questions on, for example, diet change or nutrition.

A further step would involve disaggregating confinement animal feeding operations from the regional level to the gridded level in order to better capture related air and water pollution. Given data limitations, it is unlikely that modelers will be able to track the movement of crop production from specific locations to these facilities. Rather, the feedstuffs would draw on a more aggregated regional supply pool for crop inputs.

2.4 Extension to Other Factors of Crop Production

As shown in Chap. 13, identifying key factors of production—in this case, labor—can be critical for understanding system behavior and the distributional consequences of sustainability policies. In that application, labor market rigidities are shown to limit the effectiveness of the conservation policy and generate significant losses for workers. However, labor is an extremely heterogeneous input, with significant differences in both labor supply and demand characteristics between, for example, family labor, hired labor, and migrant labor. Further disaggregation of this factor of production will be important for some applications. In general, further disaggregation of production factors is warranted when: (1) supply and demand characteristics vary significantly across factors of production, (2) there is empirical evidence supporting the specification of these supply/demand responses, and (3) the production factor in question plays a significant role in the policy under consideration.

2.5 Extension to Multiple Crops

Another challenge comes from SIMPLE-G's crop aggregation approach. In the current SIMPLE-G model, we first aggregate all crops to a single commodity (e.g., corn-equivalent output) with price-based weights and use the single crop to represent the response of crop production to shocks, assuming that the crop composition in this aggregate does not change much. While this approach simulates the general response of crop production to various policies, it is difficult to model the shock and response to a specific crop or the change in crop composition in the face of new socioeconomic drivers. Considering the example of soybean trade from above, another factor that influences China's soybean import in addition to the tariff shock is the domestic consumption of cooking oil and livestock production (the latter uses soy meal as a major input). These drivers will cause crop-specific shocks to the model and change the crop composition that was imposed in the initial aggregation stage. For these reasons, disaggregating crops will be important in some cases. However, in keeping with the SIMPLE philosophy, we believe that this should be done in a parsimonious way. Our goal is not to replicate the detailed, complex models of agricultural commodity markets that have been developed over the past decades. Rather, we would seek to find an intermediate ground in which important sustainability challenges can be addressed without making the framework too demanding to allow parameterization, replication, and validation.

There are different approaches for modeling multiple crops in SIMPLE-G while using the most reliable data sources and keeping the model simple enough to facilitate analysis and provide policy insights. One issue in modeling multicrop production is the low accuracy of satellite imagery in determining the type of crops grown. While the USDA provides high-resolution cropland data layer products, information for much of the rest of the world is less reliable. For example, Brazil's *MapBiomas*, an important collaborative network of more than 70 organizations that produces annual maps of land cover and land use in Brazil, does not report maize cropland. According to the *MapBiomas Brasil* website, it does not have a maize category because maize is included in the broader class of "other temporary crops." Maize is one of the most important temporary crops in Brazil, accounting for about 17% of the total harvested area in 2022. However, it is not distinguished from other temporary crops because of the limitations of satellite imagery and the classification methods used by *MapBiomas*.

The path forward that we propose for multicrop versions of SIMPLE-G involves specifying multiple output production functions at the grid-cell level. In this way, we do not require separate inputs for each individual crop, which can be difficult to obtain at a fine spatial scale. These crops can be exported or may compete with imports from other regions. Following the aggregation of domestic and imported crop commodities, these different categories of crops substitute for one another in the same three demand categories used in the standard SIMPLE model (recall Chap. 4): direct consumption, food processing, and livestock consumption (Fig. 18.1).

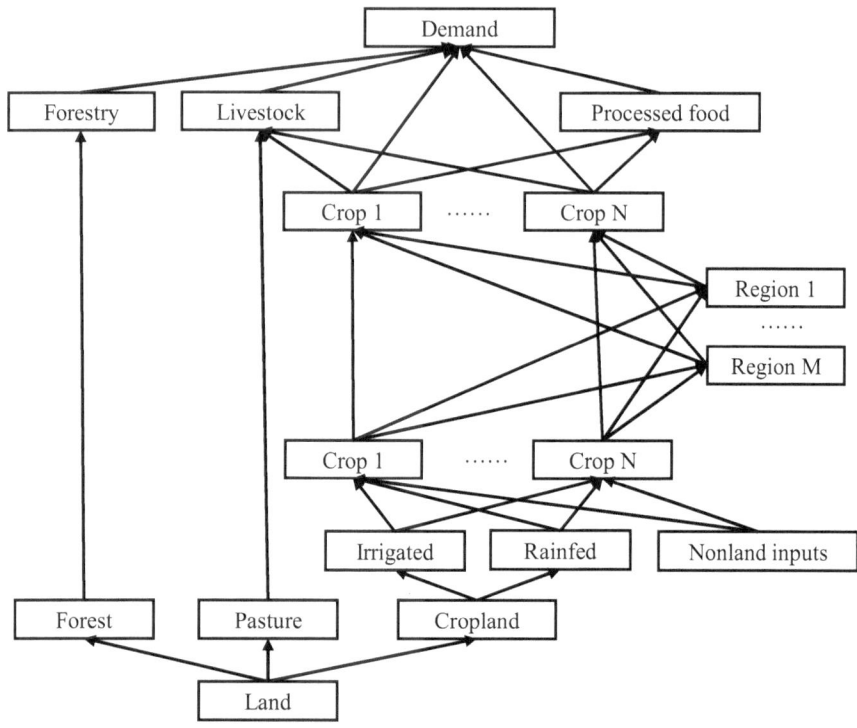

Fig. 18.1 Nesting SIMPLE-G within the GTAP-AEZ framework. Livestock and forestry demand for land are derived from a national production function, while crop production is modeled at the grid-cell level

2.6 Integration into GTAP-AEZ

Once a version of SIMPLE-G has been extended to handle bilateral trade and multiple crops, it is a relatively short step to embed this gridded framework in a general equilibrium model such as the GTAP model. Why might one wish to do so? Doesn't this violate the principle of simplicity? We believe that this should be done only when it is necessary to place this fine-scale economic analysis in a broader context. In a sense, this is really just an extension of the GTAP-AEZ model (Hertel et al. 2009, 2010), which is widely used to look at land-based sustainability policies (e.g., renewable fuels standards and other climate mitigation strategies) in an economy-wide context. More recently, the GTAP-AEZ model has been paired with fine-scale ecological modeling to uncover the macroeconomic benefits of local conservation policies (Johnson et al. 2023). A key limitation of previous work is the need to downscale economic results from the AEZs to the local level. By embedding SIMPLE-G in the GTAP model, the economic responses to conservation policies could be incorporated at the local level. Alternatively, for some purposes, it may be sufficient to link SIMPLE-G with the fine-scale ecological models.

Figure 18.1 provides an overview of how this embedding might be accomplished. As with the GTAP-AEZ model, the demand for land is derived from the forest, livestock, and crop sectors. In the case of forestry and ruminant livestock, the production functions are modeled at the national level, with demand for land services by AEZ treated as a constant elasticity of substitution nest within the national production function. This approach captures the spatial pattern of land competition across uses without modeling the full production functions for forestry and livestock at the AEZ level. These national sectors now demand land services at a finer scale (the grid cell), but nothing else has changed. However, in the case of crop production, we model the full production function at the grid cell, including the competition between rainfed and irrigated production. This analysis is more demanding, in terms of data requirements and computation, but it allows us to address the type of sustainability challenges highlighted above.

2.7 Beyond Integration: Collaborative Research for Understanding Human–Environment Linkages

Multidisciplinary research—where experts from diverse fields collaborate on a common problem—is crucial for addressing complex challenges such as food security and environmental sustainability. While each field possesses unique expertise, attempting to tackle these intricate issues within single disciplines often leads to incomplete solutions and missed opportunities. Given the importance of natural resources and the environment in SIMPLE-G, some users might consider extending SIMPLE-G to directly incorporate detailed hydrology or agroecology relationships. However, here we draw the line. We prefer not to make SIMPLE into COMPLEX! Rather, we suggest linking SIMPLE-G to a compatible hydrologic model, such as WBM, as shown in Chap. 17, or to an agroecological model, such as Agro-IBIS, as described in Chap. 14. This approach allows each field to contribute its specific knowledge and expertise, leading to a more comprehensive and robust understanding of the interconnected systems. This collaborative approach not only leverages the strengths of each field but also avoids the dilution of expertise that can occur when one discipline attempts to encompass the entirety of a complex modeling challenge. By working together, researchers can achieve greater depth and breadth in their analysis, ultimately leading to more effective and sustainable solutions for the interconnected challenges of our world.

Significant challenges must be overcome to achieve seamless collaboration. One major obstacle lies in the inherent differences between disciplinary approaches and the models themselves. Temporal and spatial resolutions differ, with hydrological and ecological models often operating at finer scales and shorter timeframes than economic models. This disparity requires careful data aggregation and interpolation to ensure compatibility across models. Furthermore, the computational software used to develop and run different models can be incompatible, hindering communication and data exchange. Additionally, conflicting definitions and terminology across disciplines can lead to misinterpretations and inconsistencies when attempting to link models. Finally, the computational demands of running large-scale linked models can be immense, requiring access to high-performance computing (HPC) resources. The lack of readily available HPC resources for many researchers can limit the feasibility of collaborative modeling approaches.

The C3F framework (Woo et al. 2022) offers a promising solution for overcoming the challenges of collaborative linkages between human and environmental models. The authors apply the C3F framework to link SIMPLE-G to WBM for the United States. By leveraging container technology, C3F enables researchers to independently develop and package their models with their specific software and data dependencies. This modular approach facilitates communication and data exchange between models while avoiding potential conflicts due to incompatible software environments. Furthermore, containerization facilitates the standardization of temporal and spatial dimensions across models, allowing researchers to define and configure consistent timeframes and spatial resolutions within each container, regardless of the models' inherent differences. This eliminates the need for complex data manipulation and interpolation, ensuring smooth communication and data exchange between models operating at different scales.

The unique feature of C3F is that it breaks away from the traditional model-coupling approach that seeks convergence between models, which is a time-consuming and often computationally expensive process. Instead, it leverages the concept of adaptive expectations. This allows each model to learn from the outputs of other models and adjust its own inputs accordingly, creating a dynamic and iterative process in much the same way as is done by real economic decision-makers. C3F facilitates this by running models consecutively, where each model receives and integrates the latest outputs from other models before making its own runs. In other words, in SIMPLE-G, farmers in each grid cell make decisions based on past WBM simulations of climate and weather and then make decisions about the extent and intensity of irrigation. Then, the WBM generates water balances considering the farmers' decisions at the beginning of the season. This eliminates the need for a single, unified solution, allowing the models to adapt and evolve in real time, reflecting the dynamic nature of the coupled human–environment system. This adaptive approach significantly reduces the computational burden compared with traditional convergence methods, making collaborative modeling more accessible and efficient. This approach also allows for a more flexible and realistic representation of the system, where model predictions are continuously updated based on the latest information, reflecting the dynamic interplay between human and environmental processes.

2.8 SIMPLE-S: A Tool for Model Development, Calibration, and Education

While the applications in Part IV demonstrate the importance of incorporating spatial heterogeneity of shocks and responses at the gridded level in SIMPLE-G, one may wonder whether it is always better to develop and work with the gridded version than with a more aggregated version. Let us consider the following circumstances under which an alternative formulation (SIMPLE-Subregion or SIMPLE-S) might be relevant:

Model Development and Testing At the model development stage, researchers may initially seek "reasonable" results—responses that are consistent with economic theory and real-world observations in direction and relative magnitude—as opposed to obtaining more precise results with considerably greater time costs. Although favored by model users, the gridded feature of SIMPLE-G can render it unnecessarily bulky for model developers, especially when the model needs to be solved repeatedly for testing purposes.

Parameter Calibration SIMPLE-G requires parameters defined at multiple scales (regional, subregional, and gridded). When researchers are interested in calibrating certain parameters (or iteratively narrowing the range of initial values for further calibration), it may be more efficient to work with the model aggregated to a specific level of interest (e.g., county, province, state, and region).

Education On the basis of our teaching experience, we find that using the gridded version for educational purposes faces three challenges. First, it is usually more difficult to interpret simulation results at the grid level, which are highly dependent on grid-specific parameters and baseline data; therefore, we have to aggregate these results further to the state or subregional level to allow for ready interpretation. In addition, the powerful AnalyseGE software, which enables detailed analysis of model outcomes on an equation-by-equation basis (recall Chap. 11), is not very functional when there are more than 20 grid cells in a region—hence the use of a mini model (Chap. 11). Second, solving SIMPLE-G at the grid level usually requires 8–15 min, depending on the computer and experiment, which may delay the teaching activity. Finally, for education and training purposes, students usually use a free-trial version of GEMPACK to learn the model. However, the size of the SIMPLE-G model (i.e., the number of endogenous variables and equations) is too large to be solved with this free-trial version.

Considering these three areas of need, we conclude that, in addition to the aggregated version (SIMPLE) and the gridded version (SIMPLE-G), it is also helpful to develop a version of intermediate size, which we name SIMPLE-S. SIMPLE-S may be developed based on an existing SIMPLE-G model; it maintains the same model and data structure as SIMPLE-G but aggregates grids to subregional levels. For example, in SIMPLE-S-US, the 76,651 5-arcmin grids of the United States are aggregated to nine USDA Farm Resource Regions (subregions) by

Table 18.1 Gridded (SIMPLE-G) and subregional (SIMPLE-S) versions in the case of the US-focused model from Chap. 12

	SIMPLE-G-US	SIMPLE-S-US
Grids	76,651 (US) + 15 (non-US regions)	9 (US) + 15 (non-US regions)
Model size	1,893,828 endogenous variables	2658 endogenous variables
Time cost	8 min 52 s	4 s
GEMPACK License	Formal (purchased) license	Free trial license

summing the value and quantity data and calculating a weighted average of gridded parameters, while the rest of the world remains at the regional level. The conversion from SIMPLE-G to SIMPLE-S is conducted with R script, which fully automates the development of the subregional version.

Table 18.1 briefly compares the SIMPLE-G-US model from Chap. 12 and its subregion version. By aggregating US grids into subregions, SIMPLE-S reduces the model size and allows the model to be solved almost instantaneously. SIMPLE-S also preserves all the equations and database structure from SIMPLE-G-US. From a research development perspective, SIMPLE-S can be used as a "wind tunnel model" for SIMPLE-G. Much as model planes are tested in a wind tunnel prior to building full-scale versions, researchers can use SIMPLE-S to test new functional forms and modules, or calibrate regional or subregional parameters, much more efficiently. Progress made in SIMPLE-S is readily transmissible to SIMPLE-G. For educational purposes, SIMPLE-S also serves as a "training manikin" version of the gridded version. Much as medical students learn their trade by first operating on manikins, students of SIMPLE-G can explore the model with various experiments and get immediate feedback before embarking on the full-scale model. Additionally, the results at the subregional level are often easier to interpret and compare to readily accessible statistics. Once familiar with SIMPLE-S, students can readily transfer their skills and experience to SIMPLE-G for further research.

3 Vision for a SIMPLE-G Community

In addition to the SIMPLE-G framework extensions discussed above, an important step toward its continuous development and application is to establish a SIMPLE-G community that brings together research capacities and interests from a broader range of individuals to address the serious sustainability challenges facing the world today. In our vision, the SIMPLE-G community can be summarized with a reinterpretation of the *global–local–global* approach to research (Hertel et al. 2023). We envision the SIMPLE-G community as one that connects *global* researchers, educators, and students who are interested in developing and applying SIMPLE-G models; encourages them to share *local* data, knowledge, and ideas for integrative research; and enhances research impacts on *global* audience and stakeholders.

3.1 Global Network

From its origin at Purdue University, SIMPLE-G has been introduced to researchers from multiple domestic and international universities and research institutes. Currently, the most important platform for SIMPLE-G training for external researchers is the SIMPLE-G short course, held biennially at Purdue since 2019. The course consists of 3 weeks of online training on the background knowledge of the non-gridded SIMPLE model and 1 week of on-site training that focuses on SIMPLE-G applications and underlying economic theory. This short course is not restricted to passive learning: During the on-site section, participants have an opportunity to collaborate with others on a group project, laying the foundation for future research projects and long-term collaboration. Based on the basic 4-week training schedule, we have also developed a workshop version of the short course that can be taught over 2 days. This abbreviated version has been successfully implemented as part of the NSF-I-GUIDE summer school (Boulder, Colorado, United States) and the China Energy Modeling Forum's model Capacity Building Workshop (Beijing, China). By November 2023, more than 60 researchers had completed SIMPLE-G training, bringing the model from Purdue to more than 40 institutes in five countries.

The group of SIMPLE-G short course participants, together with researchers and students from Purdue's Center for Global Trade Analysis, now forms the foundation for a broader global community of SIMPLE-G users and contributors. One major challenge is that, given the requirements of time commitment and cost, as well as capacity restrictions on the number of participants, the biennial short course and occasional in-person workshops cannot satisfy the increasing interest in SIMPLE-G. This is particularly true for researchers from developing countries who may be precluded from joining the on-site short courses for budgetary reasons. With these limitations in mind, this book and its accompanying lectures and files are our attempt to provide accessible teaching and training materials for SIMPLE-G to global researchers. We welcome you, readers of the book, to join the SIMPLE-G community for broader and long-term collaboration.

3.2 Local Expertise

From our experience developing region-specific models, we have learned that building a high-quality spatially explicit model depends on strong participation from research partners with local expertise, providing data inputs, local knowledge, and research ideas. Local researchers are often aware of additional data sources that are not widely recognized by their international colleagues or are less frequently used due to language barriers. By offering these data sources, they can make important contributions to establishing gridded databases for region-specific versions of SIMPLE-G. Even if the data are already available at the global level, local knowledge can also help researchers judge whether the data correctly represent real-world situations. For example, when we developed the region-specific version of SIMPLE-G for Brazil (Chap. 15), our local collaborators indicated that an important feature of Brazilian agriculture is the rotation between crops and grazing within the same year on certain croplands. However, this land-use pattern is not accurately classified or measured by global land cover datasets. We followed the suggestion of local collaborators and used a Brazilian land cover dataset, which improved the accuracy of land-use data at the grid level. Finally, and most importantly, researchers with local expertise invariably have a better understanding of the current research and policy questions faced by decision-makers; they also have a better understanding of the political, social, and cultural factors that are critical to analyzing those questions with SIMPLE-G.

3.3   Global Impacts

Fueled by global researchers and local expertise, the SIMPLE-G community will serve as an incubator for future studies that address global and local sustainability challenges. Improving the exposure and influence of SIMPLE-G models and applications has been a continuous focus of the Purdue research team. We have organized SIMPLE-G track sessions and workshops at various academic conferences—including the annual meetings of the Agricultural and Applied Economics Association, American Geographical Union, and GTAP—and presented SIMPLE-G studies in GLASSNET and I-GUIDE webinars. These events have attracted attention from the research community, which has been particularly interested in the ease with which SIMPLE-G can facilitate multidisciplinary studies. In the future, we will continue organizing relevant sessions and workshops at academic conferences, to provide additional opportunities for members of the SIMPLE-G community to present their work to a broad audience. We will also invite representatives from global policy-making agencies, think tanks, and local stakeholders to join the SIMPLE-G community and communicate with researchers, which will promote the integration of innovative research methodology with solutions to real-world challenges.

3.4   SIMPLE-G as a Component of GLASSNET

Given the breadth of the sustainability challenges facing the world, SIMPLE-G is just one of many tools needed to inform the public debate. In an effort to mobilize a wider range of databases, tools, and interdisciplinary collaborations needed to inform the UN Sustainable Development Goals, Hertel and colleagues have initiated GLASSNET—a network of networks supported by the US National Science Foundation and focused on sustainable management of the world's land and freshwater resources (https://mygeohub.org/groups/glassnet/). The purpose of GLASSNET is to facilitate global–local–global analyses, and a wide range of disciplines and researchers are involved. In addition to the project leadership, GLASSNET has a Science Council and a Stakeholder Advisory Board drawing on a diverse set of scientists, policy advisors, and decision-makers. This network of networks can play an important role in guiding analysts and policy advisors to appropriate tools for addressing the diverse challenges posed by the UN Sustainable Development Goals. By documenting the SIMPLE-G model and pairing it with teaching materials, we hope that this framework can be employed by users from a wide range of disciplines and geographies. Readers are invited to join GLASSNET to engage with the broader community of practice seeking to address the global challenges to land and water sustainability in the coming decades.

References

Aguiar, Angel, Maksym Chepeliev, Erwin L. Corong, Robert McDougall, and Dominique van der Mensbrugghe. 2019. The GTAP data base: Version 10. *Journal of Global Economic Analysis* 4: 1–27. https://doi.org/10.21642/JGEA.040101AF.

Aguiar, Angel, Maksym Chepeliev, Erwin Corong, and Dominique Van Der Mensbrugghe. 2022. The Global Trade Analysis Project (GTAP) data base: Version 11. *Journal of Global Economic Analysis* 7: 1–37. https://doi.org/10.21642/JGEA.070201AF.

Cisneros-Pineda, Alfredo, Jeffrey S. Dukes, Justin Johnson, Sylvie Brouder, Navin Ramankutty, Erwin Corong, and Abhishek Chaudhary. 2023. The missing markets link in global-to-local-to-global analyses of biodiversity and ecosystem services. *Environmental Research Letters* 18: 041003. https://doi.org/10.1088/1748-9326/acc473.

Gilbert, Marius, Gaëlle Nicolas, Giusepina Cinardi, Thomas P. Van Boeckel, Sophie O. Vanwambeke, G.R. William Wint, and Timothy P. Robinson. 2018. Global distribution data for cattle, buffaloes, horses, sheep, goats, pigs, chickens and ducks in 2010. *Scientific Data* 5: 180227. https://doi.org/10.1038/sdata.2018.227.

Hertel, Thomas, Huey-Lin Lee, Steven Rose, and Brent Sohngen. 2009. Chapter 6: Modeling land-use related greenhouse gas sources and sinks and their mitigation potential. In *Economic analysis of land use in global climate change policy*, ed. Thomas W. Hertel, Steven Rose, and Richard S.J. Tol. Abingdon: Routledge.

Hertel, Thomas W., Alla A. Golub, Andrew D. Jones, Michael O'Hare, Richard J. Plevin, and Daniel M. Kammen. 2010. Effects of US maize ethanol on global land use and greenhouse gas emissions: Estimating market-mediated responses. *Bioscience* 60: 223–231. https://doi.org/10.1525/bio.2010.60.3.8.

Hertel, Thomas W., Elena Irwin, Stephen Polasky, and Navin Ramankutty. 2023. Focus on global–local–global analysis of sustainability. *Environmental Research Letters* 18: 100201. https://doi.org/10.1088/1748-9326/acf8da.

Johnson, Justin Andrew, Uris Lantz Baldos, Erwin Corong, Thomas Hertel, Stephen Polasky, Raffaello Cervigni, Toby Roxburgh, Giovanni Ruta, Colette Salemi, and Sumil Thakrar. 2023. Investing in nature can improve equity and economic returns. *Proceedings of the National Academy of Sciences* 120: e2220401120. https://doi.org/10.1073/pnas.2220401120.

Lopez Barrera, Emiliano, and Thomas Hertel. 2023. Solutions to the double burden of malnutrition also generate health and environmental benefits. *Nature Food* 4: 616–624. https://doi.org/10.1038/s43016-023-00798-7.

Woo, Jungha, Lan Zhao, Danielle S. Grogan, Iman Haqiqi, Richard Lammers, and Carol X. Song. 2022. C3F: Collaborative container-based model coupling framework. In *Practice and experience in advanced research computing*, 1–8. Boston: ACM. https://doi.org/10.1145/3491418.3530298.

Zuidema, Shan, Jing Liu, Maksym G. Chepeliev, David R. Johnson, Uris Lantz C. Baldos, Steve Frolking, Christopher J. Kucharik, Wilfred M. Wollheim, and Thomas W. Hertel. 2023. US climate policy yields water quality cobenefits in the Mississippi Basin and Gulf of Mexico. *Proceedings of the National Academy of Sciences* 120: e2302087120. https://doi.org/10.1073/pnas.2302087120.

Index

A

Adaptations, 4, 38, 67, 174, 212, 264–266, 274, 276, 277, 284, 286, 290–300
Agricultural labor markets, 200–202, 204, 208–210, 212, 213
Agricultural markets, 19, 36, 152, 173, 200, 267, 286
Agricultural policies, 310
Agricultural production, 16, 24, 25, 35, 36, 38, 41, 46, 64, 76, 81–83, 88, 93, 96, 115, 129, 136–138, 142, 144, 147–149, 154, 179, 181, 200–202, 207, 212, 238, 244, 247, 253, 254, 258, 265, 276, 277, 284–286, 289–291, 293, 311
AnalyseGE, 6, 56, 57, 319

B

Biodiversity, 51, 52, 160, 192, 217, 277, 310
Boundary conditions, 161, 308
Brazil, 5, 41, 85, 94, 115–120, 140, 236, 237, 239–242, 245–247, 249, 250, 272, 273, 276, 288, 292, 312, 313, 315, 322

C

Climate change, 3, 7, 20, 35, 38, 40, 51, 93, 107–109, 135, 136, 141, 176–178, 184, 283, 287, 295, 298, 299, 311
Commuting zones (CZs), 6, 44, 61, 204
Compound extremes, 286
Computational efficiency, 110

Conservation policies, 15, 16, 24–27, 29–31, 36, 57, 62, 103, 123, 160, 163, 200, 201, 203, 211–213, 227, 237, 240, 242–243, 253, 255, 277, 309, 314, 316
Controlled drainage, 221–224, 226, 227, 229
Cost shares, 25–27, 53, 58, 82–84, 89, 106, 161, 162, 169, 240, 241
Crop exports, 236, 238, 313
Crop sales, 82, 90, 117, 124

D

Data-driven analysis, 81
Demand elasticities, 17, 19, 94, 95
Distributional impacts, 200, 309

E

Economic decisions, 7, 36, 93–100, 113, 114, 161, 176, 278, 286
Economic theories, 4, 5, 11–20, 30, 31, 160, 165, 200, 202–204, 319, 321
Elasticity of substitution, 25, 26, 28, 38–40, 46, 47, 52, 55, 70, 75, 119, 162, 169, 205, 219, 258, 317
Environmental policies, 13, 16, 136, 140, 141, 144–150, 152–156, 160, 174, 192, 226, 228
Environmental sustainability, 77, 93, 103, 171, 175, 307, 308, 310, 312, 317
Equilibrium models, 35, 104, 141, 154, 277, 286, 316
Equity, 14, 191, 213, 228, 309

F
Factor mobility, 59–61, 250
Fertilizer tax, 309
Food production, 7, 35, 36, 46, 74, 83, 107,
        108, 256, 277, 284, 290, 294, 299, 310,
        312
Food security, 5, 20, 46, 61, 77, 82, 83, 95, 103,
        136, 141, 147, 148, 156, 171, 175, 176,
        193, 254, 256, 283, 284, 286, 291, 292,
        294–296, 298–300, 307, 310–312, 317

G
Geospatial data, 51, 120
Geospatial economic models, 35–48, 114, 130,
        311
Global changes, 23, 24, 108, 159, 168, 170,
        173, 176, 179, 181, 192, 206–207, 269,
        275, 278, 296
Global land use, 3, 11, 14, 16, 18
Greenhouse gas (GHG) emissions, 5, 83–84,
        95, 136–138, 140–142, 145, 148, 149,
        152–156, 160, 249, 283–286, 291, 293,
        297, 298, 310, 314
Greenhouse gas emissions intensity, 150
Gridded production, 43, 53, 54, 56, 76, 259
Groundwater conservation policies, 5, 180, 181,
        203, 210, 211, 267
Groundwater scarcity, 207, 260

I
Income elasticities, 13, 38, 44, 45, 47, 74, 94
Input supply response, 26
Integrated assessment modeling, 15
International trade, 16–18, 36, 38, 46, 47, 67,
        69, 75, 95, 116, 175, 258, 259, 265, 277,
        290, 293, 294, 299, 313
Irrigation, 12, 24, 38, 52, 84, 95, 105, 124, 173,
        219, 240, 255, 286, 318

L
Labor markets, 5, 30, 40, 52, 61, 64, 192,
        199–213, 260, 274, 279, 308, 309, 314
Land rents, 27, 30, 44, 58, 60, 64, 66, 88, 89,
        96, 150, 155, 163, 265
Land use, 3, 13, 30, 36, 64, 86, 107, 114, 136,
        159, 200, 236, 255, 286, 310
Land-use changes, 11, 13–18, 40, 82–84, 93,
        122, 123, 127, 129, 135–138, 140,
        148–150, 152, 155, 176, 178, 249, 311
Local sustainability, 174, 175, 180–181, 188,
        200, 322

M
Market equilibrium, 17, 42–44, 47, 51–77
Market-mediated spillovers, 228, 308
Minimodels, 6, 56, 57, 159–161, 163, 166,
        169–171
Model transparency
Model validation, 109, 113–130, 240
Multiscale analysis, 6
Multiscale modeling, 104, 191
Multiscale models, 113, 212

N
Nonpoint source water pollution, 5, 218
Nutrient management, 218, 221, 223

P
Price elasticity of demand, 18, 74
Price elasticity of supply, 13, 14, 203
Product differentiation, 19, 40, 41, 46, 47, 52,
        67–72, 75, 178, 258

Q
Quantitative models, 30, 31, 103

R
Regional economics, 207, 286, 288
Research and development (R&D), 5, 13, 136,
        144, 146, 149, 150, 160, 190, 307

S
SIMPLE-G, 4, 11, 23, 35, 51, 81, 93, 103, 114,
        160, 173, 200, 218, 236, 254, 283, 307
SIMPLE model, 5, 15, 18, 35, 69, 73, 75, 96,
        115, 136, 141, 142, 145, 159, 160, 260,
        315, 321
Spillover effects, 7, 36, 40–42, 46, 77, 160, 174,
        175, 181, 192, 203, 210, 211, 225, 226,
        228, 236, 238, 247–248, 250, 255, 264,
        307–310, 313
Spillovers, 16–18, 40–42, 136, 142, 144, 147,
        152, 154, 155, 173–193, 225, 228, 267,
        309
Supply elasticities, 14, 18, 24, 26–28, 30, 38,
        47, 55, 56, 61, 67, 72, 93, 95–97, 99,
        100, 161, 162, 166, 169, 203–205, 212,
        237, 242, 243, 246, 248, 249
Sustainability, 3, 14, 23, 36, 51, 81, 93, 103,
        128, 142, 171, 173, 200, 253, 299, 309

Sustainability policy, 5, 7, 36, 38, 40, 56, 62, 77, 109, 164, 173–193, 199–213, 253, 254, 260, 266, 269, 273, 274, 276–279, 307, 310, 314, 316

Sustainable Development Goals (SDGs), 3, 4, 36, 307, 311, 322, 323

Sustainable groundwater use, 255–257, 267, 269, 276, 277, 279

T
Technological change, 16, 19, 108, 135, 181

Total factor productivity (TFP), 6, 25, 27, 54, 75, 117, 123, 124, 137, 140, 142–146, 148, 149, 152, 154, 159–171, 174, 179–182, 190

Trade, 4, 5, 16, 19, 35, 36, 46, 56, 67, 73–76, 81, 82, 94, 95, 104, 109, 116, 119, 120, 129, 136, 140, 175, 176, 178, 191, 200, 236, 253–279, 290, 294–296, 299, 310, 312, 313, 315, 316, 320, 321

Transformation elasticity, 63, 66, 76, 95, 96

Transportation infrastructure, 61, 235–250

V
Value of water, 58, 88, 89, 97, 99, 176, 186

W
Water, 4, 15, 24, 37, 51, 81, 93, 103, 114, 160, 173, 203, 217, 254, 283, 307

Water resources, 3, 7, 35, 36, 52, 77, 93, 107, 128, 129, 174, 177, 178, 181, 182, 184, 186, 190, 199, 203, 253, 256, 258, 262, 264, 288, 289, 308

Water stress, 38, 52, 176, 192, 284, 290, 310

Water use, 4, 23, 30, 46, 77, 84, 87, 97, 99, 128, 179, 190, 192, 200, 259, 261, 262, 264, 268, 269, 274, 278, 285, 296

Wetland restoration, 221, 222, 224–226, 229, 309, 310